AF360923

Land Use Planning for Natural Hazards

Land Use Planning for Natural Hazards

Editors

George D. Bathrellos
Hariklia D. Skilodimou

MDPI • Basel • Beijing • Wuhan • Barcelona • Belgrade • Manchester • Tokyo • Cluj • Tianjin

Editors
George D. Bathrellos
University of Patras
Greece

Hariklia D. Skilodimou
National and Kapodistrian
University of Athens
Greece

Editorial Office
MDPI
St. Alban-Anlage 66
4052 Basel, Switzerland

This is a reprint of articles from the Special Issue published online in the open access journal *Land* (ISSN 2073-445X) (available at: https://www.mdpi.com/journal/land/special_issues/landuse_natural_hazards).

For citation purposes, cite each article independently as indicated on the article page online and as indicated below:

LastName, A.A.; LastName, B.B.; LastName, C.C. Article Title. *Journal Name* **Year**, *Volume Number*, Page Range.

ISBN 978-3-03943-925-6 (Hbk)
ISBN 978-3-03943-926-3 (PDF)

Contents

About the Editors

George D. Bathrellos, Associate Professor at University of Patras, Faculty of Sciences, Department of Geology, Sector of General, Marine Geology & Geodynamics. Research interests: geomorphology (pure and applied); natural hazards; urban planning; modeling; GIS. Teaching experience: Dr. Bathrellos has taught Geomorphology, Physical Geography, GIS, Urban Planning, Environmental Management and Monitoring, and Hydrology in graduate and post-graduate courses in four Greek universities. Publications: He has contributed over 100 publications in international conferences and journals.

Selected publications:

1. Tsolaki-Fiaka, S., Bathrellos, G.D., Skilodimou, H.D. (2018): Multi-Criteria Decision Analysis for an Abandoned Quarry in the Evros Region (NE Greece). *Land*, 7 (2): 43, doi: 10.3390/land7020043, MDPI Publishing. −23 citations

2. Skilodimou, H.D., Bathrellos, G.D., Chousianitis, K., Youssef, A.M., Pradhan, B. (2019): Multi-hazard assessment modeling via multi-criteria analysis and GIS: A case study. *Environmental Earth Sciences*, 78 (2): 47, doi: 10.1007/s12665-018-8003-4, Springer. −43 citations

3. Bathrellos, G.D., Skilodimou, H.D., Chousianitis, K., Youssef, A.M., Pradhan, B. (2017): Suitability estimation for urban development using multi-hazard assessment map. *Science of the Total Environment*, 575: 119 –134, doi: 10.1016/j.scitotenv.2016.10.025, Elsevier. −137 citations

4. Bathrellos, G.D., Karymbalis, E., Skilodimou, H.D., Gaki - Papanastassiou, K., Baltas, E.A. (2016): Urban flood hazard assessment in the basin of Athens Metropolitan city, Greece. *Environmental Earth Sciences*, 75 (4): 319, doi: 10.1007/s12665-015-5157-1, Springer. −53 citations

5. Chousianitis, K., Del Gaudio, V., Sabatakakis, N., Kavoura, K., Drakatos, G., Bathrellos, G.D., Skilodimou, H.D. (2016): Assessment of earthquake-induced landslide hazard in Greece: From Arias Intensity to spatial distribution of slope resistance demand. *Bulletin of the Seismological Society of America*, 106 (1): 174 –188, doi: 10.1785/0120150172, Seismological Society of America. −47 citations

6. Rozos, D., Skilodimou, H.D., Loupasakis C., Bathrellos G.D. (2013): Application of the revised universal soil loss equation model on landslide prevention. An example from N. Euboea (Evia) Island, Greece. *Environmental Earth Sciences*, 70 (7): 3255-3266, doi: 10.1007/s12665-013-2390-3, Springer. −65 citations

7. Papadopoulou-Vrynioti, K., Bathrellos, G.D., Skilodimou, H.D., Kaviris, G., Makropoulos, K. (2013): Karst collapse susceptibility mapping considering peak ground acceleration in a rapidly growing urban area. *Engineering Geology*, 158: 77-88, doi: 10.1016/ j.enggeo.2013.02.009, Elsevier. −108 citations

8. Bathrellos, G.D., Gaki-Papanastassiou, K., Skilodimou, H.D., Papanastassiou, D., Chousianitis, K.G. (2012): Potential suitability for urban planning and industry development by using natural hazard maps and geological - geomorphological parameters. *Environmental Earth Sciences*, 66 (2): 537–548, doi: 10. 1007/s12665-011-1263-x, Springer. −168 citations

Hariklia D. Skilodimou, Ph.D. Researcher and Teaching Staff at National & Kapodistrian University of Athens, Faculty of Geology & Geoenvironment, Department of Geography & Climatology Research interests: geomorphology (pure and applied); natural hazards; modeling; GIS. Teaching experience: Dr. Skilodimou has taught Geomorphology, Physical Geography, GIS, Urban Planning, Environmental Management and Monitoring, and Hydrology in graduate and post-graduate courses in two Greek universities. Publications: She has contributed over 100 publications in international conferences and journals.

Selected publications:

1. Tsolaki-Fiaka, S., Bathrellos, G.D., Skilodimou, H.D. (2018): Multi-Criteria Decision Analysis for an Abandoned Quarry in the Evros Region (NE Greece). *Land*, 7 (2): 43, doi: 10.3390/land7020043, MDPI Publishing. −23 citations

2. Skilodimou, H.D., Bathrellos, G.D., Chousianitis, K., Youssef, A.M., Pradhan, B. (2019): Multi-hazard assessment modeling via multi-criteria analysis and GIS: A case study. *Environmental Earth Sciences*, 78 (2): 47, doi: 10.1007/s12665-018-8003-4, Springer. −43 citations

3. Bathrellos, G.D., Skilodimou, H.D., Chousianitis, K., Youssef, A.M., Pradhan, B. (2017): Suitability estimation for urban development using multi-hazard assessment map. *Science of the Total Environment*, 575: 119–134, doi: 10.1016/j.scitotenv.2016.10.025, Elsevier. −137 citations

4. Bathrellos, G.D., Karymbalis, E., Skilodimou, H.D., Gaki - Papanastassiou, K., Baltas, E.A. (2016): Urban flood hazard assessment in the basin of Athens Metropolitan city, Greece. *Environmental Earth Sciences*, 75 (4): 319, doi: 10.1007/s12665-015-5157-1, Springer. −53 citations

5. Chousianitis, K., Del Gaudio, V., Sabatakakis, N., Kavoura, K., Drakatos, G., Bathrellos, G.D., Skilodimou, H.D. (2016): Assessment of earthquake-induced landslide hazard in Greece: From Arias Intensity to spatial distribution of slope resistance demand. *Bulletin of the Seismological Society of America*, 106 (1): 174–188, doi: 10.1785/0120150172, Seismological Society of America. −47 citations

6. Rozos, D., Skilodimou, H.D., Loupasakis C., Bathrellos G.D. (2013): Application of the revised universal soil loss equation model on landslide prevention. An example from N. Euboea (Evia) Island, Greece. *Environmental Earth Sciences*, 70 (7): 3255-3266, doi: 10.1007/s12665-013-2390-3, Springer. −65 citations

7. Papadopoulou-Vrynioti, K., Bathrellos, G.D., Skilodimou, H.D., Kaviris, G., Makropoulos, K. (2013): Karst collapse susceptibility mapping considering peak ground acceleration in a rapidly growing urban area. *Engineering Geology*, 158: 77-88, doi: 10.1016/ j.enggeo.2013.02.009, Elsevier. −108 citations

8. Bathrellos, G.D., Gaki-Papanastassiou, K., Skilodimou, H.D., Papanastassiou, D., Chousianitis, K.G. (2012): Potential suitability for urban planning and industry development by using natural hazard maps and geological - geomorphological parameters. *Environmental Earth Sciences*, 66 (2): 537–548, doi: 10. 1007/s12665-011-1263-x, Springer. −168 citations

 *land*

Editorial

Land Use Planning for Natural Hazards

George D. Bathrellos * and Hariklia D. Skilodimou

Department of Geography and Climatology, Faculty of Geology and Geoenvironment, National and Kapodistrian University of Athens, University Campus, Zografou, ZC 15784 Athens, Greece
* Correspondence: gbathrellos@geol.uoa.gr; Tel.: +30-2107274882

Received: 17 August 2019; Accepted: 23 August 2019; Published: 25 August 2019

The Earth's landscape has a complex evolution and is the result of the interactions involving surficial processes, climate, tectonic, and human activity. In this context, the morphological changes in landforms due to active tectonics or climate change have the potential to affect and, in some cases, even to control human activities [1–5]. On the other hand, human activity and man-made constructions have the ability to change the landscape and, in this way, have impacted on natural hazards.

Natural hazards are physical phenomena that occur worldwide and contribute to the evolution of Earth's landscape. These phenomena affected the natural environment and existing biota, even before the appearance of man on Earth. Nowadays, they are an important global problem threatening human life. Natural hazards can damage both the natural and man-made environment [6]. Their impacts differ from place to place and frequently they appear to have adverse long-term effects [7].

On a global scale, overpopulation and urban development have the ability to increase the occurrence of natural hazards and their impacts both in the developed and developing world. For instance, deforestation causes increased rates of soil erosion and sediment transport, resulting in, for example, land degradation and flooding. Generally, natural hazards occur more frequently in relation to our capability to restore the effects of past events [8–10].

The sustainability of urban development can be influenced by several factors such as economic development, socioeconomic policy, population growth, physical environment, and natural hazards [11,12]. However, during planning, development, and management of an urban environment, only the economic and social parameters are usually taken into account. Consequently, in vulnerable locations, e.g., steeply sloping areas or those with degraded soils, the natural hazards that often occur, such as mass movements, can cause extensive damage, disrupt social and economic networks and lead to the loss of human lives and property [13–17].

Therefore, in order to minimize the loss of human life and reduce the economic consequences, proper planning and management of natural hazards are essential. However, consideration of the natural hazards and their influence on landscape evolution during the land use planning stage is essential.

In many cases, land use planning for addressing natural hazards is based on the probability of an event occurring, with little or no consideration of the consequences associated with natural hazard events [18]. For instance, floodplains are fertile, level, easy to excavate, near water and, thus, are favorable sites for urban development. In several cases, urbanization of floodplains has increased the probability of flooding, thereby causing disasters. Flood damage in such environments appears to increase, despite the construction of flood control works such as dams and river channelization [19–22].

The relationship between natural hazards and land use seems to be two-way. On the one hand, natural hazards and their associated consequences have the ability to cause changes in the landscape and thus, affect land use. On the other hand, human activities and land-use changes can lead to natural hazards.

In order to avoid the aforementioned effects, it is necessary for the decision-makers, engineers, planners, and managers to take into account the physical parameters of an area, as well as susceptibility to the natural hazards. The geology and the geomorphology of an area are important in the assurance of sustainable land management and in the protection of human life in urban areas [23].

In conclusion, is important that engineers, policymakers, and planners employ land use planning based on natural hazard maps in the evaluation and selection of suitable areas for sustainable urban development with fundamental concerns for the protection of the environment and of human life.

This special issue focuses on land use planning for natural hazards. Various types of natural hazards such as land degradation and desertification, coastal hazard, floods, and landslides, as well as their interactions with human activities, are presented in this volume.

Briassoulis H. [24] examines the use of Land Use Planning (LUP) to combat Land Degradation and Desertification (LDD). Various and interdependent socio-economic, cultural, political and institutional criteria play an important role in LDD or contribute to the management of land resources [25]. The paper presents desertification and the pertinent institutional context and studies whether and how LDD concerns enter the LUP process and the issues arising at each stage. The provision of an enabling, higher-level institutional environment should be prioritized to support phronetic-strategic integrated LUP at lower levels, which future research should explore theoretically, methodologically and empirically to realize the integrative potential of LUP and foster its effectiveness in combating LDD at the local and regional levels.

Tragaki, et al. [26] assess coastal hazard vulnerability based on geomorphic, oceanographic and demographic parameters in the Peloponnese (southern Greece). Nowadays, coastal areas around Greece are susceptible to climate change-related hazards [27]. The paper assesses the physical and social vulnerability of the Peloponnese to both coastal erosion and flooding caused by climate change-related hazards. The Coastal Vulnerability Index (CVI) and the Social Vulnerability Index (SVI) were estimated. The results showed that about 20% of the shoreline along the western and northwestern coast of the study area has high and very high physical vulnerability. Moreover, high and very high social vulnerabilities characterize communities along the northwestern part of the Peloponnese. The recognition of highly vulnerable coastal areas is very useful for coastal land use planning.

Two papers apply methods that provide vital information for land use planning and flood hazard mitigation. Rijal et al. [28] examine flood hazard mapping in the rapidly urbanizing city of Birendranagar, Nepal. Natural hazards and urbanization can interact to increase land-use changes in Nepal [29], and floods have caused loss of life and property in Birendranagar. The study focuses on the underlying land-cover dynamics and flood hazards of the study area. The spatiotemporal urbanization dynamics and associated land-use and land-cover (LULC) changes of the city from 1989 to 2016 allowed areas with high flood hazard risk to be identified. The urban area expanded nearly by 700%, while the cultivated land declined simultaneously by 12% between 1989 and 2016. This, and the loss of forests contributed significantly to increased flood hazard. Steep slopes, excessive land utilization, and intense monsoonal precipitation aggravated hazards locally.

Bathrellos et al. [30] undertake spatio-temporal analysis of flooding in the drainage basin of the Pinios River (Thessaly, Central Greece). The paper identifies the flood hazard by using historical flood events which occurred between 1979 and 2010, old topographic maps and geomorphic parameters. The flood occurrences increased during the period 1990–2010, most flood events were in October. The majority of occurrences are recorded in the southern part of the study area. There is a certain amount of clustering of flood events in the areas of former marshes and lakes as well as in the lowest and flattest parts of the study area. The applied method provides valuable information for land-use planning at a regional scale leading to the determination of the safe and non-safe areas for urban activities.

Skilodimou et al. [31] examine the relation of physical and anthropogenic factors with landslide activity in a mountainous part of northern Peloponnese in southern Greece. The existing landslides, lithology, slope angle, rainfall, road network along with land use of the study area were analyzed. The results prove that Plio-Pleistocene fine-grained sediments and flysch, relatively steep slopes and a rise in the amount of rainfall are strongly associated with the occurrence of landslides. A 100m wide zone along each road increases the probability of landslides while the extensively cultivated land of the study area is strongly related to landslide activity. This procedure may be utilized in landslide hazard assessment mapping as well as to new and existing land use planning projects.

Conflicts of Interest: The authors declare no conflict of interest.

References

1. Bathrellos, G.D.; Skilodimou, H.D.; Maroukian, H. The spatial distribution of Middle and Late Pleistocene cirques in Greece. *Geogr. Ann. A* **2014**, *9*, 323–338. [CrossRef]
2. Skilodimou, H.D.; Bathrellos, G.D.; Maroukian, H.; Gaki-Papanastassiou, K. Late Quaternary evolution of the lower reaches of Ziliana stream in south Mt. Olympus (Greece). *Geogr. Fis. Din. Quat.* **2014**, *37*, 43–50.
3. Kokinou, E.; Skilodimou, H.D.; Bathrellos, G.D.; Antonarakou, A.; Kamberis, E. Morphotectonic analysis, structural evolution/pattern of a contractional ridge: Giouchtas Mt., Central Crete, Greece. *J. Earth Syst. Sci.* **2015**, *124*, 587–602. [CrossRef]
4. Bathrellos, G.D.; Skilodimou, H.D.; Maroukian, H. The significance of tectonism in the glaciations of Greece. *Geol. Soc. Spec. Publ.* **2017**, *433*, 237–250. [CrossRef]
5. Bathrellos, G.D.; Skilodimou, H.D.; Maroukian, H.; Gaki-Papanastassiou, K.; Kouli, K.; Tsourou, T.; Tsaparas, N. Pleistocene glacial and lacustrine activity in the southern part of Mount Olympus (central Greece). *Area* **2017**, *49*, 137–147. [CrossRef]
6. Bathrellos, G.D.; Skilodimou, H.D.; Chousianitis, K.; Youssef, A.M.; Pradhan, B. Suitability estimation for urban development using multi-hazard assessment map. *Sci. Total Environ.* **2017**, *575*, 119–134. [CrossRef] [PubMed]
7. Alcántara-Ayala, I. Geomorphology, natural hazards, vulnerability and prevention of natural disasters in developing countries. *Geomorphology* **2002**, *47*, 107–124. [CrossRef]
8. Guzzetti, F.; Carrara, A.; Cardinali, M.; Reichenbach, P. Landslide hazard evaluation: A review of current techniques and their application in a multi-scale study, Central Italy. *Geomorphology* **1999**, *31*, 181–216. [CrossRef]
9. Tsolaki-Fiaka, S.; Bathrellos, G.D.; Skilodimou, H.D. Multi-criteria decision analysis for abandoned quarry restoration in Evros Region (NE Greece). *Land* **2018**, *7*, 43. [CrossRef]
10. Skilodimou, H.D.; Bathrellos, G.D.; Chousianitis, K.; Youssef, A.M.; Pradhan, B. Multi-hazard assessment modeling via multi-criteria analysis and GIS: A case study. *Environ. Earth Sci.* **2019**, *78*, 47. [CrossRef]
11. Fedeski, M.; Gwilliam, J. Urban sustainability in the presence of flood and geological hazards: The development of a GIS-based vulnerability and risk assessment methodology. *Landsc. Urban Plan.* **2007**, *83*, 50–61. [CrossRef]
12. Thapa, R.B.; Murayama, Y. Drivers of urban growth in the Kathmandu Valley, Nepal: Examining the efficacy of the analytic hierarchy process. *Appl. Geogr.* **2010**, *30*, 70–83. [CrossRef]
13. Bathrellos, G.D.; Kalivas, D.P.; Skilodimou, H.D. Landslide susceptibility mapping models, applied to natural and urban planning, using G.I.S. *Estud. Geol.-Madrid* **2009**, *65*, 49–65. [CrossRef]
14. Rozos, D.; Bathrellos, G.D.; Skilodimou, H.D. Comparison of the implementation of Rock Engineering System (RES) and Analytic Hierarchy Process (AHP) methods, based on landslide susceptibility maps, compiled in GIS environment. A case study from the Eastern Achaia County of Peloponnesus, Greece. *Environ. Earth Sci.* **2011**, *63*, 49–63. [CrossRef]
15. Bathrellos, G.D.; Gaki-Papanastassiou, K.; Skilodimou, H.D.; Skianis, G.A.; Chousianitis, K.G. Assessment of rural community and agricultural development using geomorphological-geological factors and GIS in the Trikala prefecture (Central Greece). *Stoch. Environ. Res. Risk Assess.* **2013**, *27*, 573–588. [CrossRef]
16. Papadopoulou-Vrynioti, K.; Bathrellos, G.D.; Skilodimou, H.D.; Kaviris, G.; Makropoulos, K. Karst collapse susceptibility mapping considering peak ground acceleration in a rapidly growing urban area. *Eng. Geol.* **2013**, *158*, 77–88. [CrossRef]
17. Bathrellos, G.D.; Kalivas, D.P.; Skilodimou, H.D. Landslide Susceptibility Assessment Mapping: A Case Study in Central Greece. In *Remote Sensing of Hydrometeorological Hazards*; Petropoulos, G.P., Islam, T., Eds.; CRC Press, Taylor & Francis Group: London, UK, 2017; pp. 493–512, ISBN-13: 978-1498777582.

18. Saunders, W.S.; Kilvington, M. Innovative land use planning for natural hazard risk reduction: A consequence-driven approach from New Zealand. *Int. J. Disaster Risk Reduct.* **2016**, *18*, 244–255. [CrossRef]

19. Bathrellos, G.; Skilodimou, H. Geomorphic Hazards and Disasters. *Bull. Geol. Soc. Greece* **2006**, *39*, 96–103. (In Greek)

20. Goudie, S.A. *Encyclopedia of Geomorphology*; Goudie, S.A., Ed.; Taylor & Francis Group: New York, NY, USA, 2006; pp. 378–379.

21. Bathrellos, G.D. An overview in Urban Geology and Urban Geomorphology. *Bull. Geol. Soc. Greece* **2007**, *40*, 1354–1364. [CrossRef]

22. Bathrellos, G.D.; Karymbalis, E.; Skilodimou, H.D.; Gaki-Papanastassiou, K.; Baltas, E.A. Urban flood hazard assessment in the basin of Athens Metropolitan city, Greece. *Environ. Earth Sci.* **2016**, *75*, 319. [CrossRef]

23. Bathrellos, G.D.; Gaki-Papanastassiou, K.; Skilodimou, H.D.; Papanastassiou, D.; Chousianitis, K.G. Potential suitability for urban planning and industry development by using natural hazard maps and geological-geomorphological parameters. *Environ. Earth Sci.* **2012**, *66*, 537–548. [CrossRef]

24. Briassoulis, H. Combating Land Degradation and Desertification: The Land-Use Planning Quandary. *Land* **2019**, *8*, 27. [CrossRef]

25. Briassoulis, H. Desertification. In *Science and Policy. An A to Z Guide to Issues and Controversies*; Steel, B., Ed.; CQ Press: Thousand Oaks, CA, USA, 2014; pp. 138–145.

26. Tragaki, A.; Gallousi, C.; Karymbalis, E. Coastal hazard vulnerability assessment based on geomorphic, oceanographic and demographic parameters: The case of the Peloponnese (Southern Greece). *Land* **2018**, *7*, 56. [CrossRef]

27. Karymbalis, E.; Chalkias, C.; Chalkias, G.; Grigoropoulou, E.; Manthos, G.; Ferentinou, M. Assessment of the sensitivity of the southern coast of the Gulf of Corinth (Peloponnese, Greece) to sea-level rise. *Cent. Eur. J. Geosci.* **2012**, *4*, 561–577. [CrossRef]

28. Rijal, S.; Rimal, B.; Sloan, S. Flood hazard mapping of a rapidly urbanizing city in the foothills (Birendranagar, Surkhet) of Nepal. *Land* **2018**, *7*, 60. [CrossRef]

29. Rimal, B.; Zhang, L.; Keshtkar, H.; Sun, X.; Rijal, S. Quantifying the Spatiotemporal Pattern of Urban Expansion and Hazard and Risk Area Identification in the Kaski District of Nepal. *Land* **2018**, *7*, 37. [CrossRef]

30. Bathrellos, G.D.; Skilodimou, H.D.; Soukis, K.; Koskeridou, E. Temporal and spatial analysis of flood occurrences in drainage basin of Pinios River (Thessaly, central Greece). *Land* **2018**, *7*, 106. [CrossRef]

31. Skilodimou, H.D.; Bathrellos, G.D.; Koskeridou, E.; Soukis, K.; Rozos, D. Physical and anthropogenic factors related to landslide activity in the Northern Peloponnese, Greece. *Land* **2018**, *7*, 85. [CrossRef]

 *land*

Article

Combating Land Degradation and Desertification: The Land-Use Planning Quandary

Helen Briassoulis

Department of Geography, University of the Aegean, 81100 Lesvos, Greece; e.briassouli@aegean.gr;
Tel.: +30-225-103-6411

Received: 15 December 2018; Accepted: 28 January 2019; Published: 1 February 2019

Abstract: Land-use planning (LUP), an instrument of land governance, is often employed to protect land and humans against natural and human-induced hazards, strengthen the resilience of land systems, and secure their sustainability. The United Nations Convention to Combat Desertification (UNCCD) underlines the critical role of appropriate local action to address the global threat of land degradation and desertification (LDD) and calls for the use of local and regional LUP to combat LDD and achieve land degradation neutrality. The paper explores the challenges of putting this call into practice. After presenting desertification and the pertinent institutional context, the paper examines whether and how LDD concerns enter the stages of the LUP process and the issues arising at each stage. LDD problem complexity, the prevailing mode of governance, and the planning style endorsed, combined with LDD awareness, knowledge and perception, value priorities, geographic particularities and historical circumstances, underlie the main challenges confronting LUP; namely, adequate representation of LDD at each stage of LUP, conflict resolution between LDD-related and development goals, need for cooperation, collaboration and coordination of numerous and diverse actors, sectors, institutions and policy domains from multiple spatial/organizational levels and uncertainty regarding present and future environmental and socio-economic change. In order to realize the integrative potential of LUP and foster its effectiveness in combating LDD at the local and regional levels, the provision of an enabling, higher-level institutional environment should be prioritized to support phronetic-strategic integrated LUP at lower levels, which future research should explore theoretically, methodologically and empirically.

Keywords: integrated land-use planning; land degradation; desertification; policy; phronetic approach

1. Introduction

Land[1] mediates all interactions between the natural environment, society and the economy [2,3]. Land resources provide ecosystem services but also pose constraints on human activity which, if violated, generate important unwanted environmental and socio-economic consequences. The alarming pace at which land resources are degrading in recent times has been recognized at the international and subglobal levels [4–7]. Sustainable Development Goal (SDG) 15, one of the 17 SDGs decided at the Rio+20 conference in 2012, is specifically geared to "protect, restore and promote sustainable use of terrestrial ecosystems, sustainably manage forests, combat desertification, and halt and reverse land degradation and halt biodiversity loss" [8].

---

[1] "Land means the terrestrial bio-productive system that comprises soil, vegetation, other biota, and the ecological and hydrological processes that operate within the system" [1].

The wise governance of land resources in coupled human-environment systems, or land systems[2], is key to strengthening their resilience against natural hazards and numerous other environmental, technological and socio-economic disturbances [10–13]. 'Land systems science' is a term coined to denote the contemporary, interdisciplinary scientific domain concerned with the theories, approaches, analytical tools and instruments related to the analysis and resolution of land-use problems [12,14]. Land-use planning (LUP) is an instrument of land governance that has been used since ancient times to protect land and humans against natural hazards and to address pertinent land use-related issues in order to secure the sustainability of land systems [15–19].

Desertification, an extreme form of natural and human-induced land degradation, and drought, a natural hazard, threaten the smooth functioning of land systems [7]. Desertification differs from other natural hazards, such as fires, floods and earthquakes, because it is a wide-net, higher level, multi-scalar and long-term phenomenon; at the local level, it is mostly experienced as (severe) land degradation. Land degradation and desertification (LDD) is the term commonly used in official (e.g., UN, European Union) and scientific quarters, and which is adopted in this paper. The causes and consequences of LDD concern several interdependent human activities, directly implicate more than one land resources (soil, water and vegetation) and involve diverse economic sectors, social groups and institutions, spanning the local to global spectrum. The incidence of LDD exposes land resources and human populations to multiple threats: loss of land productivity, food insecurity, water shortages and scarcity, economic hardship, social deprivation and health risks [20–25].

The signing of the United Nations Convention to Combat Desertification (UNCCD), one of the three Rio Multilateral Environmental Agreements (or, Conventions)[3], in 1994[4] underlines the global significance of the phenomenon as well as the critical role of appropriate local action. Since 2012, the UNCCD has embraced SDG 15 and, more specifically, Target 15.3 that sets the ambitious goal to achieve land degradation neutrality (LDN) by 2030. Among the several actions for its effective implementation, the UNCCD and other international and sub-global organizations incite the signatory parties to use local and regional land use planning to help combat desertification and mitigate the negative effects of drought in affected areas [26] and, more recently, to address the LDN target[5] [27]. Putting this straightforward call into practice, however, presents considerable challenges, several of which have been noted since the early days of the UNCCD [28].

In complex land systems, a multitude of actors interacting on and across scales continuously place diverse demands on a multitude of interconnected resources to satisfy various, often conflicting, goals, environmental protection being one of them, that differ in priority among groups [13,29]. Conflicts arise over the allocation of land among competing uses which accrue short- and long-term costs and benefits to individuals and groups. If human activities locate in resource-poor (e.g., water), hazard-prone and other high-risk areas, biophysical constraints imply high costs of protection or, if they are ignored, significant environmental and socio-economic costs result. LUP aims to arbitrate and resolve these conflicts and issues to secure the sustainability of local and regional development [30]. In LDD-prone areas, in particular, the LDN goal makes LUP an inevitable instrument in the fight against LDD. This is a demanding undertaking because LUP is called to harmonize the LDN with numerous other goals and it is, furthermore, complicated by the ever-present uncertainty regarding future human needs, goals and priorities, environmental conditions, socio-economic and technological

---

2   "Land systems constitute complex, adaptive social-ecological systems (Berkes et al., 1998) shaped by interactions between (i) the different actors and demands that act upon land, (ii) the technologies, institutions, and cultural practices through which societies shape land use, and (iii) feedbacks between land use and environmental dynamics (Millennium Ecosystem Assessment (MA), 2003; Verburg et al., 2015)." [9], (p. 53).

3   The other two are the 'sister' Conventions of the UNCCD, the UNCBD (United Nations Convention for Biodiversity Conservation) and the UNFCCC (United Nations Framework Convention for Climate Change).

4   The UNCCD came into force in December 1996.

5   "Furthermore, through Decision 2/COP.12, the UNCCD endorsed the formulation, revision and implementation of action programmes in view of the 2030 Agenda for Sustainable Development, (United Nations General Assembly, 2015) encouraging the linkage between planning and the implementation of LDN" [27], (p. 76).

change, unpredictable events and their changing constellations. Harmonization points to the need to apply *phronesis* (practical wisdom) [31] in making land-use decisions to safeguard the potential of affected areas to successfully adapt to changing conditions and, thus, secure their resilience and enhance their sustainable development prospects.

This paper delves into this land-use planning quandary aiming to show that LUP is not a straightforward but a complex endeavor, reveal the LUP challenges facing the fight against LDD and suggest avenues to handle them to foster the effectiveness of LUP efforts. The discussion is general applying to most (democratic) socio-political contexts although geographic particularities and historical circumstances determine the actual form the issues and challenges obtain. The second section briefly presents the main features of desertification and the institutional context to combat it. The third section introduces land-use change and land-use planning and explores the issues and challenges arising at each stage of the LUP process in the context of combating LDD at the local and regional levels. The concluding section suggests necessary priority actions to realize the integrative potential of LUP and, thus, improve its effectiveness in combating LDD that indicate future research directions.

2. Desertification and the Institutional Context to Combat Desertification

2.1. Desertification

Desertification has received and, with the escalation of global warming, is receiving significant political support at the international and subglobal levels [4,6,7,22]. However, it remains a politically contentious issue; the existence of more 100 definitions is telling [32]. The UNCCD definition, which is mostly used by now, states that desertification is "land degradation in arid, semiarid and subhumid tropics caused by a combination of climatic factors and human activities" [1]. Land degradation means reduction or loss of the biological or economic productivity and complexity of rainfed cropland, irrigated cropland, or range, pasture, forest and woodlands resulting from land uses or from a process or combination of processes, including those arising from human activities and habitation patterns, such as: (a) soil erosion caused by wind and/or water, (b) deterioration of the physical, chemical, biological and economic properties of soils and (c) long-term loss of natural vegetation [1,23].

This definition makes clear that (a) desertification is land degradation in the drylands[6], i.e., in areas with adverse biophysical conditions, (b) it leads to an extreme, often irreversible, state of degradation implying reduction or loss of both biological *and* economic productivity *and* complexity of land, (c) the natural resources concerned are climate, soil, water and vegetation, (d) it involves both natural and human-induced processes operating at multiple spatial and temporal scales, and (e) a variety of human activities and users of land are implicated.

Biophysical and human *driving forces* (indirect drivers), from the local to the global levels, underlie *the proximate causes* (direct drivers) of LDD; i.e., human activities, such as agriculture, animal husbandry, forestry, housing, tourism, transport, extraction and energy production. The associated land-using and potentially resource-degrading practices include intensive cultivation, monocultures, abandonment of traditional practices (e.g., terracing), poor or no maintenance of rural holdings, overgrazing, deforestation, forest fires, water overdrafts, extraction, drainage of wetlands, and large infrastructure works [4,5,22–24,33–36]. Their combined action modifies land resources and produces land use and land cover change which, under adverse biophysical conditions, set in motion processes of LDD.

The biophysical drivers of desertification include climate, geology, soil conditions, hydrology, topography and vegetative cover. Selected important characteristics of these drivers are: low and uneven annual and interannual rainfall distribution, extreme weather events and out-of-phase rainy and vegetative seasons; soil depth, structure and stability, organic content, stoniness of land,

---

[6]   i.e., "areas, other than polar and sub-polar regions, in which the ratio of annual precipitation to potential evapotranspiration falls within the range from 0.05 to 0.65" [1].

soil–water balance; slope gradient and slope aspect; surface and ground water availability; and biomass productivity[7] [37]. Slow and fast physical and/or chemical processes are involved in LDD. The former include soil erosion, compaction, salinization, alkalization and nitrification. The latter include drought and extreme weather events [25,37,38].

The diverse and interdependent human (socio-economic, cultural, political and institutional) drivers of LDD play a dual role; they either underlie the incidence of LDD or contribute to its mitigation by changing the valuation, modes of utilization and management of land resources. Important among them are: population structure and dynamics (mobility and migration), poverty and social inequality; changes in technology, modes of production, social values, consumption patterns, life styles, family structure, employment composition, market and/or public policy-induced agricultural product prices, capital availability and competition among economic activities [20,25,32,39].

The institutional drivers of desertification are particularly important. They encompass international economic and environmental regimes (e.g., trade, climate change and biodiversity) as well as supranational policies, such as the European Union (EU) Common Agricultural Policy (CAP), transport policy, the Structural Funds (SFs) and their national level counterparts [40]. Several national policies negatively affect bioclimatically sensitive regions, setting the stage for their degradation. Important national level concerns include the mode of governance, which depends on the prevailing political regime, inappropriate or inexistent environmental and spatial planning legislation, problematic plan and policy implementation, unclear, uncoordinated or inexistent systems of resource rights for critical resources, such as water and soil, administrative compartmentalization and lack of coordination [33]. At the local level, land tenure and ownership constitute critical institutional influences. Rural land rental, combined with absentee ownership, land fragmentation, and vague and incompatible resource regimes often give rise to inappropriate land management and degradation, impeding the proper implementation of formal policies [25,37].

Geographic location, accessibility and the spatial distribution of economic activities, uses of land, population and infrastructure determine the particular nonlinear interactions among biophysical and human drivers and underlie the processes, such as agricultural intensification, urbanization, industrialization, etc., that judge the incidence and magnitude of LDD in a region. The urban–rural dynamics, in particular, greatly affects the long-term prospects of the phenomenon as it consolidates the complex, multi-scale influences and pressures on land resources. Lastly, biophysical and human macro-forces and events operating within a period, e.g., wars, famines, natural disasters, new technologies, price shocks and resource crises have important off-site effects on the incidence of LDD [25].

In contrast to other natural hazards, the biophysical, socio-economic and other impacts of desertification may range from localized and short-term to large-scale and long-term, owing to the diverse biophysical and human drivers that act and interact non-linearly at different speeds (fast and slow processes) on multiple spatial and temporal scales. This is the most important source of uncertainty concerning LDD with important implications for its definition, identification, assessment and choice of proper measures to combat it.

2.2. The Institutional Context to Combat Desertification

The institutional activity pertinent to LDD and drought spans the global to local spectrum. At the international level, the UNCCD is the principal direct institutional regime that provides the broad frame of actions to combat desertification. It comprises five Annexes that correspond to five groupings of world regions and related region-specific desertification problems. Annex IV (Northern Mediterranean) includes 16 signatories, of which 10 are EU member states. The organizational apparatus of the UNCCD comprises the Conference of the Parties (COP), a permanent secretariat and a Committee on Science and Technology (CST). Their mandate includes support for the elaboration,

---

[7]   Land is considered desertified when biomass productivity drops below a certain threshold value.

implementation and co-ordination of various instruments; co-ordination of the UNCCD with its sister conventions, the United Nations Convention for Biodiversity Conservation (UNCBD) and the 40United Nations Framework Convention on Climate Change (UNFCCC); research and development, technology transfer, acquisition and adaptation; capacity building, education, awareness raising and provision of financial resources and mechanisms to facilitate implementation[8]. The UNCCD Secretariat provides regular monitoring and reporting on the implementation of proposed actions [33,37].

The UNCCD, acknowledging the complexity of desertification and recognizing the vital role of local level responses in sustainable resource management, offers guiding principles for an integrated 'top-down' and 'bottom-up' approach to designing interventions in affected areas and encourages multi-level communication and collaboration [33]. It prescribes general and specific obligations of the signatory parties, underlining the institutional conditions required to facilitate effective implementation. The most important obligation is the preparation of Regional Action Programmes (RAPs) and National Action Programmes (NAP), following a UNCCD-defined template [1]. These programmes should be linked to national sustainable development programmes based on participatory processes with input from the field and the scientific community. The affected countries are urged to strengthen extant and enact new, including spatial and land-use planning, legislation.

Given the implementation problems recorded over time, COP8 [41] approved a 10-year strategy to improve UNCCD implementation and secure adequate financing that prioritizes the integration of desertification concerns into development planning and policies. The most important development since that time is perhaps the agreement of the parties to endorse Target 15.3 of SDG 15 that states "By 2030, combat desertification, restore degraded land and soil, including land affected by desertification, drought and floods, and strive to achieve a land degradation-neutral world" [3][9]. Land Degradation Neutrality (LDN) was defined as "A state whereby the amount and quality of land resources, necessary to support ecosystem functions and services and enhance food security, remains stable or increases within specified temporal and spatial scales and ecosystems" [27], (p. 33).

LDN is considered a hybrid lay-scientific concept that "aims to maintain and increase the amount of healthy and productive land resources, *in line with national development priorities*" (emphasis added) [42], (p. 2). As such, it represents "a flexible target that can be implemented at local, regional or national scales. *It recognizes the sovereignty of nations to manage the trade-offs* and to capitalize on the synergies between biological and economic productivity (emphasis added) [42], (p. 2). It has been hailed as "a paradigm shift in land management policies and practices … a unique approach that counterbalances the expected loss of productive land with the recovery of degraded areas. *It squarely places the measures to conserve, sustainably manage and restore land in the context of land use planning*" (emphasis added) [43]. Practically, it could be achieved by: "(a) managing land more sustainably, which would reduce the rate of degradation; and (b) increasing the rate of restoration of degraded land, so that the two trends converge to give a zero net rate of land degradation" [44], (p. 12).

It is widely recognized that the implementation of this ambitious target requires multi-stakeholder engagement, cross-scale and intersectoral planning and strong national-scale coordination that should serve to manage and streamline diverse local, regional and sectoral governance structures[10]. If successful, LDN might reinforce the implementation of the convention and contribute to the achievement of other goals such as climate change mitigation and adaptation, biodiversity conservation, ecosystem restoration, food and water security, disaster risk reduction, and poverty eradication. In other words, it may serve as a vehicle to affect the coordination and integration of the three

---

⁸ Lacking autonomous financing, the UNCCD, under the guidance of the COP, employs the Global Mechanism as a brokering body to facilitate the effective and efficient channeling of financial resources and mechanisms to affected countries. COP6 designated the Global Environment Facility as a UNCCD financial mechanism. Developing synergies with the Conventions on Biological Diversity (CBD) and Climate Change (UNFCCC) is expected to improve its financing prospects [41].

⁹ Available online: https://www.unccd.int/actions/ldn-target-setting-programme; (accessed on 28 September 2018).

¹⁰ See, [45] for a discussion of open challenges.

conventions (UNCCD, UNCBD and UNFCCCC) that variously materialize, in one way or another, in the process of using land.

The European Union has supported the fight against desertification through research funding (Framework Programmes since 1989), specific projects (e.g., INTERREG, LIFE), research at the Joint Research Centre (Ispra, Italy), technical and information support provided by the European Environment Agency, and specific measures included in the CAP (agri-environmental measures) and the Structural Funds [20]. The Thematic Strategy for Soil Protection [46] remains the only direct institutional response to the issue as efforts to institute an EU Soil Framework directive have failed, the main reason being that planning the uses of land is considered a matter of national sovereignty and not of EU competence[11]. Indirectly, provisions included in horizontal and sectoral policies, such as the EIA, SEA, Habitats and Water Framework directives, provide instruments to address LDD at national and subnational levels.

At the national level, direct desertification policies do not exist besides the NAPs. The co-ordination of their implementation has been assigned to ministries of agriculture or the environment. Most NAPs emphasize measures targeting the proximate causes, such as agriculture, forestry and animal husbandry, and not the driving forces of LDD [20]. Spatial planning activities are supposed to observe the provisions of the NAPs in affected areas. How close this requirement is being followed in practice remains an open question. Indirectly, several pieces of national horizontal environmental and sectoral legislation and policies[12] concerning soil protection, afforestation, fire protection, water resources conservation, nature conservation and other issues may contribute to mitigating the longer term occurrence of LDD. National development frameworks often prescribe the integration of pertinent measures in land use planning strategies [30].

Summarizing, LDD is a process during which extreme, irreversible states of land degradation may emerge at a higher level in the long run. Desertification constitutes an aggregate, macro-feature of the state of the land. It involves complex, non-linear, context-, scale- and path-dependent interactions between natural resources and human activities driven by numerous, multi-scalar, interacting biophysical and human driving forces, among which nature–society institutions (international environmental regimes, policies, customary land management regimes, etc.) play a pivotal role. Considerable controversy still surrounds its definition and causality [25,47,48], making it difficult to disentangle the biophysical from the anthropogenic causes, to accurately assess the land affected and/or at risk, and to predict its consequences and reversibility. Hence, it cannot be stated with certainty whether and when an area will be 'locked' in an irreversibly desertified state.

The complexity of LDD and the associated scientific uncertainty carry over to and combine with a similarly complex world of practice where numerous individual and collective actors as well as formal and informal institutions are implicated on and across multiple spatial and temporal scales in making land use decisions. Land-use planning, functioning within this milieu, is called to assist in coping with a contentious, wicked socio-ecological problem, LDD.

3. Land-Use Planning to Combat Land Degradation and Desertification: Uncovering the Complexities

3.1. Land-Use Change and Land-Use Planning

Land-use patterns and their evolution over time, i.e., land-use change, result from the constant interaction between demand and supply factors from the local to the global level, which are mediated and regulated by human decision making. The demand factors are distinguished into driving forces

---

[11] After eight years of deliberations, the proposal for an EU Soil directive was withdrawn in April 2014 because five countries (UK, Germany, Austria, France, the Netherlands) blocked the process.
[12] In EU member states, in particular, the transposition of community regulations and directives to the national legislation has provided a considerable range of measures toward this purpose.

and proximate causes. The driving forces (indirect drivers) underlie the generation of demand for various economic activities and, consequently, the demand for land of particular characteristics. Higher level factors, such as population, affluence, technology, socio-economic organization, culture, international and subglobal institutions, political systems and their change, combine with lower level factors, such as demographic and socio-cultural traits, activity-specific costs, benefits and profits, value systems, formal and customary resource regimes and rights, and resource use practices, to shape the choice, extent and intensity of actual land use [45]. Certain driving forces, such as international agreements, subglobal and national policies, and resource-conserving practices, often act as mitigating forces that moderate the unwanted impacts of land-use change. Finally, the proximate causes (direct drivers) of land-use change include various human activities and their locational requirements, preferences and impacts [45].

The supply factors encompass the biophysical and anthropogenic resources of an area, also referred to as 'capitals'—natural, physical, landesque, technological, human, social, economic, institutional, political [49–51]. The biophysical resources include climate and weather, geology and geomorphology (topography), soils, water, vegetation, fauna and flora. The anthropogenic resources comprise past and present manmade structures (buildings, settlements, archaeological sites, etc.), social infrastructure (schools, hospitals, etc.), physical-technical infrastructure (transport, communication, energy and other networks), human and social capital, institutions, land tenure, land ownership, culture and value systems. The interplay of the features, spatial distribution and dynamics of these factors influences the suitability of land resources for particular activities, their sensitivity to various uses, their accessibility and their activity-specific economic and socio-cultural value that, eventually, determine the relative priority of land for various uses. Land-use conflicts commonly arise as several activities may have similar locational preferences but their co-existence is either impossible or results in unwanted environmental and socio-economic impacts. The actual land-use pattern is determined by past land use trajectories and historic contingencies that favor the domination of one or more uses of land over other competing uses in particular areas and contexts.

Land-use change, either autonomous or planned, takes two forms; *modification* of an existing land use type (e.g., change between crops, change in forest use) or *conversion* from one land use type to another (e.g., from agricultural to urban or industrial land). Land-use planning is the main instrument of planned land-use change. It is the purposeful, anticipatory activity of intervening in an existing, always fluid and dynamic, state of affairs with the aim to modify the land-use pattern in order to achieve certain goals and reduce uncertainty about the future impacts and consequences of human activities [19,52,53]. Given differences in the suitability of land resources for particular activities, the presence of natural constraints on activity location[13] and the competition among activities for the same resources, LUP broadly aims to (a) guide the choice of activities and land uses that can make the best (socio-economically efficient) use of land resources, i.e., cause the least environmental and socio-economic harm, avoid waste and generate the highest present and future environmental and socio-economic benefits, (b) resolve land-use conflicts, thus, alleviating current negative impacts and achieving positive environmental and socio-economic outcomes and (c) distribute equitably the costs and benefits of proposed interventions among the parties involved [3,52,54–57]. Essentially, LUP strives to match the demand for land resources by human activities with their supply to optimize general and place-specific goals and the overarching goal of sustainable local and regional development.

The numerous definitions encountered in the literature over time concur on the quintessence of LUP: (a) it is a process during which the actors involved set goals related to a problem (or, problems), develop alternative courses of action, choose available or devise new means to implement a chosen alternative to achieve the goals, evaluate the outcomes and make necessary modifications during

---

[13] e.g., geologic faults, seismic activity, unstable slopes, floodplains, habitats of endangered species, desertification-sensitive areas, erosion-prone areas, etc.

implementation; and (b) it is not a one-off operation that produces blueprints but a continuous, iterative, future-oriented decision-making process that comprises certain basic stages, serves certain key functions and is practiced following different styles [58,59].

The LUP process comprises certain basic stages: (a) problem definition and goal setting; (b) problem description and analysis; specification of planning objectives; (c) alternative plan formulation; (d) plan evaluation and choice of the preferred alternative; (e) plan implementation; (f) plan monitoring, evaluation and modification [52,54,60]. Figure 1 schematically presents this process and briefly describes the stages. These stages are not discrete and clearly delineated, they do not necessarily follow a linear succession in practice and they may overlap as some planning activities take place contemporaneously (e.g., problem definition and description, plan formulation and evaluation). Their boundaries are blurred, continuous feedbacks occur from one to another as demand and supply-related environmental and socio-economic conditions change, new actors get involved, new information is gathered, thus, continuously modifying the definition of the planning problem and the subsequent stages.

Land-use planning, like planning in general, is both a technical and a political process [61,62] that serves three functions: an *intelligence function* (data collection and analysis to build the information base concerning the demand and supply side of the planning problem), *advance plan-making* and *action/implementation* [52]. Depending on how LUP and the elements of the process are conceived, and on whether a proactive or a reactive decision making culture prevails, different planning styles (or, modes) are followed in practice. These include the classical (and by now mostly defunct) comprehensive or synoptic planning, incremental, transactive, advocacy, mixed scanning, contingency, adaptive and participatory planning [56,61,63,64]. The last three styles are particularly important in the case of natural hazards in general and LDD in particular.

In contemporary democratic, but increasingly complex societies, desirable features of LUP necessary to enhance its performance and effectiveness are strategic orientation, coordination and integration with development planning, wide participation (bottom up decision making), flexibility and adaptation [58,65,66]. These require well developed social and institutional capital and, above all, political will to devise and implement scientifically sound and socially acceptable solutions to socio-ecological problems that are characterized by considerable uncertainty [56]. In this context, the coordinating role of the advance plan-making function is critical in bridging the technical (intelligence) with the political (action/implementation) functions of LUP, thus, underlining the role of both science and politics in judging the outcomes of the process.

On the technical front, present-day LUP employs a variety of approaches, such as the landscape, the ecosystem and the multifunctional land-use approach [67–70], quantitative and qualitative analytical methods and techniques originating in the natural and the social sciences, such as land suitability analysis, environmental impact and risk analysis, needs assessment, survey research, environmental and social impact analysis, scenario analysis, integrated environmental-economic modeling, socio-ecological systems analysis [64,71–73], and planning support systems as well as traditional and contemporary data collection techniques (e.g., Earth observation systems, censuses, surveys).

On the political front, LUP combines a broad variety of technical/technological, physical, economic, financial, social, institutional and educational means and instruments, originating in environmental and socio-economic policy sectors and in various spatial/organizational levels, to produce and support the implementation of land use plans. Formal (institutionalized) and informal negotiation, mediation, bargaining and conflict resolution mechanisms and procedures are utilized toward this purpose [74,75].

The effectiveness of land-use planning depends critically on the availability of all types of resources (or, capitals) in the right combinations, in the right place and at the right time to successfully implement the land-use plan and meet the LUP goals and objectives. These requirements are rarely met in practice due to several reasons as discussed in the context of combating LDD below.

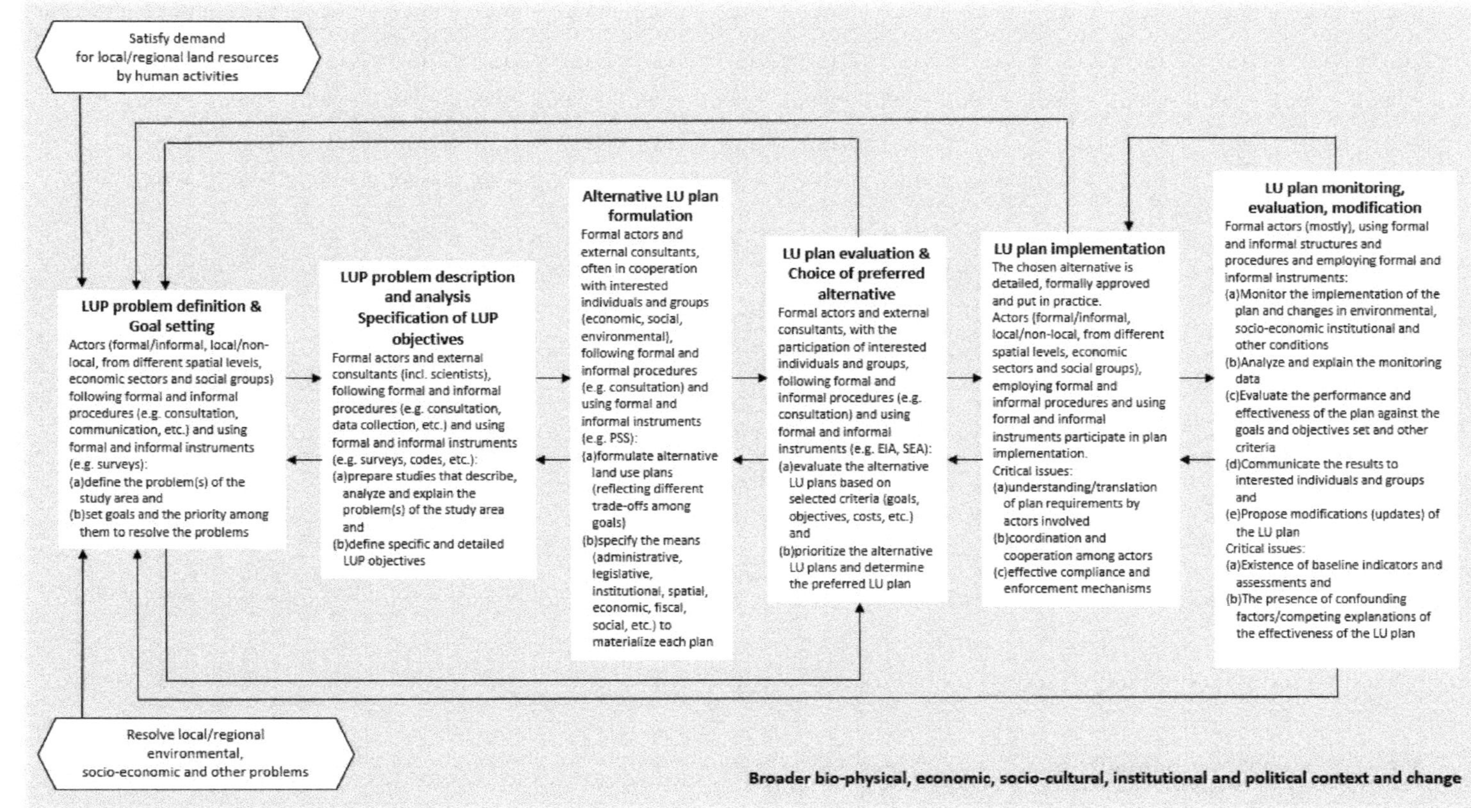

Figure 1. Schematic presentation of the land-use planning (LUP) process and description of the stages.

3.2. Land-Use Planning to Combat Land Degradation and Desertification

In affected areas, the general land-use planning problem is to determine those uses of land that will protect their land resources against LDD and will secure their socio-economic vitality (economic welfare, quality of life and social equity)—i.e., promote their sustainable development. The LDD-specific goal is to stop, reverse or moderate the degradation of land resources by properly selecting, siting and monitoring human activities. The land degradation neutrality (LDN) target gives more concrete, although not necessarily precise, direction: (a) reduce the rate of degradation by managing land more sustainably; and (b) increase the rate of restoration of degraded land, so that the two eventually produce zero net land degradation.

In order to assess the prospects of land-use planning successfully meeting the overarching goal of combating LDD and mitigating the impacts of drought at the local and the regional level, this section undertakes a reality check of the ideal LUP process; more specifically, it inquires whether and how LDD concerns enter, or may enter, the stages of the process. For each stage, the interplay of demand and supply factors is examined with regard to the actors involved, the organizational apparatus (decision structures and procedures) available and the instruments used. The use of stages is made for the purposes of analysis only because, as mentioned before, they intermingle and do not follow a given order in practice.

3.3. Problem Definition and Goal Setting

This stage concerns the question of *who* defines *what* problem and sets the LUP goals for the study area. It is the most critical stage of the LUP process and influences all subsequent stages. It takes place both formally and informally depending on the institutional status of planning and the administrative organization and culture that range from formal/centralized to informal/decentralized. A variety of mixed real world situations are encompassed in between. It is both demand- and supply-driven as the need for LUP in an area may arise from external or internal demand for its resources by economic activities and/or from the need to resolve environmental, socio-economic and other problems.

Various kinds of actors are implicated in LUP problem definition and goal setting. Some of them represent the demand and some other the supply side of the problem. Who and how can legitimately participate is determined by state laws and the associated procedures and instruments (e.g., consultation, committee membership).

Formal public actors (individuals and agencies) from various spatial/organizational levels represent spatial planning, regional and rural development, economic (agriculture, forestry, industry, extraction, tourism, energy production) and environmental sectors (water, soil, forests, nature protection, etc.). The latter potentially stand for LDD concerns. Planning consultants are also formally involved as states usually commission the preparation of LUP studies to specialized firms.

Other individual and collective actors may participate, as in decentralized systems following inclusionary, participatory processes, or they variously seek to influence formal actors and shift the LUP problem definition and the goals set to their advantage. These include economic interests placing demands on land [59], related intermediaries (e.g., construction and real estate sectors), social and environmental groups (e.g., NGOs, civic groups, etc.). In recent times, actors are increasingly non-locals (e.g., multinational firms, the EU, the World Bank) as in the case of tourism and second home development, infrastructure projects, energy production, industrial agriculture and forestry, etc. LDD concerns may be represented depending on the mandate and interests of these actors. Scientists may participate formally or they indirectly influence problem definition by disseminating scientific evidence concerning the magnitude and severity of LDD (studies, maps, indicators, etc.) among other problems.

Informal actors always co-exist with formal actors, act autonomously and in parallel with them, placing demands on land resources, making and carrying out their 'plans' outside the official apparatus in affected areas. They often create *de facto* land uses (e.g., informal agriculture, housing) and produce environmental, LDD-related problems (e.g., deforestation, fires, overgrazing, water overdrafting), that formal LUP is called to handle. Informal residential and tourism development in coastal LDD-sensitive areas in the Mediterranean and elsewhere and illegal deforestation in the Amazon are cases in point.

The effective legislation mandates the competences of and linkages among formal actors. The diverse linkages developing between formal and informal actors as well as among informal actors take two basic forms; coordination/cooperation or conflict. The perception and knowledge of the problems of the area, the vision regarding its future, the congruence of aspirations and the common interests among actors as well as the presence of externalities of the respective activities judge whether cooperation or conflict will result. In the case of conflicts, both formal and informal, customary conflict resolution procedures and prevailing power balances determine the outcome. Obviously, communication, common understanding and coordination among actors are necessary to produce a coherent LUP problem definition that encompasses LDD concerns. However, scientific uncertainty regarding LDD and its causes, together with lack of hard local and regional evidence, leaves ample room for multiple interpretations and adaptation to local circumstances.

The spatial, sectoral and administrative diversity and plurality of formal and informal actors, possessing differing mandates and knowledge of LDD, the compartmentalization of administrative structures even for the same resource (e.g., land, water, etc.) and turf politics [37] often result in vague and incoherent definitions that show up in goal setting. Formal LUP goal setting follows either in a top-down process, in the case of centralized systems and hierarchical modes of governance, or some form of participatory process in more decentralized systems. General national goals, often incorporating international obligations (e.g., climate change), include sustainable development, economic development, resource protection and conservation, as well as specific goals such as agricultural specialization, tourism development and energy production. Strategic goals concerning the protection of critical resources (water, soil, air, etc.), food and energy security, and others may be also specified. LDD-specific goals are included in the case of duly constituted NAPs. Agreement over general goals is usually easy to achieve especially if all actors share a common development vision for the area. Conflicts arise later when the more detailed planning objectives formulated reveal the critical trade-offs.

The priorities set among goals reveal the influence of dominant actors. The decisions regarding the uses of land are made on the basis of economic considerations mainly (demand side), often by nonlocal interests, and rarely on the basis of environmental considerations (supply side) except for symbolic purposes and extreme cases of environmental degradation, including LDD. Therefore, the protection of land resources is one of several socio-economic and environmental goals of LUP in affected areas, which depend on the specifics of each individual case but are, most likely, intricately related to LDD. The perception and knowledge of LDD which the formal and informal actors hold, that depend on scale, sector and social group, and the prevailing proactive or reactive decision culture critically determine the sense of urgency and the priority it is given compared to other problems. The possibilities range from total ignorance (theoretically possible but practically rather unlikely) to complete knowledge and concern for its impacts. At the local level, LDD is perceived as land degradation, not as desertification, and it is of direct concern to agricultural interests and environmental groups. LDD-related goals may be congruent and synergistic with certain goals, such as environmental protection, resource conservation and food security, or incompatible and antagonistic with some others, such as large-scale industrial, energy and tourism development. When goals conflict, the need for trade-offs among them and for specific LUP objectives that support consensus land use patterns arises.

LUP problem definition and goal-setting do not remain constant. They are often revised and modified especially when LUP processes are protracted. Economic, socio-cultural, environmental, administrative and political changes occurring at all levels modify environmental conditions (positively or negatively), introduce new actors, legislation and planning modes (e.g., participation), bring about changes in demand for land from new activities and different lifestyles, provide new knowledge and necessitate changes in priorities among goals. These changes create considerable uncertainty and necessitate emphasis on strategic goals, continuous top down-bottom up communication and coordination and broad actor representation and participation as well as integration, flexibility and openness as early in the LUP process as possible. However desirable these features may be, which characterize decentralized systems and are absent from hierarchical planning systems, they underline the complexity and explain the difficulties to come up with clear, coherent and consistent LUP problem definitions and goals in affected areas that prioritize combating LDD.

3.3.1. Problem Description and Analysis; Specification of Planning Objectives

Following the LUP problem definition, this stage aims, on the one hand, to set up the indispensable basis of plan making, i.e., the comprehensive and reliable description of the biophysical and human structure, dynamics and problems of the area, including LDD, and, on the other, to specify the LUP objectives.

Formal actors and external consultants are the most important actors at this stage although the role of other actors interested in the resources of the area should not be downplayed. The effective legislation may prescribe the relationships among these actors, such as formal requirements for plan development and the instruments (consultations between formal actors, consultants and the public) that have to be used. However, the most important relationships are informal, occurring during problem description, analysis and formulation of LUP objectives. New actors enter the process, such as data and information providers, new users of land and new formal actors, different from those involved in problem definition, responsible for formulating the specific LUP objectives.

Consultants take the problem definition and goals set as given but may propose modifications based on the description and analysis of the information and data collected for the affected area as well as on pertinent scientific theories, including those related to LDD. In this way, they may proactively introduce new concerns, activities and actors as well as underline the uncertainty surrounding the causes and effects of LDD and other problems of the area, thus, paving the road for feedbacks and problem redefinition and goal reformulation.

From the viewpoint of properly describing and analyzing LDD, the most important issues concern (a) the theoretical framing of the evolution and problems of the area; (b) the availability of reliable assessment techniques; and (c) the availability of suitable data. Scientific theories developed at higher spatial levels, including those contained in the NAPs, cannot satisfactorily address the causality of land use change and LDD at the local and regional level because they ignore the numerous, tangible and intangible, place-specific and contextual factors and their changes. Study area-specific theoretical frameworks are necessary to guide the situated description and analysis of the affected area for the purposes of LUP especially with respect to combating LDD [25,27,48,76].

Numerous LDD assessment methodologies, indicators and indices have been developed in the last decades to describe the extent and severity of LDD as well as to support LDD abatement decisions at the global, national and local/regional level [77–89]. The Environmental Sensitivity Assessment (ESA) methodology [79], in particular, concerns the assessment of desertification risk on the basis of selected indicators of climate, soil, vegetation and land management at lower levels (landscape/watershed, regional, national). It is continuously being used widely up to the present [90–96] and it has been refined through the application of Remote Sensing and GIS techniques [97–100].

From the viewpoint of their utility for LUP purposes at the local/regional level, the assumptions and limitations of these assessment methodologies and indicators should be kept in mind. Most of them use variables (or, indicators) for which published, regularly collected data are readily available. Issues related to the data used arise as discussed below. The operationalization of several factors (e.g., policy, land management) is based on expert judgment (for obvious reasons). Their suitability and transferability to lower levels of analysis and particular resources (point and nonpoint)[14], the assessment and mapping of LDD-related issues, such as the rate of degradation and the rate of restoration of degraded land, carrying capacity and many more remain open issues that should be treated judiciously [25,28,48,101].

The theoretical and methodological issues briefly discussed above raise important issues with respect to the suitability of existing data bases for the spatial scale and unit of analysis[15] and the time period for a LUP problem at hand. Existing data may concern units of analysis (e.g., mapping units, land units) that do not correspond to the actual decision making units in the study area [102] and may not be available for long time periods for all variables (environmental, economic, social, etc.) of interest. Aggregate land-use types (used in current data bases) may conceal more detailed uses of land at the level of the study area as human activities are multi-purpose and nonhomogeneous and more than one activities may occupy the same tract of land (the case of multifunctional landscapes) or co-exist vertically (the case of high-rise urban areas)[16]. Spatio-temporally consistent environmental, socio-economic and other data may be also lacking [102]. Lack of suitable data do not favor valid assessments and mapping of baseline values of pertinent LDD indicators (see, e.g., [38,87]), impact assessment of past interventions (e.g., policy measures) and of land-use change as well as projections of future demand (population, income, tourism, etc.) and supply of local resources (i.e., carrying capacity, potential, sensitivity and thresholds of critical biophysical and other factors). In sum, for LUP purposes the assessment methodologies and indicators should be used with caution as helpful but indicative and not the only guides to describe and assess, but not explain, the incidence and severity of LDD, to set LUP LDD-specific objectives, to operationalize the LDN target and to make land use decisions.

Numerous other factors complicate the description and analysis of affected areas such as the incidence of unexpected changes and the long time horizons that introduce uncertainty in predicting future land-use change, lack of appropriate environmental and socio-economic monitoring, and so on. Most importantly perhaps, lack of time, human and financial resources, data and expertise may preclude the use of suitable analytical techniques in area-specific analyses in LUP studies. These issues deem necessary detailed data collection and monitoring as well as analytical studies that are indispensable for formulating reliable study area-specific planning objectives in support of LUP to combat LDD, among other goals. Recent trends towards participatory data collection and analysis approaches [103,104] may partially ease the resolution of these issues.

3.3.2. Alternative Plan Formulation

Alternative land-use plans may be formulated to suggest different courses of action to match the present and future demand for land resources with the present and future supply of these resources to achieve the goals and the area-specific planning objectives set, a task critically hinging on problem description and analysis. In affected areas, the most critical question is which land use configuration can satisfy the goal of combatting LDD and the LDN target (or else, how the LDN target is integrated in LU plans) while meeting all other planning goals and objectives.

Plan formulation aiming at environmental protection and resource conservation should be proactive, guided, on the one hand, by strategic choices regarding critical land resources and, on the other, by the prevention, the precautionary and other principles [65,105]. The plans thus produced are more likely to safeguard the constant functioning of natural systems and the provision of ecosystem

---

[14] Such as erosion models, agro-ecological zoning, statistical models, impact assessment and future projection models (climate, population, demand, supply, etc.).

[15] As [3], (p. 117) has noted: " ... proper analysis and meaningful resolution of environmental problems can take place only within natural spatial units, such as watersheds (Manning, 1988). However, practical considerations (data availability and ease of implementation mainly) dictate the use of administrative units most of the time, the result being fragmented, partial, and often ineffective solutions to these problems. Efforts to combine the two types of entities into geopolitical units, such as the 'environomic units' being developed in Canada (Gelinas, 1988), should be encouraged, as well as the formation of inter-jurisdictional bodies ... "

[16] The problem is more acute for small study areas that may suffer from severe LDD.

services that constitute the enabling condition of sustainable development [106]. In this context, an important consideration is the property status of land and other resources, such as water; i.e., whether they are private, state or common property [107,108] that significantly influences land-use allocation decisions[17] [109,110]. It is noted, however, that land resources, irrespective of property status, are common pool resources[18] that should be preserved for the benefit of all present and future users [111]. Given the fundamental role of soil and water resources in the LDD context and not only that, it follows that their protection should constitute a strategic LUP goal. While water resources enjoy a more or less satisfactory legislative protection and status in plan formulation, this not the case with soil resources. Land-use plans often prescribe uses of land that give rise to several soil threats—erosion, decline in organic matter, sealing, contamination, compaction, decline in biodiversity, salinization, landslides [46]—that increase the risk of LDD especially in sensitive affected areas.

Alternative plan formulation is usually carried out by planning consultants (including scientists) in cooperation and consultation with formal actors but also with diverse other interested actors, some of them entering at this stage of making critical land use allocation decisions. Guidance regarding how land-use plans should meet LDD-specific objectives such as improving the rate and extent of land restoration and reducing the rate and extent of land degradation is missing. The UNCCD sets broad goals and lacks operational guidelines and satisfactory indicators to measure progress at lower levels [32]. The NAPs usually focus on land uses related to the proximate causes of LDD—agriculture, animal husbandry, forests and, to a lesser extent, on tourism, extraction, transport, energy (e.g., renewable energy sources) and large infrastructure works (transport and energy networks, airports, etc.). They propose general activity-specific measures to minimize LDD because pertinent policy measures and instruments, such as environmental regulations, sustainable land management (SLM) practices, subsidies, etc., may be already in place and usually agree with dominant sectoral preferences. This is the case of the EU CAP agri-environmental and cross-compliance measures [33].

The focus on known proximate causes of LDD agrees with the common LUP practice where the uses of land in an area are determined by the present or future demand for land by economic activities. Activity-specific obligations are formally specified in order to avoid resource degradation and other predictable externalities. This practice of coupling demand with supply overlooks two important issues. First, the status of land resources is affected by numerous and diverse indirect drivers that modify, in one way or another, the demand for and thus, the pressures on land resources. These include outmigration, land abandonment, socio-economic restructuring, changes in urban-rural dynamics, changes in life styles (e.g., trends towards eco-living, eco-tourism, etc.), land and other taxation, and so on. This implies that, second, changes in these drivers may preclude the realization of chosen land-use combinations or, if realized, drastically modify their anticipated impacts, leading to either deterioration or effective protection of land resources. Therefore, LUP instruments, such as zoning (of hazard-prone and sensitive landscape areas), natural and cultural reserves, multifunctional land use schemes, should be coupled with economic, fiscal and other instruments, e.g., land and general taxation, incentive schemes, voluntary measures, to attain the LDN and related objectives [68,110].

This broadened perspective necessarily leads to the quest for integrated land-use planning that has been voiced since the early discussions of the issue [43,48] and more recently in discussing the materialization of the LDN goal [27,38]. The need for integrated and situated land-use plan formulation becomes evident at this stage of plan formulation as the effectiveness of any land-use plan depends on the spatial coordination of several activities, policy instruments, etc. at the level of the study area and across levels to promote individual and collective goals.

---

[17] Private interests exert pressures on the LUP process and significantly affect the choices made.

[18] Common Pool Resources (CPRs) are characterized by (a) non-excludability—they are indivisible and, thus, nobody can be ethically or practically be excluded from using them and (b) subtractability—they are finite and, thus use by one user reduces the amount of resource available to others [107,108].

The alternative land-use plans reflect, to different degrees, the trade-offs among different goals and objectives and, consequently, among the associated economic activities. This is a highly-charged political stage that, combined with uncertainty and land inertia, renders integrated land-use planning a challenging endeavor. Participatory decision making and planning is proposed, and occasionally practiced, in an effort to strike balances and arrive at land-use plans with higher chances of implementation [18,19,27,56,112]. Cross-compliance measures, that require the satisfaction of environmental goals as a condition for financial assistance, are practical, although partial, integrative instruments that may be used to integrate LDD concerns into LUP.

3.3.3. Plan Evaluation and Choice of Preferred Alternative

This stage, theoretically at least, concerns the evaluation of the alternative land-use plans on the basis of criteria flowing from the goals and objectives set in order to arrive at the preferred alternative that will be implemented. The priorities among criteria, which usually favor economic over environmental considerations, determine the final evaluation outcome. In practice, two alternatives are commonly considered: the existing land use pattern and a preferred land use pattern that meets the approval of interested actors. In this case, the evaluation essentially dissolves into a formal approval process of the preferred (and only) plan. Notice, however, that several alternatives may have been considered before the preferred one was chosen.

The actors involved in this stage are more or less the same with those participating at the previous stage. Formal evaluation instruments, such as the Environmental Impact Assessment (EIA) and Strategic Environmental Assessment (SEA), are commonly used, offering the opportunity to laypersons and other individuals who have not participated in the previous stages to express their opinion regarding the proposed plan or plans (in the rare case that there are more than one) through public consultation. These procedures, although they are proactive in nature, face several theoretical, methodological (analytical) and practical (data) challenges and their results are fraught with uncertainty [113]. Consequently, their potential to resolve essential questions regarding differences among alternative land use plans, in terms of goal achievement and distributive impacts, is limited. One issue of particular relevance in the LUP context is the requirement built into the EIA and SEA to evaluate alternatives for the same problem. Studies have shown that this requirement is rarely met [114]; alternatives are eliminated using qualitative reasoning mostly in support of the preferred alternative.

LDD concerns do or may enter this stage if LDD-related (or LDN-related) criteria are included in these formal procedures; they are often raised in the context of public consultations if they have not been integrated in the proposed plan(s). Their meaningful and effective application at the local and the regional levels is conditioned by the availability of data and baseline environmental and economic[19] assessments and the suitability of assessment techniques mentioned before.

3.3.4. Plan Implementation

Plan implementation remains the most critical stage of the LUP process because it constitutes the real test of whether the chosen land-use plan can materialize under actual, dynamic and fluid conditions. When land-use plans are put to practice a host of issues inevitably surface. These include omissions during previous stages (actors, local characteristics, LDD assessments, etc.), past planning and other legacies, pressures from interest groups, perception and understanding of the planning problems and prescriptions by the actors involved, conflicts between proposed/new and customary practices, endogenous and plan-induced change (externalities) and unexpected events and contingencies (natural and technological

---

[19] Economic assessments, and the respective methodologies, of the costs of LDD and the costs of various abatement options do exist but they are still not widely used in local and regional LUP practice [115–117]; (see, also, the ELD site at: http://www.eld-initiative.org/).

disasters, political and socio-economic events) in the study area and the rest of the world [118,119]. The end result is that the implemented land-use plan deviates from the chosen, legislated plan, implying that its actual environmental and socio-economic impacts will also differ from those originally assessed.

Plan implementation implicates a wide variety of formal and informal actors from various sectors, jurisdictions and spatial/organizational levels who variously contribute to the materialization of the proposed uses of land. Examples are formal local and regional administrators of various ranks, the courts, land owners/renters, farmers and their associations, real estate and building groups, banks, professionals (individuals and groups), the media, local and supralocal environmental and civic groups, residents, and so on [66]. This variety of actors engenders differences in decision culture (reactive vs. proactive), knowledge and perception of planning problems (including LDD), multiple local understandings and interpretations of planning goals, objectives and requirements of the land-use plan, different priorities among goals and interests, competences, reactions to change and local pressures, and power imbalances when conflicts arise. During implementation, the LUP problem is essentially redefined several times; issues that were not originally considered are included (e.g., safety and soil protection issues following severe floods, landslides or earthquakes). LDD may, thus, obtain significance if interest groups raise concerns during implementation that may lead to suspension or change if the approved land-use plan has potentially negative effects on soil and water resources.

The implementation of land-use plans requires cooperation and coordination among actors from various policy domains, such as agricultural, development, spatial, economic, social and environmental, the existence of formal procedures which guide their interactions as well as informal procedures based on mutual trust; i.e., on social capital. In the case of LDD, a strong commitment to the cause of combating LDD is indispensable for proper implementation of resource (soil, water, biodiversity) conservation requirements and pertinent plan provisions.

The most critical issue in plan implementation is compliance with the plan's requirements and the existence, application and effectiveness of formal and informal enforcement mechanisms (fines, penalties, etc.). Because of the inherent uncertainty of plans and the changes occurring during plan implementation, conflicts will always arise that necessitate the presence of effective conflict-resolution mechanisms. Compliance, enforcement and related mechanisms depend on local culture, tradition, familiarity with planning and on the broader mode of governance. Participatory land-use planning approaches aim at securing compliance by involving all actors early in the process. These are increasingly applied in coping with environmental problems, such as LDD, as a way to avoid non-implementation or significant deviations that will ultimately nullify the original plan. Their success is not guaranteed, however, because on the one hand socio-cultural, institutional and several practical factors condition participation and, on the other, power balances modify the outcomes that may not always favor combating LDD (e.g., development of water-intensive activities in water-deficient and degraded areas).

The less-than-satisfactory implementation record of several land-use plans reveals that the above requirements are difficult to meet in practice, the end result being implementation delays and uncertain outcomes. LDD may worsen, under unfavorable biophysical conditions and strong development pressures or, it may be reversed when pressures are reduced for other reasons (e.g., lack of development interest in the area, local resistance to development plans, etc.).

3.3.5. Plan Monitoring, Evaluation, Modification

Monitoring the implementation of land use plans aims at identifying and evaluating the issues arising and taking corrective action. Monitoring may be prescribed in the legislation and employed using suitable monitoring mechanisms (e.g., water metering, indicators, etc.) or it takes place voluntarily (see, e.g., [84]) or spontaneously when problems arise and necessitate resolution.

Monitoring is not a singular but a multiple operation involving old and new actors who, most likely, may not have complete knowledge of the land use plan and operate within narrow sectoral domains with specific mandates. Assuming that the required apparatus is in place, feeding back monitoring results (e.g., population change, land use change, soil conditions, water consumption, production levels, etc.) to formal and other actors requires proper transmission channels for the variety of actors and settings involved. Even with proper feedback, the monitoring results represent new information that is variously processed, translated and evaluated under conditions different from those prevailing when the initial land-use problem was posed and defined, planning goals were set, the land-use plan was formulated and implementation commenced [27].

Assessing and evaluating the monitoring results requires the existence of reliable local/regional level baseline indicators and assessment models, previous assessments, etc., a condition that is rarely met in practice. Evaluating the changes in the rate of LDD and of land restoration, in particular, is limited by the incomplete local/regional level assessments. Explaining the evaluation results, i.e., answering the question "what has caused the change", requires consideration of numerous endogenous and exogenous factors that may have contributed to change, in addition to the interventions associated with the land use plan. A single indicator is often difficult to describe and assess the latter because interventions are composite and complex actions. Attributing the changes in the rate of LDD and of land restoration to the land-use plan is, thus, a task that cannot yield dependable results to support plan modifications necessary to achieve the LDN and other LDD-related goals. Careful analysis of the monitoring results is needed accounting for the co-presence of several factors and concurrent changes that take place at all levels [27].

Land-use plan revisions may be prescribed in the legislation and they are usually undertaken every 10–15 years. They involve a comprehensive revision of the conditions of the study area since the official approval of the initial plan and propose required changes that reflect the changed conditions and the new demands for local and regional resources that arise with the passage of time. Whether plan revisions are prompted by or take into account LDD-related monitoring results, for shorter time intervals, is an open question that only thorough knowledge of the study area can answer.

4. Facing the Land-Use Planning Quandary

The reality check carried out in the preceding section suggests that the effectiveness of LUP as an instrument of land governance in combating LDD at the local and regional level depends on whether and how LDD concerns enter, or may enter, the stages of the LUP process. LDD problem complexity and severity, the state of scientific knowledge, market forces driving demand for land, the prevailing mode of governance (cf. [27]), the planning style followed, combined with local awareness, knowledge and perception of LDD, value systems and planning goal priorities, power balances, geographic particularities and historical circumstances are important catalyzing influences in this respect [33]. The complex nature of LDD implies that no single state agency can deal with it exclusively, as happens with simpler and less complex hazards[20]. Multiple, diverse, formal and informal, individual and collective actors are involved and numerous, disparate, formal and informal institutions from multiple organizational levels are implicated, interacting non-linearly on and across spatial and temporal scales [33,120]. All these forces generate a fluid and uncertain context within which LUP confronts important challenges. This concluding section first summarizes the main challenges and outlines necessary priority actions to realize the integrative potential of LUP and, thus, improve its effectiveness in combating LDD. Figure 2 schematically summarizes the rationale of the preceding analysis, the challenges identified and priority actions proposed.

---

[20] Even in this case, of, e.g., earthquakes, the responsibility for dealing with the issue may be divided among several competent state agencies.

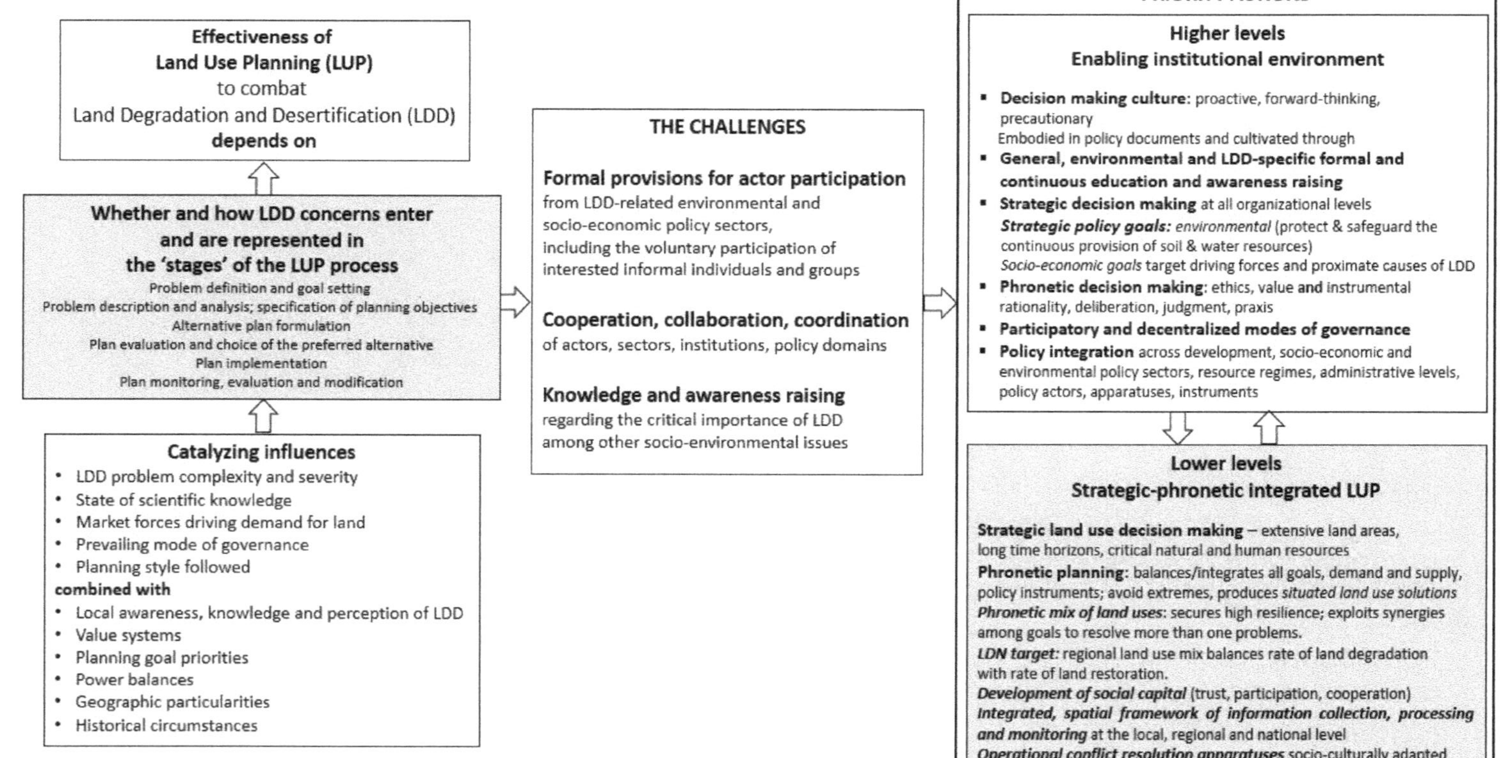

Figure 2. Enhancing the effectiveness of LUP to combat land degradation and desertification (LDD): challenges and priority actions.

The representation of LDD concerns at each stage of the LUP process depends on the existence of pertinent formal provisions for actor participation from LDD-related environmental and socio-economic policy sectors, including the voluntary participation of interested informal individuals and groups. Strong conflicts between LDD-related and socio-economic development goals are common, however, and create an adversarial context. The need thus arises for cooperation, collaboration and coordination of the numerous actors, sectors, institutions and policy domains involved to effectively resolve LUP conflicts and manage present and future environmental and socio-economic change. Hierarchical or mixed (hierarchical and market) governance systems, reactive planning styles (e.g., disjointed incrementalism), a myopic decision culture lacking a preparedness mentality, and the usual administrative and organizational compartmentalization, impede the satisfactory manipulation of these challenges as several real-world situations attest [30]. The attention and priority given to LDD in the LUP context depends on knowledge and awareness of its importance among other issues especially when strong development pressures and pro-development value systems dominate. Greater attention and concern is encountered in areas experiencing extreme LDD and lacking immediate development options and pressures. Moreover, climate change seems to have reinforced interest in effective sustainable land management to safeguard precious soil and water resources.

In order to improve the chances of LUP effectively coping with these challenges, certain priority actions are necessary at both higher and lower levels (cf. [27,118]). At higher levels, an enabling institutional environment should be provided[21] to support lower level actions. Featuring high among its crucial features is a proactive, forward-thinking and precautionary decision culture embodied in pertinent policy documents and cultivated through suitable general, environmental and LDD-specific formal and continuous education and awareness raising. Instituting strategic decision making, which is inherently long-term, should concern all spatial/organizational levels and be supported by suitable instruments. Strategic policy goals should encompass environmental, namely, protecting and safeguarding the continuous provision of soil and water resources, in addition to socio-economic goals. Socio-economic goals should target the driving forces of LDD[22] and not only the proximate causes as is current practice.

Phronesis (or, practical wisdom)[23], the most important of the principal virtues according to Aristotle, should be the guiding principle cutting across LUP decision making. The phronetic approach, first introduced by Flyvbjerg [122], underlines the situatedness of knowledge, decisions and action and emphasizes ethics, value rationality to balance instrumental rationality, deliberation, interpretation, judgment, participation, power relations and praxis. The possession and application of *phronesis* ensures the ethical employment of science (*episteme*), i.e., of universal rules, and technology (*techne*) in choosing the means in concrete, specific cases and adapting action to context [123].

Where hierarchical modes of governance and centralized planning traditions dominate, a gradual shift towards more participatory modes of governance and decentralized planning might be encouraged (and institutionalized) to benefit from greater participation of interested parties and flexibility in making decisions. However, care should be taken to ensure that participation is working for the common good and in the long run, it is based on a common understanding of LDD, and effective communication among the parties is involved as well as on cooperation around mutually agreed and shared goals and objectives. Lastly, the integrative potential of LUP can be realized only in the context of policy integration at higher spatial/organizational levels. This translates into the need for integration across development, socio-economic and environmental policy sectors, resource regimes, administrative levels, policy actors, apparatuses and instruments [39].

This enabling institutional environment is indispensable for framing and supporting the application of the widely commended and cited integrated LUP and lends it particular features

---

21   As [1] has underlined already.
22   That may be common to other socio-ecological problems.
23   "(knowing) how to exercise judgment in particular cases" (MacIntyre, 1985: 154 cited in [121], (p. 381)).

that are necessary to achieve the integration of LDD concerns within the LUP process and address the LDN target. First, integrated LUP should develop around strategic land-use decision making, which concerns extensive land areas, long time horizons and critical natural and human resources (cf. [123]), and, thus, it is particularly relevant to LDD and to policy integration.

Second, integrated LUP should adopt the phronetic approach that is suitable when dealing with highly contentious, uncertain and 'wicked' LUP problems [31] that demand flexibility and adaptation to changing conditions, thus, requiring preparedness to cope with the impacts of LDD, LUP interventions and unanticipated events. The phronetic choice of land uses avoids prioritizing certain (usually narrow) goals only, often associated with monocultures and bound to cause serious adverse impacts; instead, it favors those that balance all goals, do not compromise the strategic ones, and support a mix of uses that secures high local and regional resilience. To achieve the LDN target specifically, the regional mix of land uses should balance the rate of land degradation with the rate of land restoration. Moreover, because the direct and indirect LDD drivers usually underlie other socio-ecological problems, phronetic land-use solutions may exploit synergies among the respective goals (e.g., food security, biodiversity conservation) to resolve more than one problems, including combating other natural hazards.

Third, integrated LUP should foster participation, communication and cooperation, i.e., the development of social capital [27], as the most important precondition for generating situated, ecologically sound and socially responsive integrated land-use solutions with high implementation potential. This echoes with certain NAPs[24] that prescribe detailed analyses of affected areas and case-by-case decision making.

Strategic-phronetic integrated LUP, as it might be called, targets both the demand (driving forces and proximate causes) and the supply side (resource conservation and protection) of LDD. It echoes the adaptive co-management approach [125] and combines a variety of policy instruments to support the development and implementation of situated land use solutions. These include SLM, spatial management instruments (impact zoning, transfer or purchase of development rights, land reserves, land banking, etc.), and fiscal and economic instruments, such as subsidies, tax breaks and resource use fees. Its effective implementation hinges on the existence of an integrated, spatial framework of information collection, processing and monitoring[25] at the local and regional levels [31] and of operational, socio-culturally adapted conflict resolution apparatuses. Lastly, it is essential that issues arising during implementation should be fed back to higher level institutions to introduce necessary changes in the enabling institutional environment and the decision-making apparatuses and instruments to facilitate adaptation to changing conditions and needs that local level action has revealed. Future research is required to explore the theoretical, methodological and empirical issues that should be addressed in order to provide operational guidance for its implementation to combat LDD in the context of other local and regional socio-ecological matters.

Funding: This research received no external funding.

Conflicts of Interest: The author declare no conflict of interest.

References

1. UNCCD. *United Nations Convention to Combat Desertification*; UNCCD: Paris, France, 1994.
2. Manning, E.W. *The Analysis of Land-Use Determinants in Support of Sustainable Development*; International Institute for Applied Systems Analysis: Laxenburg, Austria, 1988.
3. Briassoulis, H. Pollution prevention for sustainable development: The land-use question. *Int. J. Sustain. Dev. World* **1994**, *1*, 110–120. [CrossRef]

---

24 Such as the Greek NAP [124].
25 Preferably based on hybrid, administrative-environmental spatial units.

4. MEA. *Ecosystems and Human Well-Being: Desertification Synthesis*; Millennium Ecosystem Assessment; World Resources Institute: Washington, DC, USA, 2005.

5. FAO. *The State of the World's Land and Water Resources for Food and Agriculture (SOLAW)—Managing Systems at Risk*; Food and Agriculture Organization of the United Nations: Rome, Italy; Earthscan: London, UK, 2011.

6. UNCCD. *The Global Land Outlook*, 1st ed.; United Nations Convention to Combat Desertification: Bonn, Germany, 2017.

7. IPBES. *Summary for Policymakers of the Assessment Report on Land Degradation and Restoration of the Intergovernmental Science-Policy Platform on Biodiversity and Ecosystem Services*; Scholes, R., Montanarella, L., Brainich, A., Barger, N., ten Brink, B., Cantele, M., Erasmus, B., Fisher, B., Gardner, T., Holland, T.J., et al., Eds.; IPBES Secretariat: Bonn, Germany, 2018.

8. United Nations. *Transforming Our World: The 2030 Agenda for Sustainable Development*; Resolution adopted by the General Assembly on 25 September 2015; A/RES/70/1; 4th Plenary Meeting; United Nations: New York, NY, USA, 2015.

9. Meyfroidt, P.; Chowdhury, R.; de Bremond, A.; Ellis, E.C.; Erb, K.H.; Filatova, T.; Garrett, R.D.; Grove, J.M.; Heinimann, A.; Kuemmerle, I.T.; et al. Middle-range theories of land system change. *Glob. Environ. Chang.* **2018**, *53*, 52–67. [CrossRef]

10. Global Land Programme. *Science Plan and Implementation Strategy*; IGBP Report No. 53/IHDP Report No. 19; IGBP Secretariat: Stockholm, Sweden, 2005; 64p.

11. Verburg, P.H.; Erb, K.-H.; Mertz, O.; Espindola, G. Land System Science: Between global challenges and local realities. *Curr. Opin. Environ. Sustain.* **2013**, *5*, 433–437. [CrossRef] [PubMed]

12. Verburg, P.H.; Crossman, N.; Ellis, E.C.; Heinimann, A.; Hostert, P.; Mertz, O.; Nagendra, H.; Sikor, T.; Erb, K.-H.; Golubiewski, N.; et al. Land system science and sustainable development of the earth system: A global land project perspective. *Anthropocene* **2015**, *12*, 29–41.

13. Dearing, J.A.; Braimoh, A.K.; Reenberg, A.; Turner, B.L.; van der Leeuw, S. Complex land systems: The need for long time perspectives to assess their future. *Ecol. Soc.* **2010**, *15*, 21. [CrossRef]

14. Land. *Land System Science*, Special Issue. 2017; 6.

15. Marsh, G.P. *Man and Nature; or, the Earth as Modified by Human Action*; The Belknap Press of Harvard University Press: Cambridge, MA, USA, 1965; (Originally published in 1864).

16. Meyer, W.B.; Turner, B.L., II (Eds.) *Changes in Land Use and Land Cover: A Global Perspective*; Cambridge University Press: Cambridge, UK, 1994.

17. Turner, B.L., II; Clark, C.; Kates, R.W.; Richards, J.F.; Mathews, J.T.; Meyer, W.B. (Eds.) *The Earth as Transformed by Human Action: Global and Regional Changes in the Biosphere over the Past 300 Years*; Cambridge University Press: Cambridge, UK, 1990.

18. Burby, R.J. *Cooperating with Nature: Confronting Natural Hazards with Land-Use Planning for Sustainable Communities*; Joseph Henry Press: Washington, DC, USA, 1998.

19. Burby, R.J.; Deyle, R.E.; Godschalk, D.R.; Olshansky, R.B. Creating hazard resilient communities through land-use planning. *Nat. Hazards Rev.* **2000**, *1*, 99–106. [CrossRef]

20. Bowyer, C.; Withana, S.; Fenn, I.; Bassi, S.; Lewis, M.; Cooper, T.; Benito, P.; Mudgal, S. *Land Degradation and Desertification*; European Parliament; Policy Department A, Economic and Scientific Policy: Brussels, Belgium, 2009.

21. Low, P.S. (Ed.) Economic and Social Impacts of Desertification, Land Degradation and Drought. White Paper I. UNCCD 2nd Scientific Conference; Prepared with the Contributions of an International Group of Scientists. 2013. Available online: http://2sc.unccd.int (accessed on 2 November 2018).

22. Geist, H. *The Causes and Progression of Desertification*; Routledge: London, UK, 2017.

23. Imeson, A. *Desertification, Land Degradation and Sustainability*; Wiley: London, UK, 2011.

24. D' Odorico, P.; Battachan, A.; Davis, K.F.; Ravi, S.; Runyan, C.W. Global desertification: Drivers and feedbacks. *Adv. Water Res.* **2013**, *51*, 326–344. [CrossRef]

25. Reynolds, J.F.; Stafford-Smith, M. *Global Desertification: Do Humans Cause Deserts?* Dahlem University Press: Berlin, Germany, 2002.

26. Enne, G.; Zanolla, C.; Peter, D. (Eds.) *Desertification in Europe; Mitigation Strategies, Land-Use Planning*; EC, DG for Research Environment and Climate, Office of the Official Publications of the European Communities: Luxembourg, 2000.

27. Orr, B.J.; Cowie, A.; Castillo Sanchez, V.M.; Chasek, P.; Crossman, N.D.; Erlewein, A.; Louwagie, G.; Maron, M.; Metternicht, G.I.; Minelli, S.; et al. *Scientific Conceptual Framework for Land Degradation Neutrality. A Report of the Science-Policy Interface*; United Nations Convention to Combat Desertification (UNCCD): Bonn, Germany, 2017; Available online: https://www.unccd.int/sites/default/files/documents/2017-08/LDN_CF_report_web-english.pdf (accessed on 2 November 2018).

28. Van der Leeuw, S.E. Some potential problems with the implementation of Annex IV of the convention to combat desertification. In *Desertification in Europe; Mitigation Strategies, Land-use Planning*; Enne, G., Zanolla, C., Peter, D., Eds.; EC, DG for Research Environment and Climate, Office of the Official Publications of the European Communities: Luxembourg, 2000; pp. 249–262.

29. Turner, B.L.I.I.; Lambin, E.F.; Reenberg, A. The emergence of land change science for global environmental change and sustainability. *Proc. Natl. Acad. Sci. USA* **2007**, *104*, 20666–20671. [CrossRef] [PubMed]

30. Metternicht, G. *Land Use Planning*; Working paper. Global Land Outlook; United Nations Convention to Combat Desertification: Bonn, Germany, 2017.

31. Flyvbjerg, B. Phronetic planning research. Theoretical and methodological reflections. *Plan. Theory Pract.* **2004**, *5*, 283–306. [CrossRef]

32. Briassoulis, H. Desertification. In *Science and Policy. An A to Z Guide to Issues and Controversies*; Steel, B., Ed.; CQ Press: Thousand Oaks, CA, USA, 2014; pp. 138–145.

33. Briassoulis, H. Governing desertification in Mediterranean Europe: The challenge of environmental policy integration in multi-level governance contexts. *Land Degrad. Dev.* **2010**, *22*, 313–325. [CrossRef]

34. Turney, D.; Fthenakis, V. Environmental impacts from the installation and operation of large-scale solar power plants. *Renew. Sustain. Energy Rev.* **2011**, *15*, 3261–3270. [CrossRef]

35. Hernandez, R.R.; Easter, S.B.; Murphy-Mariscal, M.L.; Maestre, F.T.; Tavassoli, M.; Allen, E.B.; Barrows, C.W.; Belnap, J.; Ochoa-Hueso, R.; Ravi, S.; et al. Environmental impacts of utility-scale solar energy. *Renew. Sustain. Energy Rev.* **2014**, *29*, 766–779. [CrossRef]

36. Kaoshan, D.; Bergot, A.; Liang, C.; Xiang, W.-N.; Huang, Z. Environmental issues associated with wind energy—A review. *Renew. Energy* **2015**, *75*, 911–921.

37. Briassoulis, H. Complex Environmental Problems and the Quest for Policy Integration. In *Policy Integration for Complex Environmental Problems: The Example of Mediterranean Desertification*; Briassoulis, H., Ed.; Ashgate: Cheltenham, UK, 2005; pp. 1–49.

38. Grainger, A. Is Land Degradation Neutrality feasible in dry areas? *J. Arid Environ.* **2015**, *112 Pt A*, 14–24. [CrossRef]

39. Briassoulis, H. (Ed.) *Policy Integration for Complex Environmental Problems: The Example of Mediterranean Desertification*; Ashgate: Cheltenham, UK, 2005.

40. Wilson, G.A.; Juntti, M. (Eds.) *Unraveling Desertification: Policies and Actor Networks in Southern Europe*; Wageningen Academic Publishers: Wageningen, The Netherlands, 2005.

41. COP8. *Report of the Conference of the Parties on Its Eighth Session, Held in Madrid from 3 to 14 September 2007*; United Nations Convention to Combat Desertification; ICCD/COP(8)/16/Add.1; United Nations: Bonn, Germany, 2007.

42. UNCCD. *Towards a Land Degradation Neutral World*; United Nations Convention to Combat Desertification: Bonn, Germany, 2015; Available online: https://knowledge.unccd.int/sites/default/files/inline-files/Towards%20a%20Land%20Degradation%20Neutral%20World%20A%20Sustainable%20Development%20Priority.pdf (accessed on 2 November 2018).

43. UNCCD. *Achieving Land Degradation Neutrality*; United Nations Convention to Combat Desertification: Bonn, Germany, 2015; Available online: https://www.unccd.int/actions/achieving-land-degradation-neutrality (accessed on 2 November 2018).

44. UNCCD. *Land Degradation Neutrality. Resilience at Local, National and Regional Levels*; United Nations Convention to Combat Desertification: Bonn, Germany, 2015; Available online: http://catalogue.unccd.int/858_V2_UNCCD_BRO_.pdf (accessed on 27 September 2018).

45. Briassoulis, H. Factors affecting changes in land use and land cover. In *UNESCO Encyclopedia of Life Support Systems (EOLSS)*; EOLSS Publishers: Oxford, UK, 2003.

46. CEC. *Thematic Strategy for Soil Protection*; Communication from the Commission to the Council, the European Parliament, the European Economic and Social Committee and the Committee of the Regions; COM(2006)231 Final; Commission of the European Communities: Brussels, Belgium, 2006.

47. Hermann, S.M.; Hutchinson, C.F. The Scientific Basis: Links between Land Degradation, Drought and Desertification. In *Governing Global Desertification: Linking Environmental Degradation, Poverty and Participation*; Johnson, P.M., Mayrand, K., Paquin, M., Eds.; Ashgate: Aldershot, UK, 2006; pp. 11–26.
48. Bestelmeyer, B.T.; Oki, G.S.; Duniway, M.C.; Archer, S.R.; Sayre, N.F.; Williamson, J.C.; Herrick, J.E. Desertification, land use, and the transformation of global drylands. *Front. Ecol. Environ.* **2015**, *13*, 28–36. [CrossRef]
49. Scoones, I. *Sustainable Rural Livelihoods: A Framework for Analysis*; IDS Working Paper, 72; IDS: Brighton, UK, 1998.
50. Scoones, I. Livelihoods perspectives and rural development. *J. Peasant Stud.* **2009**, *36*, 171–196. [CrossRef]
51. Blaikie, P.; Brookfield, H. *Land Degradation and Society*; Routledge: London, UK, 1987.
52. Kaiser, E.J.; Godschalk, D.J.; Chapin, F.S. *Urban Land Use Planning*, 4th ed.; University of Illinois Press: Urbana, IL, USA, 1995.
53. Van Lier, H. The role of land use planning in sustainable rural systems. *Landsc. Urban Plan.* **1998**, *41*, 83–91. [CrossRef]
54. Food and Agriculture Organization (FAO). *Guidelines for Land-Use Planning*; FAO: Rome, Italy, 1996.
55. Godschalk, D.R. Land Use Planning Challenges: Coping with Conflicts in Visions of Sustainable Development and Livable Communities. *J. Am. Plan. Assoc.* **2004**, *70*, 5–13. [CrossRef]
56. Randolph, J. *Environmental Land Use Planning and Management*, 2nd ed.; Island Press: Washington, DC, USA, 2004.
57. Godschalk, D.R.; Beatley, T.; Berke, P.; Brower, D.J.; Kaiser, E.J. *Natural Hazard Mitigation: Recasting Disaster Policy and Planning*; Island Press: Washington, DC, USA, 1999.
58. Healey, P. *Collaborative Planning*; Shaping Places in Fragmented Societies; UBC Press: Vancouver, BC, Canada, 1997.
59. Healey, P. *Urban Complexity and Spatial Strategies*; Routledge: London, UK, 2007.
60. Food and Agriculture Organization (FAO). *Planning for Sustainable Use of Land Resources*; FAO Land and Water Bulletin 2; FAO: Rome, Italy, 1995.
61. Faludi, A. *A Reader in Planning Theory*; Pergamon: Oxford, UK, 1973.
62. Howe, E.; Kaufmann, J. The ethics of contemporary American planners. *J. Am. Plan. Assoc.* **1979**, *45*, 243–255. [CrossRef]
63. Bolan, R. Emerging views of planning. *J. Am. I. Plan.* **1967**, *33*, 233–245. [CrossRef]
64. Briassoulis, H. Theoretical Orientations in Environmental Planning: An Inquiry into Alternative Styles. *Environ. Manag.* **1989**, *13*, 381–392. [CrossRef]
65. ESDP. *European Spatial Development Perspective: Towards Balanced and Sustainable Development of the Territory of the EU*; European Commission, Office for Official Publications of the European Communities: Luxembourg, 1999.
66. Albrechts, L. Strategic (spatial) planning reexamined. *Environ. Plan. B* **2004**, *31*, 743–758. [CrossRef]
67. White, R.P.; Tunstall, D.; Henninger, N. *An Ecosystem Approach to Drylands: Building Support for New Development Policies*; Information Policy Brief No.1; World Resources Institute: Washington, DC, USA, 2002.
68. Vreeker, R.; de Groot, H.I.F.; Verhoef, E.T. Urban Multifunctional Land Use: Theoretical and Empirical Insights on Economies of Scale, Scope and Diversity. *Built Environ.* **2004**, *30*, 289–307. [CrossRef]
69. Vasishth, A. A scale-hierarchic ecosystem approach to integrative ecological planning. *Prog. Plan.* **2008**, *70*, 99–132. [CrossRef]
70. Sayer, J.; Sunderland, T.; Ghazoul, J.; Pfund, J.-L.; Sheil, D.; Meijaard, E.; Venter, M.; Klintuni Boedhihartono, A.; Day, M.; Garcia, C.; et al. Ten principles for a landscape approach to reconciling agriculture, conservation, and other competing land uses. *Proc. Natl. Acad. Sci. USA* **2013**, *110*, 8349–8356. [CrossRef] [PubMed]
71. McHarg, I. *Design with Nature*; Natural History Press: Garden City, NY, USA, 1969.
72. Food and Agriculture Organization (FAO). *Land Evaluation: Towards a Revised Framework*; Land and Water Discussion Paper 6; TC/D/A1080E/1/04.07; FAO: Rome, Italy, 2007; ISSN 1729-0554. Available online: http://www.fao.org/nr/lman/docs/lman_070601_en.pdf (accessed on 2 November 2018).
73. Binder, C.R.; Hinkel, J.; Bots, P.W.G.; Pahl-Wostl, C. Comparison of frameworks for analyzing social-ecological systems. *Ecol. Soc.* **2013**, *18*, 26. [CrossRef]
74. Susskind, L.; Ozawa, C. Mediated Negotiation in the Public Sector: The Planner as Mediator. *J. Plan. Educ. Res.* **1984**, *4*, 5–15. [CrossRef]
75. Forester, J. Planning in the face of Conflict. *J. Am. Plan. Assoc.* **1987**, *53*, 434–446. [CrossRef]
76. Geist, H.J.; Lambin, E.F. Dynamic causal patterns of desertification. *BioScience* **2004**, *54*, 817–829. [CrossRef]

77. Dharumarajan, S.; Bishop, T.F.A.; Hedge, R.; Singh, S.K. Desertification vulnerability index—An effective approach to assess desertification processes: A case study in Anantapur District, Andhra Pradesh, India. *Land Degrad. Dev.* **2018**, *29*, 150–161. [CrossRef]

78. Helldén, U.; Tottrup, C. Regional desertification: A global synthesis. *Glob. Planet. Chang.* **2008**, *64*, 169–176. [CrossRef]

79. Kosmas, C.; Ferrara, A.; Briasouli, H.; Imeson, A. Methodology for mapping Environmentally Sensitive Areas (ESAs) to Desertification. In *The Medalus Project: Mediterranean Desertification and Land Use. Manual on Key Indicators of Desertification and Mapping Environmentally Sensitive Areas to Desertification*; Kosmas, C., Kirkby, M., Geeson, N., Eds.; European Union: Brussels, Belgium, 1999; pp. 31–47. ISBN 92-828-6349-2.

80. Kosmas, C.; Poesen, J.; Briassouli, H. Key indicators of desertification at the ESA scale. In *The Medalus Project: Mediterranean Desertification and Land Use. Manual on Key Indicators of Desertification and Mapping Environmentally Sensitive Areas to Desertification*; Kosmas, C., Kirkby, M., Geeson, N., Eds.; European Union: Brussels, Belgium, 1999; pp. 11–30. ISBN 92-828-6349-2.

81. Kosmas, K.; Tsara, M.; Moustakas, N.; Karavitis, C. Identification of indicators for desertification. *Ann. Arid Zones* **2003**, *42*, 393–416.

82. Kosmas, C.; Kairis, O.; Karavitis, C.; Ritsema, C.; Salvati, L.; Acikalin, S.; Alcalá, M.; Alfama, P.; Atlhopheng, J.; Barrera, J.; et al. Evaluation and Selection of Indicators for Land Degradation and Desertification Monitoring: Methodological Approach. *Environ. Manag.* **2013**, *54*, 951–970. [CrossRef] [PubMed]

83. Panagiotis, T.N.; Nadia, P.; Kapsomenakis, J. Spatial and temporal variability of the Aridity Index in Greece. *Atmos. Res.* **2013**, *119*, 140–152.

84. Reed, M.S.; Dougill, A.J. Linking degradation assessment to sustainable land management: A decision support system for Kalahari pastoralists. *J. Arid Environ.* **2010**, *74*, 149–155. [CrossRef]

85. Salvati, L.; Mancino, G.; De Zuliani, E.; Sateriano, A.; Zitti, M.; Ferrara, A. An expert system to evaluate environmental sensitivity: A local scale approach to desertification risk. *Appl. Ecol. Environ. Res.* **2013**, *11*, 611–627. [CrossRef]

86. Sommer, S.; Zucca, C.; Grainger, A.; Cherlet, M.; Zougmore, R.; Sokona, Y.; Hill, J. Application of indicator systems for monitoring and assessment of desertification from national to global scales. *Land Degrad. Dev.* **2011**, *22*, 184–197. [CrossRef]

87. Verón, S.R.; Paruelo, J.M.; Oesterheld, M. Assessing desertification. *J. Arid Environ.* **2006**, *66*, 751–763. [CrossRef]

88. Xu, D.; You, X.; Xia, C. Assessing the spatial-temporal pattern and evolution of areas sensitive to land desertification in North China. *Ecol. Indic.* **2019**, *97*, 150–158. [CrossRef]

89. Zdruli, P.; Pagliai, M.; Kapur, S.; Cano, A.F. (Eds.) *Land Degradation and Desertification: Assessment, Mitigation and Remediation*; Springer: Berlin, Germany, 2010.

90. Capozzi, F.; Di Palma, A.; De Paola, F.; Giugni, M.; Iavazzo, P.; Topa, M.E.; Adamo, P.; Giordano, S. Assessing desertification in sub-Saharan peri-urban areas: Case study applications in Burkina Faso and Senegal. *J. Geochem. Explor.* **2018**, *190*, 281–291. [CrossRef]

91. Contador, J.F.L.; Schnabel, S.; Gutiérrez-Gómez, A.; Fernández, M.P. Mapping sensitivity to land degradation in Extremadura. SW Spain. *Land Degrad. Dev.* **2009**, *20*, 129–144. [CrossRef]

92. Jiang, L.; Bao, A.; Jiaper, C.; Guo, H.; Zheng, G.; Gafforov, K.; Kurban, A.; De Mayer, P. Monitoring land sensitivity to desertification in Central Asia: Convergence or divergence? *Sci. Total Environ.* **2019**, *658*, 669–683. [CrossRef] [PubMed]

93. Labbaci, A.; Kabbachi, B.; Ezaidi, A.; Thorne, J. An Assessment of Sensitivity to Desertification in Western High Atlas of Morocco: An Application to Ain Asmama Site. In *Recent Advances in Geo-Environmental Engineering, Geomechanics and Geotechnics, and Geohazard*; Kallel, A., Erguler, Z.A., Cui, Z.D., Karrech, A., Karakus, M., Kulatilake, P., Shukla, S.K., Eds.; Advances in Science, Technology & Innovation (IEREK Interdisciplinary Series for Sustainable Development); Springer: Cham, Switzerland, 2019.

94. Lahlaoi, H.; Rhinane, H.; Hilali, A.; Lahssini, S.; Moukrim, S. Desertification assessment using MEDALUS model in watershed Oued El Maleh, Morocco. *Geosciences*. **2017**, *7*, 50. [CrossRef]

95. Prăvălie, R.; Săvulescu, I.; Patriche, C.; Dumitraşcu, M.; Bandoc, G. Spatial assessment of land degradation sensitive areas in south-western Romania using modified MEDALUS method. *Catena* **2017**, *153*, 114–130. [CrossRef]

96. Tsesmelis, D.E.; Karavitis, C.A.; Oikonomou, P.D.; Alexandris, S.; Kosmas, C. Assessment of the Vulnerability to Drought and Desertification Characteristics Using the Standardized Drought Vulnerability Index (SDVI) and the Environmentally Sensitive Areas Index (ESAI). *Resources* **2019**, *8*, 6. [CrossRef]

97. Boudjemline, F.; Semar, A. Assessment and mapping of desertification sensitivity with MEDALUS model and GIS—Case study: Basin of Hodna, Algeria. *J. Water Land Dev.* **2018**, *36*, 17–26. [CrossRef]

98. Leman, N.; Ramli, M.F.; Khirotdin, P.K. GIS-based integrated evaluation of environmentally sensitive areas (ESAs) for land use planning in Langkawi, Malaysia. *Ecol. Indic.* **2016**, *61*, 293–308. [CrossRef]

99. Rabah, B.; Aida, B. Adaptation of MEDALUS Method for the Analysis Depicting Land Degradation in Oued Labiod Valley (Eastern Algeria). In *Advances in Remote Sensing and Geo Informatics Applications*; El-Askary, H., Lee, S., Heggy, E., Pradhan, B., Eds.; Advances in Science, Technology & Innovation (IEREK Interdisciplinary Series for Sustainable Development); Springer: Cham, Switzerland, 2019.

100. Symeonakis, E.; Karathanasis, N.; Koukoulas, S.; Panagopoulos, G. Monitoring sensitivity to land degradation and desertification with the Environmentally Sensitive Area Index: The case of Lesvos island. *Land Degrad. Dev.* **2016**, *27*, 1562–1573. [CrossRef]

101. Vogt, J.V.; Safriel, U.; Von Maltitz, G.; Sokona, Y.; Zougmore, R.; Bastin, G.; Hill, J. Monitoring and assessment of land degradation and desertification: Towards new conceptual and integrated approaches. *Land Degrad. Dev.* **2011**, *22*, 150–165. [CrossRef]

102. Briassoulis, H. Policy-Oriented Integrated Analysis of Land-Use Change: An Analysis of Data Needs. *Environ. Manag.* **2001**, *27*, 1–11. [CrossRef]

103. Whitfield, S.; Reed, M.S. Participatory environmental assessment in drylands: Introducing a new approach. *J. Arid Environ.* **2012**, *77*, 1–10. [CrossRef]

104. Stringer, L.C.; Fleskens, L.; Reed, M.S.; de Vente, J.; Zengin, M. Participatory Evaluation of Monitoring and Modeling of Sustainable Land Management Technologies in Areas Prone to Land Degradation. *Environ. Manag.* **2014**, *54*, 1022–1042. [CrossRef]

105. Buxton, M.; Haynes, R.; Mercer, D.; Butt, A. Vulnerability to Bushfire Risk at Melbourne's Urban Fringe: The Failure of Regulatory Land Use Planning. *Geogr. Res.* **2011**, *49*, 1–12. [CrossRef]

106. Sadler, B.; Jacobs, P. A key to tomorrow: On the relationship of environmental assessment and sustainable development. In *Sustainable Development and Environmental Assessment: Perspectives on Planning for a Common Future*; Jacobs., P., Sadler, B., Eds.; Canadian Environmental Assessment Research Council: Ottawa, ON, Canada, 1991; pp. 3–32.

107. Ostrom, E. *Governing the Commons: The Evolution of Institutions of Collective Action*; Cambridge University Press: Cambridge, UK, 1990.

108. Bromley, D. *Environment and Economy*; Blackwell Publishers: Cambridge, UK, 1991.

109. Harou, P.A. What is the role of markets in altering the sensitivity of arid land systems to perturbation? In *Governing Desertification*; Do Humans Cause Deserts; Reynolds, J.F., Stafford-Smith, D.M., Eds.; Dahlem University Press: Berlin, Germany, 2002; pp. 253–274.

110. Deininger, K. *Land Policies for Growth and Poverty Reduction*; The World Bank: Washington, DC, USA, 2003.

111. National Research Council. *The Drama of the Commons*; The National Academies Press: Washington, DC, USA, 2002.

112. Hessel, R.; van den Berg, J.; Kaboré, O.; van Kekem, A.; Verzandvoort, S.; van Kekem, A.; Verzandvoort, S.; Dipama, J.-M. Linking participatory and GIS-based land use planning methods: A case study from Burkina Faso. *Land Use Policy* **2009**, *26*, 1162–1172. [CrossRef]

113. Morgan, R.K. Environmental impact assessment: The state of the art. *Impact Assess. Proj. A* **2012**, *30*, 5–14. [CrossRef]

114. CEC. *On the Application and Effectiveness of the EIA Directive*; Directive 85/337/EEC, as amended by Directives 97/11/EC and 2003/35/EC; COM(2009)378 Final; Commission of the European Communities: Brussels, Belgium, 2009.

115. Nkonya, E.; Gerber, N.; Baumgartner, P.; von Braun, J.; De Pinto, A.; Graw, V.; Kato, E.; Kloos, J.; Walter, T. *The Economics of Desertification, Land Degradation, and Drought toward an Integrated Global Assessment*; IFPRI Discussion Paper 01086; International Food Policy Research Institute and ZEF: Bonn, Germany, 2011.

116. UNCCD. The Economics of Desertification, Land Degradation and Drought: Methodologies and Analysis for Decision-Making. Background Document; United Nations Convention to Combat Desertification 2nd Scientific Conference. 2013. Available online: http://2sc.unccd.int (accessed on 2 November 2018).

117. Giger, M.; Liniger, H.; Sauter, C.; Schwilch, G. Economic Benefits and Costs of Sustainable Land Management Technologies: An Analysis of WOCAT's Global Data. *Land Degrad. Dev.* **2018**, *29*, 962–974. [CrossRef]

118. Glavovic, B.C.; Saunders, W.S.A.; Becker, J.S. Land-use planning for natural hazards in New Zealand: The setting, barriers, 'burning issues' and priority actions. *Nat. Hazards* **2010**, *54*, 679–706. [CrossRef]

119. Sapountzaki, K.; Wanczura, S.; Casertano, G.; Greiving, S.; Xanthopoulos, G.; Ferrara, F.F. Disconnected policies and actors and the missing role of spatial planning throughout the risk management cycle. *Nat. Hazards* **2011**, *59*, 1445–1474. [CrossRef]

120. Briassoulis, H. The Institutional Complexity of Environmental Policy and Planning Problems: The Example of Mediterranean Desertification. *J. Environ. Plan. Manag.* **2004**, *47*, 115–135. [CrossRef]

121. Shotter, J.; Tsoukas, H. Performing phronesis: On the way to engaged judgment. *Manag. Learn.* **2014**, *45*, 377–396. [CrossRef]

122. Flyvbjerg, B. *Making Social Science Matter*; Cambridge University Press: Cambridge, UK, 2001.

123. Healey, P. In Search of the "Strategic" in Spatial Strategy Making. *Plan. Theory Pract.* **2009**, *10*, 439–457. [CrossRef]

124. GNCCD. *Greek National Action Plan for Combating Desertification*; Greek National Committee for Combating Desertification: Athens, Greece, 2001.

125. Plummer, R.; Crona, B.; Armitage, D.R.; Olsson, R.; Tengö, M.; Yudina, O. Adaptive comanagement: A systematic review and analysis. *Ecol. Soc.* **2012**, *17*, 11. [CrossRef]

Article

Coastal Hazard Vulnerability Assessment Based on Geomorphic, Oceanographic and Demographic Parameters: The Case of the Peloponnese (Southern Greece)

Alexandra Tragaki, Christina Gallousi and Efthimios Karymbalis *

Department of Geography, School of Environment, Geography and Applied Economics, Harokopio University, 70 El. Venizelou Av., 17671 Kallithea, Athens, Greece; atragaki@hua.gr (A.T.); xgallousi@gmail.com (C.G.)
* Correspondence: karymbalis@hua.gr; Tel.: +30-210-9543159

Received: 5 April 2018; Accepted: 26 April 2018; Published: 1 May 2018

Abstract: Today low-lying coastal areas around the world are threatened by climate change-related hazards. The identification of highly vulnerable coastal areas is of great importance for the development of coastal management plans. The purpose of this study is to assess the physical and social vulnerability of the Peloponnese (Greece) to coastal hazards. Two indices were estimated: The Coastal Vulnerability Index (CVI) and the Social Vulnerability Index (SVI). CVI allows six physical variablesto be related in a quantitative manner whilethe proposed SVI in this studycontains mainly demographic variables and was calculated for 73 coastal municipal communities. The results reveal that 17.2% of the shoreline (254.8 km) along the western and northwestern coast of the Peloponnese, as well as at the inner Messiniakos and Lakonikos Gulfs, is of high and very high physical vulnerability. High and very high social vulnerabilities characterize communities along the northwestern part of the study area, along the coasts of the Messinian and Cape Malea peninsulas, as well as at the western coast of Saronikos Gulf.

Keywords: sea-level rise; storm surge; physical vulnerability; social vulnerability; Peloponnese; Greece

---

1. Introduction

Coastal areas have always been attractive settling grounds for human populations. They constitute the transitional zone between land and the marine environment; a particular area with unique natural and socioeconomic characteristics thatencourage the concentration of human activities [1]. Land cover change is considered an important element of recent environmental change at a global level. The rate of land-cover alteration in the coastal regions is increasing dramatically worldwide due to the increasing and intensifying human use of the land. These changes have also been analyzed in semi-arid and dry Mediterranean areas, concentrating specificallyon the consequences of farmland abandonment and reforestation carried out at different scales [2]. The Mediterranean coastal zone has a history of a millennia of, more or less, intensive human use. During the last century, the population along the Mediterranean coasts has grown impressively. Changes in the coastal landscapes are mainly due to changes in human land use that in turn are the result of changes in the wider socio-economic environment. In the era of globalization, coastal rural space takes up new content due to the changes related to the fact that rural locales are no longer agricultural sites, as in the past [3]. Moreover, plenty of rural places in coastal zone have now become "peri-urban" areas. At the present moment, besides the regions surrounding the largest cities, population increase has slowed in the remaining areas. Demographic dynamics have highlighted the gap between populations living in urban centers as opposed to the surrounding rural areas. Since the 1980s, however, several Mediterranean cities have undergone a rapid transition from the traditional "compact growth" model to the more "dispersed" ones, characterized by huge expansions of the built up area around the core [4,5]. Following economic growth, the most recent challenge was a drastic low-density sprawl coupled with impressive deconcentration processes in inner coastal cities. The mobility of the urban population to coastal rural/peri-urban areas has brought to the fore new issues related to land use/land cover change and to the changing perceptions of rural place, of local needs, and of priorities for rural development [6]. These changes in land use and land cover increase the exposure of coastal communities to a range of natural hazards [7]. Hence today low-lying coastal areas around the world are threatened by climate change-related hazards, such as the accelerated global mean sea-level rise and extreme storm surge events [8–13].

Global mean sea-level rise is caused by an increase in the volume of the oceans. This in turn is caused by thermal expansion (due to the warming of the oceans), loss of ice by glaciers and ice sheets, and a reduction of liquid water storage on land [14]. Although the exact rates of present and future global mean sea-level rise due to global warming are uncertain, it is predicted to reach approximately 53–98 cm by the year 2100 [15]. The most adverse projections are reported by Pfeffer et al. [16], who project asea-level rise likely to reach 0.8 m to 2 m. According to this study, the IPCC has not successfully modeled the dynamic development (decline) of the Greenland and Antarctic glaciers, a view also supported by other researchers (e.g., [17,18]).

Extreme storm surge events constitute an additional hazard. Storm surges, also referred to as meteorological residuals or meteorological tides, constitute along with the waves and the tidal oscillations the main components of extreme water levels along the coastal zone [9,19]. Storm surges are forced by wind driven water circulation towards or away from the coast and by atmospheric pressure driven changes of the water level; i.e., the inverse barometric effect. Some studies report an increased intensity and frequency of extreme water levels along several coastal regions in the world [20–23]. Hence the anticipated increase in extreme total water levels due to relative sea-level rise can be further enforced by an increase of the extreme storm surge level, which can exceed 30% of the relative sea-level rise [24].

Present sea-level trends in the Mediterranean basin have been estimated by analyzing available tide-gauge records longer than 35 years anda sea-level trend between 1.2 and 1.5 ± 0.1 mm/year has been calculated for the longest records [25]. Decadal sea-level trends in the Mediterranean are not always consistent with global values. In particular, during the 1990s the Mediterranean has shown enhanced sea level rise of up to 5 mm/year compared to the global average (mostly attributed to higher warming). The sea-level trend for the period between 1944 and 1989, observed in Alexandria (Egypt), at the only long-term gauge station in the Eastern Mediterranean, is 1.9 ± 0.2 mm/year and is the highest of the other stations in the basin [25]. According to the recent reports [15], mean sea-level in the Mediterranean is expected to rise at the rate of 5 cm/decade during the 21st century.

Regional projections of storm surge levels have been generated along the Mediterranean [26–28]. These projections show a general decreasing trend in the storminess of storm surge extremes in the Mediterranean Sea for a period of 150 years (1951–2100) under most of the considered climate change scenarios. This decreasing trend is mostly related to the frequency of local peaks and the duration and spatial coverage of the storm surges. However, the magnitudes of sea surface elevation extremes may increase in several Mediterranean sub-regions during the 21st century. There are clear distinctions in the contributions of winds and pressure fields to the sea-level height for various regions of the Mediterranean Sea, as well as on the seasonal variability of extreme values. The Aegean and Adriatic Seas are characteristic examples, where high surges are predicted to be mainly induced by low pressure systems and favorable winds, respectively [29].

Among the significant negative effects of the climate change-related coastal hazards are: coastal erosion with consequent loss of land of great economic, social, and environmental valueand infrastructure, frequent flooding of the low-lying coastal plains, inundation of ecologically significant wetlands, and threats to cultural and historical resources. Hence the identification of "sensitive" sections of coastline as well as the assessment of social vulnerability of the coastal communities is necessary for the development of coastal management plans. The understanding of a coastlines' response to both long-term and short-term sea-level rise as well as the assessment of the vulnerability of the coastal communities have become an important issue in recent years. Various approaches (mainly in the form of indices) have been proposed to predict the evolution of the coastal zone under the influence of sea-level rise. The relative physical vulnerability of different coastal environments to sea-level rise may be quantified by considering information regarding the important variables that contribute to coastal evolution in a given area, such as coastal geomorphology and slope, shoreline displacement, rate of relative sea-level change, tide range, wave height, and other related factors. Indices based mainly on these physical variables have been used to assess the physical vulnerability of coasts in the USA, Europe, Brazil, India, and Greece [30–37].

Recently, research on hazards in coastal areas has highlighted the need to incorporate other variable than physical variables in an attempt to capture the so-called social vulnerability [33,38–40]. Social vulnerability is described by various characteristics that condition a community's ability to respond to, cope with, recover from, and adapt to environmental hazards. According to a steadily growing literature, the socio-economic and demographic features are the main factors influencing social vulnerability [41]. In this context, investigations have been made relating to the evaluation of risk in coastal zones by taking into account not only natural but also socio-economic variables [39,42–45].

The aim of this study is to assess the overall vulnerability (physical and social) of the Peloponnese (southern Greek mainland) to both coastal erosion and flooding caused by climate change-related hazards. The assessment relies on the calculation of a Coastal Vulnerability Index (CVI) and a Social Vulnerability Index (SVI). The use of a quantitatively derived social vulnerability index, such as the proposed SVI, is important for two reasons. First, the method provides a useful tool for comparing the spatial variability in social vulnerability using a single value derived from multivariate characteristics. Second, SVI can be linked (statistically and spatially) to more physically based indices in calculating the overall vulnerability of a specific place. Not only does this index significantly contribute to the methods and metrics used in vulnerability science, but it also provides important comparative information for policy makers and emergency managers.

2. Study Area

The Peloponnese is a peninsula that covers an area of some 21,549.6 km^2 and constitutes the southernmost part of mainland Greece (Figure 1). It is actually an island separated fromthe central part of the country by theGulf of Corinth, a restricted marine embayment, with a nearly 105 km longitudinal axis lying in the E-W direction. At the western end of the Gulf, the Peloponnese is connected to the mainland of Greece by the bridge of Rio-Antirio. To the east the Peloponnese is separated from the Greek mainland by the Corinth Canal (an artificially dredged channel, 8 m deep and 21 m wide) which links the Gulf of Corinth with the Saronikos Gulf. The study area has mountainous interior and deeply indented coasts, especially along its eastern coastline (Figure 1). It possesses four south-pointing peninsulas: the Messinian, the Mani, the Cape Malea, and the Argolid in the far northeast, which are separated by the NW-SE trending Gulfs of Messinakos, Lakonikos, and Argolikos.

The Peloponnese consists of 152 municipal communities (a municipal community is the lowest level of government within the organizational structure of Greece), 73 of which are coastal (Figure 1). It has a population of 1,047,000, which corresponds to less than 10 % of the total population of Greece. Most of its population is concentrated in the coastal zone. The average population density in the coastal municipal communities is 77.8 inhabitants/km^2, whereas the mean value for Greece is 82.29 inhabitants/km^2. Its GDP contributes less than 7% of total GDP in Greece. Unemployment rates are higher than the national rates and three times higher than the average unemployment rate in the EU-28. The economy of the study area is distributed between the primary sector (mostly agricultural activities) and tourism related servicesand a small yet gradually growing industrial activityover the last years. Despite the dominance of the mountainous terrain, the Peloponnese hosts some of the most fertile lands in the country, producing renowned wine labels and top quality olive oil. Tourism is the region's heavy industry with high quality services and infrastructure facilities mainly along the coastline. Nowadays, the infrastructure works in transportation create new challenges for the area. The national road that connects Athens with Patras (the capital of the Peloponnese and the third largest city in Greece), extends along the north shoreline of the Peloponnese, while numerous cities and settlements, such as Patras, Kalamata, Korinthos, Nafplio, etc., are coastal (Figure 1).

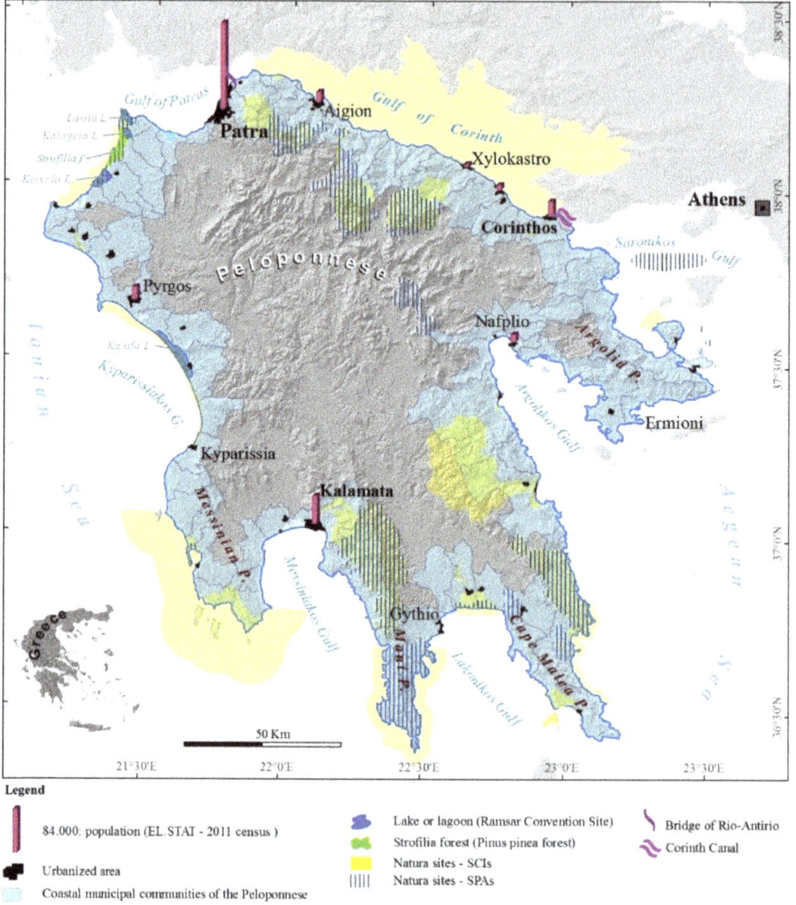

Figure 1. Hill-shaded map of the Peloponnese. The map also shows the 73 coastal municipal communities of the study area.

The Peloponnese is also of great environmental significance since it hosts ecologically important areas. Along its northwestern coast there is a 28 km zone with sandy beaches, dunes, lagoons (Kalogria, Strofilia, Lamia, and Kotychi, which is a Ramsar Convention site and the most significant lagoon in the Peloponnese), marshy areas, salt and freshwater wetlands, and the most extensive Pinuspinea forest in Greece (Strofilia forest) (Figure 1). Along the shore of the Kyparissiakos Gulf exists the significant wetland of Kaiafa Lake and a unique sand dune ecosystem of great ecological value.

From the above description it is more than evident that the Peloponnese deserves an assessment vulnerability study to spot regions of very high or high physical and social vulnerability, in order to promote relevant integrated management planning.

3. Materials and Methods

3.1. Physical Vulnerability

For the assessment of the physical vulnerability of the Peloponnese to climate change-related coastal hazards we applied the index proposed by Thieler and Hammar-Klose [31], which modified the initial CVI [38]. The CVI allows six physical variables: coastal geomorphology, shoreline shifting rate, coastal slope, relative sea-level rise rate, mean wave height, and mean tidal range to be related in a quantitative manner. In each variable a relative risk value is assigned based on the potential magnitude of its contribution to the physical changes on the coast as sea-level rises. The variables are ranked from 1 to 5 according to Table 1, with rank 1 indicating very low vulnerability and rank 5 indicating very high vulnerability. The theoretical minimum and maximum values of the CVI are 0.4 and 51.0 respectively. The ranges of vulnerability ranking, used in this study, are those proposed for the coastal environment of Greece [36]. The CVI is calculated as the square root of the product of the six variables divided by their total number.

$$\text{CVI} = \sqrt{\frac{a \cdot b \cdot c \cdot d \cdot e \cdot f}{6}}, \tag{1}$$

where, a: geomorphology, b: shoreline erosion/accretion rate, c: coastal slope, d: relative sea-level rise rate, e: mean significant wave height and f: mean tide range.

Categorization of coastal geomorphology classes in this study (Table 1) was undertaken using recent ortho-rectified aerial photographs (taken in 2009). Since this variable also represents the bedrock outcropping along the shoreline, data for the rock types were interpreted from the geological maps of Greece (at 1:50,000 scale),published by the Greek Institute of Geology and Mineral Exploration (IGME).

Table 1. Ranges for vulnerability ranking of the six CVI variables. The ranges of vulnerability ranking, used in this study, are those proposed for the coastal environment of Greece [36].

Variables	Vulnerability Categories				
	1	**2**	**3**	**4**	**5**
Geomorphology	Rocky, cliffed coasts	Medium cliffs, indented coasts	Low cliffs, alluvial plains	Cobble Beaches, Lagoons	Sandy beaches, deltas
Shoreline Erosion (−)/ Accretion (+) rate (m/year)	>(+1.5)	(+1.5)–(+0.5)	(+0.5)–(−0.5)	(−0.5)–(−1.5)	<(−1.5)
Coastal Slope (%)	>12	12–9	9–6	6–3	<3
Relative Sea-Level rise (mm/year)	<1.8	1.8–2.5	2.5–3.0	3.0–3.4	>3.4
Mean Wave Height (m)	<0.3	0.3–0.6	0.6–0.9	0.9–1.2	>1.2
Mean Tide Range (m)	<0.2	0.2–0.4	0.4–0.6	0.6–0.8	>0.8
Physical Vulnerability	very low	low	moderate	high	very high

Shoreline change rates were derived from ortho-rectified aerial photographs taken in 1969 and 2009, obtained from the Hellenic Military Geographical Service and the Hellenic Cadastre (Ktimatologio S.A.), respectively. The photomosaics of these photographs were manipulated within the GIS environment to digitize the shorelines of 1969 and 2009. The thematic layers of the 1969 and 2009 shorelines (in vector format) were overlaid, and with the use of GIS-based distance analysis functions, the final shoreline change map for the 40-year time period was obtained with estimated accretion and erosion rates.

To estimate the coastal slope values, a slope map of the coastal zone of the Peloponnese has been created with the use of the 1:5000 scale, topographic maps. With these maps as the main elevation source, the 5 m resolution Digital Elevation Model (DEM) of the coastal belt with an elevation from 0 to 50 m was created for the study area. Next, for this zone, the slope map was implemented within the ArcGIS spatial analysis extension environment, and a map of slope zones (according to Table 1) was constructed. Finally, for the assignment of the proper slope categorization to each coastline segment, an intersection of slope zones with the coastline was performed.

The coast of the Peloponnese relative sea-level change is the sum of the eustatism component and the local long-term tectonic vertical land movements. Published information concerning the late Holocene relative sea-level trends at the broader area of the Peloponnese was considered [46].

Mean annual values of the significant wave height were abstracted from the "Wave and Wind Atlas of the Hellenic Seas" [47], which are based on offshore measurements for the period between 1999 and 2007 (POSEIDON program), whereas tidal range was deduced from published information [48]. Every section of the coastline was assigned a risk value based on each specific data variable and the CVI was calculated.

GIS software ArcGIS (ver. 10.2), provided the platform for the coastal mapping and the calculation of the CVI. For each variable, the entire coastline of the study area is segmented into five sensitivity classes, and a sensitivity rank number is assigned to each segment of the coast (indicating the vulnerability level in terms of the given variable). The method of computing the CVI in the present study is similar to that applied by Pendleton et al., [49] Thieller and Hammar-Klose [31] and Abuodha and Woodroffe [50]. The difference is that instead of the "raster" approach, input parameters and final CVI values were estimated in coastline segments.This modified approach seems appropriate for medium scale. The size of each segment was 50 m. As mentioned above, for each of the six variables, a ranking on a scale of 1–5 was assigned to each segment (with rank 1 representing very low vulnerability and rank 5 indicating very high vulnerability) following the classification scheme outlined in Table 1. The final CVI map was generated by combining all of the variables. This map contains nearly 3500 segments of the coastline, each of which has a unique identity in its corresponding attribute table. Another column was added to this attribute table for the CVI formula so that the system generated the CVI values for all of the coastline segments of the Peloponnese. The combined CVI value was added as an attribute value for each coastline segment. Subsequently, the "natural breaks" classification method was used to categorize coastline segments according to their CVI magnitude for the construction of the final CVI zonation maps.

3.2. Social Vulnerability

Six variables have been used as indicators to assess social vulnerability of the coastal municipal communities of the Peloponnese: population density, share of women in total population, share of persons above 65 in total population, share of children below 5 in total population, share of foreign-born in total population, and share of low educated in total population. Demographic and socio-economic data were collected for all 1033 Greek municipal communities. Data were provided by the Hellenic Statistical Authority (EL.STAT) based on the 2011 Population and Household Census.

Since the six different variables used for the SVI calculation are measured in different units and scales, data must be standardized in order to be suitable for multivariate analysis and correlation tests. As most of the relevant data are not normally distributed, the method chosen is that of range standardization given by the following formula:

$$x'_{A,i} = \frac{x_{A,i} - x_{A,min}}{x_{A,max} - x_{A,min}}, \tag{2}$$

where $x'_{A,i}$: the standardized value of the variable A referring to the municipal community i, $x_{A,i}$: the observed value of the variable A referring to the municipal community i, $x_{A,min}$: the minimum observed value of the variable A, and $x_{A,max}$: the maximum observed value of the variable A.

All observation values are therefore between 0 (the case with the lowest value) and 1 (the case with the highest value). The six new standardized values were placed in an additive model to compose the SVI for each municipal community, making the assumption of an equal contribution of each variable to the community's overall vulnerability. The produced indicator is a relative measure of the overall vulnerability for each municipal community. The SVI_i for the municipal community i is therefore given by the following formula:

$$SVI_i = \sum_{A=1}^{6} x'_{A,i},$$ (3)

At a second stage, SVI_i scores are classified based on standard deviations from the mean into five categories, ranging from less than -1σ on the lower end to more than $+1\sigma$ on the upper end. Value 1 is given to the less vulnerable municipal communities while value 5 is given to the most vulnerable ones.

For each index (CVI and SVI) maps were produced depicting the geographic distribution of physical and social vulnerability along the coastal zone of the Peloponnese.

4. Results and Discussion

4.1. Coastal Vulnerability Index

The "coastal geomorphology" variable is non-numerical and expresses the relative response of different types of coastal landforms to sea-level rise. Geomorphology is ranked qualitatively according to the relative strength of the coastal landforms and rocks outcropping along the coast. The coastal zone of the Peloponnese is made up of a wide variety of features. The predominant coastal landform is steep marine cliffs made up of rocks highly resistant to coastal erosion, such as limestone and metamorphic rocks (592.9 km—40.0%) followed by sandy and shingle beaches, located along the coastal alluvial plains and river deltas (343.2 km, which is 23.2 % of the coastline), cobble pocket beaches (217.1 km—14.6%), medium rocky cliffs and indented coasts (180.6 km—12.2%) and low rocky cliffs or cliffs made up of less resistant to erosion formations (147.9 km—10%) (Table 2, Figure 2). Steep cliffs made up of hard geological formations (limestone and metamorphic rocks) offer maximum resistance and were classed with a rank of 1, whereas medium rocky coastal cliffs and indented coasts or cliffs of medium slope made up of weaker formations were assigned sensitivity ranks of 2 and 3, respectively. Cobble pocket beaches are susceptible to the effects of both natural marine processes and relative sea-level rise and are given a rank of 4 (high vulnerability). Sandy beaches, river deltas and fan-deltas were assigned a sensitivity rank of 5, while stabilized dune formations are considered that provide a kind of protection to the land behind them and were assigned a vulnerability rank of 4 (Table 2).

Table 2. The coastline length (as percentage % of the total shoreline length) in each vulnerability category (1–5) for the physical variables (a–f) of the CVI.

	Variables	Physical Vulnerability Classes				
		1 **Very Low**	**2** **Low**	**3** **Moderate**	**4** **High**	**5** **Very High**
a.	Geomorphology	40.0	12.2	10.0	14.6	23.2
b.	Shoreline Change	0.2	2.8	91.2	5.3	0.5
c.	Coastal Slope	55.5	8.3	7.9	14.1	14.3
d.	Relative Sea-level Rise	0.0	100.0	0.0	0.0	0.0
e.	Significant Wave Height	33.4	39.4	23.4	3.8	0.0
f.	Tidal Range	100.0	0.0	0.0	0.0	0.0
	CVI values	<1.2	1.2–3.2	3.2–5.2	5.2–7.1	>7.1
	CVI (%)	46.5	18.7	17.5	12.0	5.2

The shoreline change variable attempts to capture the historical trend of shoreline movement by determining the overall patterns of erosion or accretion. Shoreline change is one of the more complex parameters because the trend is typically variable over time [49]. The ranking of the shoreline change rate is based on the range of change in beach width values. A shoreline length of 1350.9 km, which corresponds to 91.2% of the coastline, is relatively stable (mean shoreline shifting rate within

±0.5 m/year), between 1969 and 2009. Nearly 78.2 km (5.3%) of the coastline is retreating with a mean rate between −0.5 and −1.5 m/year and a very small segment of the shoreline (7.8 km—0.5%), which mainly consists of sandy beaches at the aprons of the Alfios and Evrotas River deltas, have undergone fast erosion (with rates > −1.5 m/year). Nearly 41.5 km (2.8 %) of the shoreline has been prograded with a mean accretion rate between +0.5 and +1.5 for the period between 1969 and 2009, while a very small percentage of the coastline (0.2%—3.3 km) is prograding faster with a mean accretion rate higher than +1.5 m/year (Table 2). Accretion occurs at the mouth of some streams (Figure 2) due to increased sediment supply, especially during the rainy period of the year [51,52].

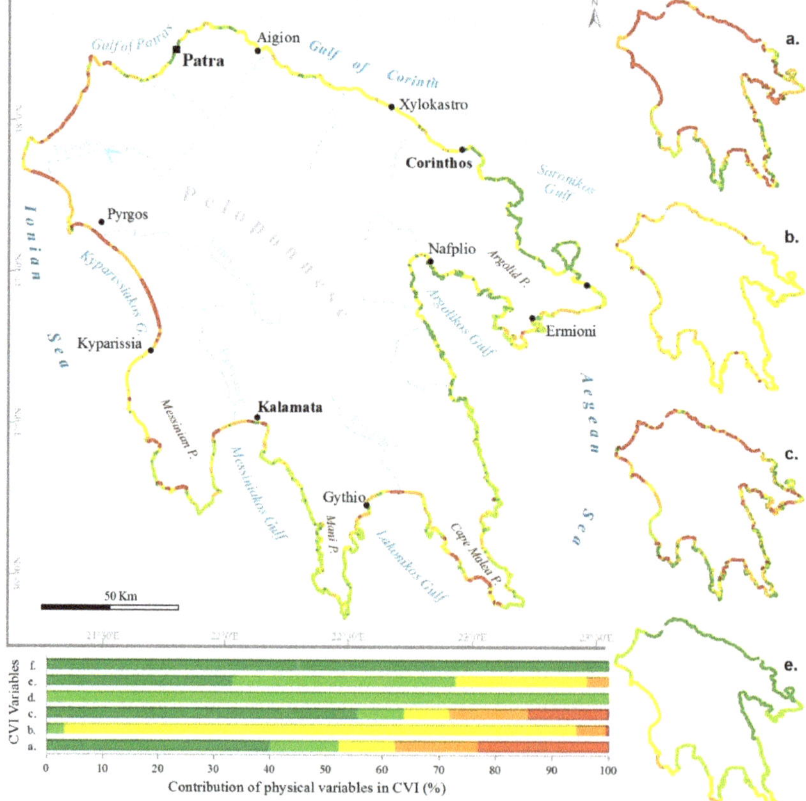

Figure 2. Maps of classification of the Peloponnese coastline into five physical vulnerability classes according to the variables of the Coastal Vulnerability Index (CVI) values. Inset maps (**a–c,e**) indicate the vulnerability ranking of various segments of the coast based on the CVI variables ((**a**) geomorphology, (**b**) shoreline change rate, (**c**) coastal slope and (**e**) mean significant wave height). Since the variables of relative sea-level rise and tidal range are considered to have the same value along the entire coastline of the Peloponnese, the figure does not include vulnerability ranking maps for these variables. The bar diagram shows the length (in percentage %) of shoreline in each vulnerability category for the CVI variables.

Determination of the regional coastal slope identifies the relative sensitivity of inundation and the potential rapidity of shoreline retreat because low-sloping coastal regions are thought to retreat faster than steeper regions [51]. The ranges for vulnerability ranking of the coastal slope variable are the same as those used in similar studies from around the world [31,49]. Regions with coastal slopes lower than 3% were characterized with very high vulnerability, whereas coastal cliffs with slopes higher than 12% were classified as areas of very low vulnerability (Table 1). Nearly 56.0% of the coastal zone of the study area (which corresponds to 822.9 km) belongs to the very low vulnerability class. Very low vulnerability area lies mainly along the western coast of the Saronikos Gulf as well as along the east and west coasts of the four NW-SE trending peninsulas: the Messinian, the Mani, the Cape Malea, and the Argolid. On the other hand,14.3% of the study coastal zone (which corresponds to 211.4 km) is low-lying and is characterized as very highly susceptible to inundation. This very highly vulnerable area lies partly along the southern coastline of the Gulf of Corinth as well as along part of the western coast of the Peloponneseand at the coast of the inner parts of the Messinakos, Lakonikos and Argolikos Gulfs (Figure 2). Low-lying coasts are occupied by beaches developed along the fronts of alluvial plains and river deltas. Nearly 14.0% of the coast (208.4 km) is comprisedof slope between 3 and 6% and is considered highly vulnerable, whereas 116.4 km (7.9%) of the coastal zone is of moderate vulnerability, having a slope between 6% and 9% (Table 2).

Due to the lack of recent accurate long-term sea-level measurements, the values for the variable of relative sea-level rise were estimated coupling the effects of eustatism and local land level movements caused by tectonics. The relative sea-level rise is considered to have the same value along the coastline of the Peloponnese (Table 2, Figure 2) and took a value < 2.0 mm/year based on estimations from studies relevant to eustatic sea-level rise in Greece [46]. This ranking is in agreement with the mean eustatic global sea-level rise rate for the time period between 1850 and 1950 [9], as well as with the trend of the sea-level for the period 1985–2001 at the station of Kalamata (0.6 $\pm$ 0.1 mm/year) [19].

Wave heights are proportional to the square root of wave energy, which is a measure of the capacity for erosion.The wave climate of the Peloponnese coastline is affected by offshore significant wave heights between 0.1 and 1.0 m [47], according to the output of the wave model (POSEIDON program), which has been calibrated with the use of offshore field measurements. A significant part of the coastline of the Gulf of Patras, as well as the entire coast of the Gulfs of Corinth and Argolikos is dominated by offshore significant wave heights <0.3 m. Thus, 494.8 km (33.4%) of the Peloponnese coastline is considered to have very low vulnerability (rank 1). Nearly 584.4 km (39.4%) of the coastline, located mainly along the northwestern and eastern shore of the Peloponnese, including also the coasts of Messiniakos and Lakonikos Gulfs, is characterized by offshore significant wave heights between 0.3 and 0.6 m (rank 2). Approximately 23.0% of the study area (which corresponds to 346.2 km), that lies mostly along the western coast of the Peloponnese and partly the NW-SE facing coasts of the Messinian, the Mani, and the Cape Malea peninsulas are of moderate vulnerability since offshore wave heights range between 0.6 and 0.9 m. Finally, small segments (56.3 km—3.8%) of the southwestern shores of the Messinian and Mani peninsulas are dominated by significant wave heights between 0.9 and 1.2 m and are ranked as highly vulnerable (Figure 2e).

The tidal range is linked to both inundation and erosion hazards [30]. For the vulnerability ranking of the tidal range variable for the Peloponnese, the ranges proposed for Greece were considered (Table 1). These ranges were proposed taking into account the tide range variations for the Greek seas, which are generally less than 10 cm [48]. However, the overall fluctuation of sea-level exceeds 0.5 m due to meteorological forcing (differences in barometric pressure, wind, and wave setup) [53]. In this study, tidal range is ranked as such that extremely microtidal (tidal range < 0.2 m) coasts are at low risk and less microtidal (tidal range > 0.8 m) coasts are at high risk. The reasoning is that although a large tidal range dissipates wave energy, limiting beach or cliff erosion to a brief period of high tide, it also delineates a broad zone of intertidal area that will be most susceptible to inundation following long-term sea-level rise [30]. Furthermore, the velocity of tidal currents depends partially on the tidal range. High tidal range is associated with stronger tidal currents that are capable of eroding and transporting sediment [30]. The coast of the Peloponnese is a microtidal environment with tidal (astronomical) range <4 cm. As such, the tidal range variable is ranked with the value 1 (very low vulnerability) (Table 2, Figure 2).

The CVI values alongside the Peloponnese range between 0.82 and 11.18. The median value of the index for the study area is 4.11 and the standard deviation is 2.17. The geographical distribution of the vulnerability of the Peloponnese coast to climate change-related coastal hazards is presented schematically in Figure 2. CVI values above 7.1 are classified as having very high vulnerability. Nearly 77.6 km, corresponding to 5.2% of the total coastline length, was assigned to this category. A total length of 177.2 km (12.0%) of the coastline is classified as having high vulnerability (CVI values between 5.2 and 7.1). The deltaic plains of the most extensive deltas (those of Alfios and Pinios Rivers) of the western Peloponnese as well as the coastal plains of the inner Messiniakos and Lakonikos Gulfs are characterized by very high and high levels of vulnerability, primarily due to the low regional coastal slope, the high erodible of the coastal landforms, the high rates of erosion and the relatively high offshore mean significant wave heights. About 17.5% (259.5 km) of the shoreline, mainly along the Northern Peloponnese is moderately vulnerable (with CVI values between 3.2 and 5.2) while 18.7% (277.8 km) is of low vulnerability (CVI values between 1.2 and 3.2). Finally, values below 1.2 are assigned to the very low vulnerability category with 46.5% (689.7 km) of the shoreline belonging to this class (Table 2). The low and very low vulnerability categories are primarily located at the western rocky coasts of the Peloponnese, as well as along the NW-SE oriented cliffy coasts of the Messinian, Mani, and Cape Malea peninsulas (Figure 2).

4.2. Social Vulnerability Index (SVI)

Densely populated areas are thought to be more vulnerable, as the total amount of people and assets per km^2 poses a higher vulnerability of total damage in case of a disaster. Additionally, high population density increases the evacuation time as well as the consequences of a disaster (injuries and fatalities). In our case-study, population density ranges from 5.8 to 1352.7 inhabitants/km^2 with a mean score of 109.9 (standard deviation: 196.99). In respect to this criterion, 25 municipal communities (which correspond to 34.2% of the 73 total communities) are classified as low vulnerable, whereas the other 48 communities belong to the moderate vulnerability class (Table 3). Higher population densities are concentrated primarily along the southern coast of the Gulf of Corinth and at the western part of the Peloponnese (Figure 3a).

Table 3. Number of coastal communities (as percentage% of the total coastal municipal communities of the Peloponnese) in each vulnerability category (1–5) for the social variables (a–f) of the SVI.

	Variables	Social Vulnerability Classes				
		1 Very Low	2 Low	3 Moderate	4 High	5 Very High
a.	Population Density	0.0	34.2	65.8	0.0	0.0
b.	Share of Women in t.p.	16.4	5.5	65.8	12.3	0.0
c.	Share of Persons above 65 in t.p.	13.7	26.0	50.7	5.5	4.1
d.	Share of Children below 5 in t.p.	4.1	11.0	57.5	16.4	11.0
e.	Share of foreign-born in t.p.	0.0	8.2	39.7	17.8	34.2
f.	Share of Low Educated in t.p.	8.2	31.5	45.2	4.1	11.0
SVI	Number of Communities	8	15	22	14	14
	(%)	11.0	20.5	30.1	19.2	19.2

Women, especially those living in less developed economies or in low status households, have limited access to knowledge and resources and are therefore less efficient in protecting or helping themselves out of a disaster [54]. Their care-giving role is an additional aggravating factor. Thus, relevant literature considers high shares of women in a population a high risk factor. In our case-study, the share of women in the total population ranges from 28.9% to 51.6%, with amean value of 48.9% (standard deviation: 0.028). Most of the communities (48 out of 73) are moderately vulnerable, 16 municipal communities (21.9%) belong to the very low and low vulnerability classes, while 9 (12.3%) of them are highly vulnerable (Table 3 and Figure 3b).

Elders, especially those living alone, often have mobility limitations and special needs that may require the assistance of others. Hence high values of "share of persons above 65 in total population" indicate high vulnerability. The shares of elders within the communities populations range from 10.7% to 38.9%, with a mean score of 23.55% (standard deviation: 0.058). Most of the communities (37 corresponding to 50.7%) fall into the moderate vulnerability category, whereas 29 (39.7%) show low and very low vulnerabilities (Table 3). Regarding this particular variable, four highly and three very highly vulnerable coastal communities are located in the southern part of the Peloponnese (Figure 3c).

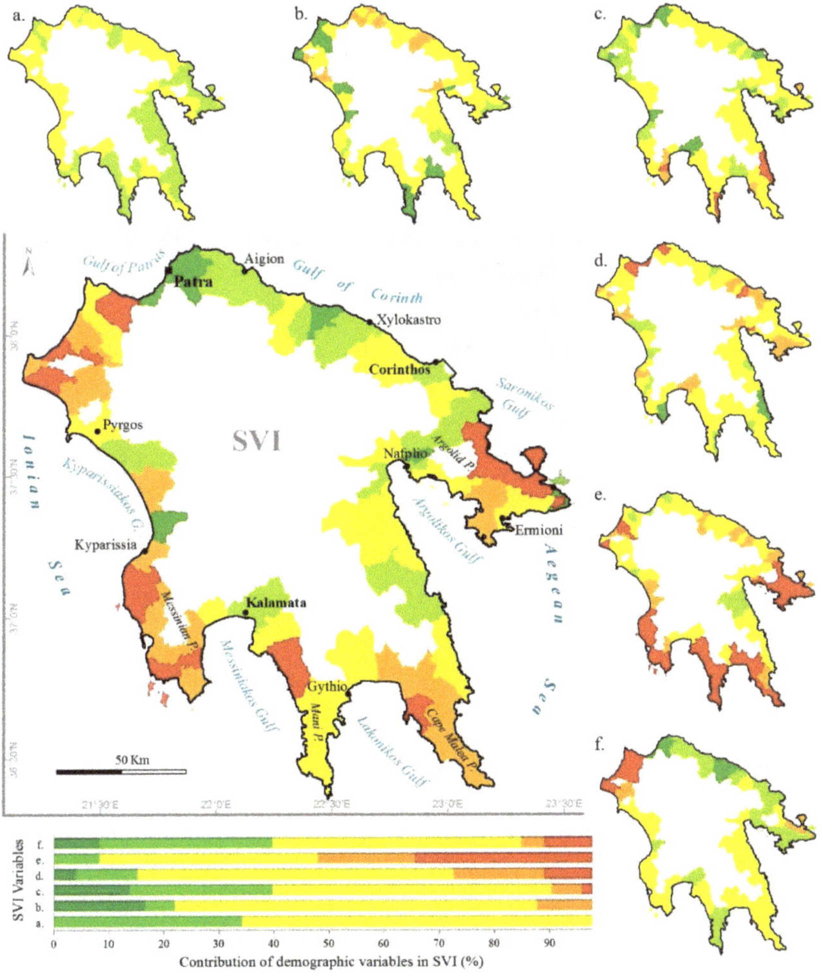

Figure 3. Maps of classification of the coastal municipal communities of the Peloponnese into five social vulnerability classes according to the variables of the Social Vulnerability Index (SVI) values. Inset maps (**a–f**) indicate the social vulnerability ranking of the communities based on the categorization of the SVI variables' values ((**a**) population density, (**b**) share of women in total population, (**c**) share of persons above 65 in total population, (**d**) share of children below 5 in total population, (**e**) share of foreign-born in total population and (**f**) share of low educated in total population). Inset bar diagram show the length (in percentage %) of shoreline in each vulnerability category for the SVI variables.

Children (<5 years) can hardly protect themselves during a disaster for they lack the necessary resources, knowledge, as well as life experience to survive. The values of the variable "share of children below 5 in total population" range between 2.9% and 9.8%, with a mean score of 5.34% (standard deviation: 0.057). Vulnerability increases with the values of this parameter. The majority of the coastal municipal communities (42–57.5%) show moderate vulnerability, 20 out of 73 (27.4%), primarily located at the north and northwest part of the Peloponnese, belong to the high and very high vulnerability classes, while the rest show low and very low vulnerabilities (Table 3, Figure 3d).

Racial and ethnic minorities are supposed to be more vulnerable to hazards for they are more likely to be poor [55]. Since we lack detailed ethnic data at this administrative level, we usethe "share of foreign-born in total population". Non-natives are thought to be more vulnerable for different reasons: (i) disaster communication is made more difficult due to limited language skills (ii) the impact may be greater due to higher social and economic marginalization as well as to cultural differences. Across the coastal municipalities of Peloponnese, the shares of foreignersrange from 2.9% to 35.1% of total population with mean score of 11.8% (standard deviation: 0.057). A significant number of communities (38 out of 73—52%), especially those of the Argolid, Cape Malea, Mani, and Messinian peninsulas are characterized by high and very high vulnerability (Figure 3e). A concentration of highly and very highly vulnerable communities is also observed at the northwestern part of the Peloponnese.

"Share of low educated in total population" is a variable that reflects the socio-economic status of an area. Low education implies limited access to information, followed by poor prevention measures and is often related to social and economic deprivation. The share of low-educated within the municipalities populationgoes from a minimum value of 9.5% to a maximum of 21.6% with mean score of 14.9% (standard deviation: 0.033). In respect to this demographic feature, most of the coastal municipal communities of the Peloponnese belong to the moderate, low, and very low vulnerability categories (Table 3). The geographic distribution of this parameter shows the presence of highly and very highly vulnerable communities at the northwestern part of the study area. Some communities along the southwestern shores of Saronikos Gulf are also of high vulnerability (Figure 3f).

In order to compose the SVI and due to the different scales of measures used for each variable, the values have been standardized [56]. The SVI ranges from a minimum value of 0.91 (lowest social vulnerability) to 2.67 (highest social vulnerability) with a mean score of 2.11 (standard deviation: 0.25). The geographical distribution of the social vulnerability to climate change-related coastal hazards for the Peloponnese is presented schematically in Figure 3. Most communities exhibit low or moderate levels of social vulnerablity since more than 6 out of 10 communities are classified in the first three groups. However, the share of highly and very highly vulnerable communities goes up to almost 38.4% (Table 3). High and very high levels of social vulnerability are clustered in communities along the coasts of the Messinian and Cape Malea peninsulas, as well as along the western coast of Saronikos Gulf and at the northwestern part of the Peloponnese. In contrast, the communities along the southern coast of the Gulf of Corinth, and most of the communities along the east coast of the Gulf of Patras, as well as most of the Argolikos Gulf coastal communities to the east, have moderate to very low social vulnerability levels (Figure 3).

The comparison among the final two CVI and SVI maps shows that seven coastal municipal communities of the Peloponnese (Vouprasia, Lechena, Vartholomio, Gastouni, Amaliada, Zacharo, and Elos) show high and very high levels of overall (both physical and social) vulnerability to costal hazards. The high overall vulnerability of these communities is a function of high or very high values of SVI while more than 80% of their coastal zone is also of high and very high physical vulnerability. The highly vulnerable (physically and socially) communities are concentrated at the northwestern coastal zone of the Peloponnese, while there is one highly vulnerable community at the inner part of the Lakonikos Gulf (Figure 4).

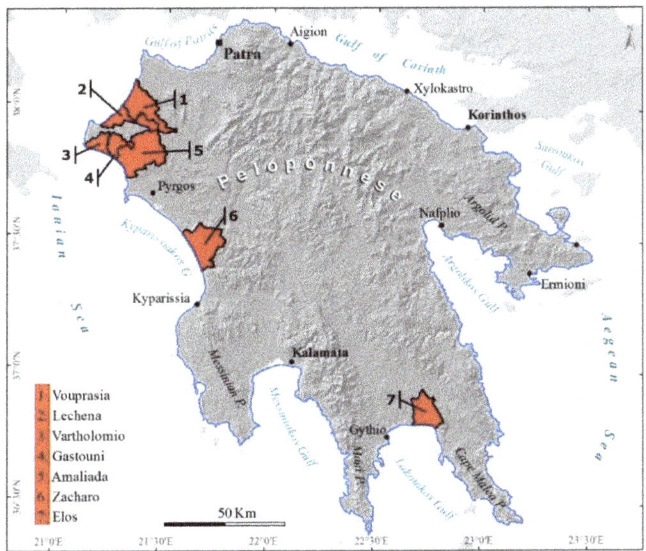

Figure 4. Location map of the seven more vulnerable (both physically and socially) municipal communities of the Peloponnese. These municipal communities have high or very high values of SVI while more than 80% of their coastal zone is also of high and very high physical vulnerability.

5. Conclusions

Of the six physical vulnerability variables, "geomorphology", "regional coastal slope", and "mean significant wave height" introduce the greatest variability to the CVI values. Among the other three parameters, "shoreline change rate" shows a small variation, while "tidal range" and "relative sea-level rise rate" have the same values along the entire coastline. On the other hand, all six variables used as indicators to assess social vulnerability, show a great variability. Among the social vulnerability variables,"share of foreign-born in total population", "share of children below 5 in total population" and "share of low educated in total population" are those that increase the SVI values within the study area.

The application of the proposed methodology reveals that there are differences among the geographical distribution of physical and social vulnerability. Nearly 17.2% of the shoreline (254.8 km) consisting primarily of low-lying sandy beaches and deltaic plains, has high and very high levels of physical vulnerability. Coastline segments of high physical vulnerability level are geographically concentrated at the western and northwestern coast of the Peloponnese, as well as at the inner coastal areas of Messiniakos, and the Lakonikos Gulfs. The highly and very highly social vulnerable communities are concentrated along the northwestern coast of the Peloponnese, along the coasts of the Messinian and Cape Malea peninsulas, as well as at the western coast of Saronikos Gulf. There are seven high risk coastal municipal communities in the study area (Vouprasia, Lechena, Vartholomio, Gastouni, Amaliada, Zacharo and Elos) andthey present high and very high values of social vulnerability. Accordingly, more than 80% of their coastal zone is also of high and very high physical vulnerability. Six of them are located in the northwestern Peloponnese and host ecologically important sites.

This study provided a comprehensive and detailed spatial GIS digital database of topographic, geological, and physio-geographical characteristicsfor 1481.7 km of shoreline, as well as population and demographic data for the coastal municipal communities of the Peloponnese, which can be renewed and expanded further to incorporate newly available data (e.g., storm surge, socio-economic status, land values), including new variables (e.g., sediment budget, environmental parameters) in the future for better results of physical and social vulnerability assessment.

The implementation of the indices used in this study, at the national scale could assist in a preliminary identification of the hazardous and vulnerable coastal areas of the country. CVI and SVI can be also integrated and used in the policies defined by Agenda 2030 for Sustainable Development, which is a plan of action for the people, the planet and for prosperity that also seeks to strengthen universal peace and freedom. The main target of Goal 13 is to take urgent action to combat climate change and its impacts. After the initial assessment of the most physically vulnerable segments of the coastlines to coastal hazards, focusing on regions with specific socio-economic and environmental interest is required, in order to re-examine the study area on a larger, more detailed, scale. The results of the categorization of the coasts in different risk categories can be a useful tool for managers of coastal areas and for those responsible for carrying out national protection policies, strategies and planning of the coastal zone against climate change driven natural hazards. Hence the application of CVI and SVI along the Greek coast would strengthen resilience and adaptive capacity to climate-related hazards and coastal natural disasters in the country.

Author Contributions: A.T. analyzed the demographic data and developed the Social Vulnerability Index; C.G. created and organized the GIS spatial database and created the maps of the geographic distribution of the vulnerability variables (physical and social); E.K. analyzed the data regarding the Coastal Vulnerability Index; A.T, C.G. and E.K. wrote the paper.

Acknowledgments: We would like to thank G.B. and H.S., Guest Editors of this Special Issue, for their helpful suggestions as well as the anonymous reviewers for their comments and corrections that significantly improved the paper.

Conflicts of Interest: The authors declare no conflict of interest.

References

1. Costanza, R. The ecological, economic, and social importance of the oceans. *Ecol. Econ.* **1999**, *31*, 199–213. [CrossRef]
2. Bellot, J.; Bonet, A.; Pena, J.; Sánchez, J. Human impacts on land cover and water balances in a Coastal Mediterranean county. *Environ. Manag.* **2007**, *39*, 412–422. [CrossRef] [PubMed]
3. Chalkias, C.; Papadopoulos, A.; Ouils, A.; Karymbalis, E.; Detsis, V. Land cover changes in the coastal peri-urban zone of Corinth, Greece. In Proceedings of the Tenth International Conference on the Mediterranean Coastal Environment, Rhodes, Greece, 25–29 October 2011; Volume 2, pp. 913–923.
4. Salvati, L.; Munafò, M.; Morelli, V.G.; Sabbi, A. Lowdensitysettlementsand land use changes in a Mediterranean urban region. *Landsc. Urban Plan.* **2012**, *105*, 43–52. [CrossRef]
5. Catalán, B.; Saurí, D.; Serra, P. Urban sprawl in the Mediterranean? Patterns of growth and change in the Barcelona metropolitan region 1993–2000. *Landsc. Urban Plan.* **2008**, *85*, 174–184. [CrossRef]
6. Romano, B.; Zullo, F. The urban transformation of Italy's Adriatic Coast Strip: Fifty years of unsustainability. *Land Use Policy* **2014**, *38*, 26–36. [CrossRef]
7. Satta, A.; Puddu, M.; Venturini, S.; Giupponi, C. Assessment of coastal risks to climate change related impacts at the regional scale: The case of the Mediterranean region. *Int. J. Disaster Risk Reduct.* **2017**, *24*, 284–296. [CrossRef]
8. Church, J.; White, N. Sea-level rise from the late 19th to the early 21st century. *Surv. Geophys.* **2011**, *32*, 585–602. [CrossRef]
9. Losada, I.J.; Reguero, B.G.; Méndez, F.J.; Castanedo, S.; Abascal, A.J.; Mínguez, R. Long-term changes in sea-level components in Latin America and the Caribbean. *Glob. Planet. Chang.* **2013**, *104*, 34–50. [CrossRef]
10. Hinkel, J.; Lincke, D.; Vafeidis, A.T.; Perrette, M.; Nicholls, R.J.; Tol, R.S.J.; Marzeion, B.; Fettweis, X.; Ionescu, C.; Levermann, A. Coastal flood damage and adaptation costs under 21st century sea-level rise. *Proc. Natl. Acad. Sci. USA* **2014**, *111*, 3292–3297. [CrossRef] [PubMed]
11. Hogarth, P. Preliminary analysis of acceleration of sea level rise through the twentieth century using extended tide gauge data sets (August 2014). *J. Geophys. Res. Oceans* **2014**, *119*, 7645–7659. [CrossRef]
12. Hoggart, S.P.G.; Hanley, M.E.; Parker, D.J.; Simmonds, D.J.; Bilton, D.T.; Filipova-Marinova, M.; Franklin, E.L.; Kotsev, I.; Penning-Rowsell, E.C.; Rundle, S.D.; et al. The consequences of doing nothing: The effects of seawater flooding on coastal zones. *Coast. Eng.* **2014**, *87*, 169–182. [CrossRef]

13. Jevrejeva, S.; Moore, J.C.; Grinsted, A.; Matthews, A.P.; Spada, G. Trends and acceleration in global and regional sea levels since 1807. *Glob. Planet. Chang.* **2014**, *113*, 11–22. [CrossRef]
14. Allen, J.C.; Komar, P.D. Climate controls on US west coast erosion processes. *J. Coast. Res.* **2006**, *22*, 511–529. [CrossRef]
15. IPCC. Summary for policy makers. In *Climate Change 2013: The Physical Science Basis*; Stocker, T.F., Qin, D., Plattner, G.K., Tigno, M., Allen, S.K., Boschung, J., Nauels, A., Xia, Y., Bex, V., Midgley, P.M., Eds.; Cambridge University Press: Cambridge, UK, 2013.
16. Pfeffer, W.T.; Harperand, J.T.; O'Neel, S. Kinematic constraints on Glacier Contributions to 21st Century Sea-Level Rise. *Science* **2008**, *321*, 1340–1343. [CrossRef] [PubMed]
17. Rohling, E.J.; Grant, K.; Bolshaw, M.; Roberts, A.P.; Siddall, M.; Hemleben, C.H.; Kucera, M. Antarctic temperature and global sea-level closely coupled over the past five glacial cycles. *Nat. Geosci.* **2009**, *2*, 500–504. [CrossRef]
18. Grinsted, A.; Moore, J.C.; Jevrejeva, S. Reconstructing sea-level from paleo and projected temperatures 200 to 2100 AD. *Clim.Dyn.* **2010**, *34*, 461–472. [CrossRef]
19. Lowe, J.A.; Woodworth, P.L.; Knutson, T.; McDonald, R.E.; McInnes, K.L.; Woth, K.; von Storch, H.; Wolf, J.; Swail, V.; Bernier, N.B.; et al. *Past and Future Changes in Extreme Sea Levels and Waves. Understanding Sea-Level Rise and Variability*; Wiley-Blackwell: London, UK, 2010.
20. Ullmann, A.; Monbaliu, J. Changes in atmospheric circulation over the North Atlantic and sea-surge variations along the Belgian coast during the twentieth century. *Int. J. Climatol.* **2010**, *30*, 558–568. [CrossRef]
21. Izaguirre, C.; Méndez, F.J.; Espejo, A.; Losada, I.J.; Reguero, B.G. Extreme wave climate changes in Central-South America. *Clim. Chang.* **2013**, *119*, 277–290. [CrossRef]
22. Wang, X.L.; Feng, Y.; Swail, V.R. Changes in global ocean wave heights as projected using multimodel CMIP5 simulations. *Geophys. Res. Lett.* **2014**, *41*, 1026–1034. [CrossRef]
23. Weisse, R.; Bellafiore, D.; Menéndez, M.; Méndez, F.; Nicholls, R.J.; Umgiesser, G.; Willems, P. Changing extreme sea levels along European coasts. *Coast. Eng.* **2014**, *87*, 4–14. [CrossRef]
24. Vousdoukas, M.; Voukouvalas, E.; Annunziato, A.; Giardino, A.; Feyen, L. Projections of extreme storm surge levels along Europe. *Clim. Dyn.* **2016**, *47*, 3171–3190. [CrossRef]
25. Marcos, M.; Tsimplis, M.N. Coastal sea-level trends in Southern Europe. *Geophys. J. Int.* **2008**, *175*, 70–82. [CrossRef]
26. Marcos, M.; Jordà, G.; Gomis, D.; Pérez, B. Changes in storm surges in southern Europe from a regional model under climate change scenarios. *Glob. Planet. Chang.* **2011**, *77*, 116–128. [CrossRef]
27. Jordà, G.; Gomis, D.; Álvarez-Fanjul, E.; Somot, S. Atmospheric contribution to Mediterranean and nearby Atlantic sea level variability under different climate change scenarios. *Glob. Planet. Chang.* **2012**, *80–81*, 198–214. [CrossRef]
28. Conte, D.; Lionello, P. Characteristics of large positive and negative surges in the Mediterranean Sea and their attenuation in future climate scenarios. *Glob. Planet. Chang.* **2013**, *111*, 159–173. [CrossRef]
29. Androulidakis, Y.; Kombiadou, K.; Makris, C.; Baltikas, V.; Krestenitis, Y. Storm surges in the Mediterranean Sea: Variability and trends under future climatic conditions. *Dyn. Atmos. Oceans* **2015**, *71*, 56–82. [CrossRef]
30. Gornitz, V. Global coastal hazards from future sea-level rise. *Glob.Planet. Chang.* **1991**, *89*, 379–398. [CrossRef]
31. Thieler, E.R.; Hammar-Klose, E.S. *National Assessment of Coastal Vulnerability to Sea-Level Rise*; U.S. Atlantic Coast; U.S. Geological Survey: Reston, VA, USA, 1999.
32. Pendleton, E.A.; Thieler, E.R.; Williams, S.J. *Coastal Vulnerability Assessment of Cape Hatteras National Seashore (CAHA) to Sea-Level Rise*; USGS Open File Report; US Geological Society: Reston, VA, USA, 2008.
33. Diez, P.G.; Perillo, G.M.E.; Piccolo, C.M. Vulnerability to sea-level rise on the coast of the Buenos Aires Province. *J. Coast. Res.* **2007**, *23*, 19–126. [CrossRef]
34. Rao, K.N.; Subraelu, P.; Rao, T.V.; Malini, B.H.; Ratheesh, R.; Bhattacharya, S.; Rajawat, A.S. Sea-level rise and coastal vulnerability: An assessment of Andhra Pradesh coast, India through remote sensing and GIS. *J. Coast. Conserv.* **2008**, *12*, 195–207. [CrossRef]
35. Gaki-Papanastassiou, K.; Karymbalis, E.; Poulos, S.; Seni, A.; Zouva, C. Coastal vulnerability assessment to sea-level rise based on geomorphological and oceanographical parameters: The case of Argolikos Gulf, Peloponnese, Greece. *Hell. J. Geosci.* **2011**, *45*, 109–121.

36. Karymbalis, E.; Chalkias, C.; Chalkias, G.; Grigoropoulou, E.; Manthos, G.; Ferentinou, M. Assessment of the sensitivity of the southern coast of the Gulf of Corinth (Peloponnese, Greece) to sea-level rise. *Cent. Eur. J. Geosci.* **2012**, *4*, 561–577. [CrossRef]

37. Karymbalis, E.; Chalkias, C.; Ferentinou, M.; Chalkias, G.; Magklara, M. Assessment of the Sensitivity of Salamina and Elafonissos islands to Sea-level Rise. *J. Coast. Res.* **2014**, *70*, 378–384. [CrossRef]

38. Gornitz, V.; Daniels, R.C.; White, T.W.; Birdwell, K.R. The development of a coastal vulnerability assessment database: Vulnerability to sea-level rise in the U.S. southeast. *J. Coast. Res.* **1994**, *12*, 327–338.

39. Boruff, B.; Emrich, C.; Cutter, S.L. Erosion hazard vulnerability of US coastal countries. *J. Coast. Res.* **2005**, *21*, 932–942. [CrossRef]

40. Lichter, M.; Felsenstein, D. Assessing the costs of sea-level rise and extreme flooding at the local level: A GIS-based approach. *Ocean Coast. Manag.* **2012**, *59*, 47–62. [CrossRef]

41. Flanagan, B.E.; Gregory, E.W.; Hallisey, E.J.; Heitgerd, J.L.; Lewis, B. A Social Vulnerability Index for Disaster Management. *J. Homel. Secur. Emerg. Manag.* **2011**, *8*. [CrossRef]

42. Szlafsztein, C.; Sterr, H. A GIS-based vulnerability assessment of coastal natural hazards, state of Pará, Brazil. *J. Coast. Conserv.* **2007**, *11*, 53–66. [CrossRef]

43. Gorokhovich, Y.; Leiserowitz, A.; Dugan, D. Integrating Coastal Vulnerability and Community-Based Subsistence Resource Mapping in Northwest Alaska. *J. Coast. Res.* **2014**, *30*, 158–169. [CrossRef]

44. Satta, A.; Puddu, M.; Firth, J.; Lafitte, A. *Climate Risk Management Tools: Towards a Multi-Scale Coastal Risk Index for the Mediterranean*; Plan Bleu Report; Plan Bleu: Valbonne, France, 2015.

45. Mavromatidi, A.; Briche, E.; Claeys, C. Mapping and analyzing socio-environmentalvulnerability to coastal hazards induced by climate change: An application to coastal Mediterranean cities in France. *Cities* **2018**, *72*, 189–200. [CrossRef]

46. Lambeck, K. Sea-level change and shore-line evolution in Aegean Greece since Upper Palaeolithic time. *Antiquity* **1996**, *70*, 588–611. [CrossRef]

47. Soukisian, T.; Hatzinaki, M.; Korres, G.; Papadopoulos, A.; Kallos, G.; Anadranistakis, E. *Wave and Wind Atlas of the Hellenic Seas*; Hellenic Centre for Marine Research Publication: Anavyssos, Greece, 2007.

48. Tsimplis, M.N. Tidal oscillations in the Aegean and Ionian Seas. *Estuar. Coast. Shelf Sci.* **1994**, *39*, 201–208. [CrossRef]

49. Pendleton, E.A.; Thieler, E.R.; Williams, S.J. Importance of coastal change variables in determining vulnerability to sea- and lake-level change. *J. Coast. Res.* **2010**, *26*, 176–183. [CrossRef]

50. Abuodha, P.A.O.; Woodroffe, C.D. Assessing vulnerability to sea-level rise using a coastal sensitivity index: A case study from southeast Australia. *J. Coast. Conserv.* **2010**, *14*, 189–205. [CrossRef]

51. Dwarakish, G.S.; Vinay, S.A.; Natesan, U.; Asano, T.; Kakinuma, T.; Venkataramana, K.; Jagedeesha, B.; Badita, M.K. Coastal vulnerability assessment of the future sea-level rise in Udupi coastal zone of Karnataka state, west coast of India. *Ocean Coast. Manag.* **2009**, *52*, 467–478. [CrossRef]

52. Parcharidis, I.; Kourkouli, P.; Karymbalis, E.; Foumelis, M.; Karathanassi, V. Time Series Synthetic Aperture Radar Interferometry for Ground Deformation Monitoring over a Small Scale Tectonically Active Deltaic Environment (Mornos, Central Greece). *J. Coast. Res.* **2013**, *29*, 325–338. [CrossRef]

53. Alexandrakis, G.; Karditsa, A.; Poulos, S.; Ghionis, G.; Kampanis, N.A. Vulnerability assessment for to erosion of the coastal zone to a potential sea-level rise: The case of the Aegean Hellenic coast. In *Environmental Systems in Encyclopedia of Life Support Systems (EOLSS)*; Sydow, A., Ed.; Eolss Publisher: Oxford, UK, 2009.

54. Enarson, E.; Morrow, B.H. A gendered perspective: The voices of women. In *Hurricane Andrew: Ethnicity, Gender, and the Sociology of Disasters*; Peacock, W.G., Morrow, B.H., Gladwin, H., Eds.; International Hurricane Center, Laboratory for Social and Behavioral Research: Miami, FL, USA, 1997; pp. 116–140.

55. Rygel, L.; O'Sullivan, D.; Yarnal, B. A method for constructing a Social Vulnerability Index: An application to hurricane storm surges in a developed country. *Mitig. Adapt. Strateg. Glob. Chang.* **2006**, *11*, 741–764. [CrossRef]

56. Cutter, S.L.; Mitchell, J.T.; Scott, M.S. Revealing the vulnerability of people and places: A case study of Georgetown county, South Carolina. *Ann. Assoc. Am. Geogr.* **2000**, *90*, 713–737. [CrossRef]

Article

Temporal and Spatial Analysis of Flood Occurrences in the Drainage Basin of Pinios River (Thessaly, Central Greece)

George D. Bathrellos [1,*], Hariklia D. Skilodimou [1], Konstantinos Soukis [2] and Efterpi Koskeridou [3]

[1] Department of Geography and Climatology, Faculty of Geology and Geoenvironment, National and Kapodistrian University of Athens, University Campus, ZC. 15784 Zografou, Athens, Greece; hskilodimou@geol.uoa.gr

[2] Department of Dynamic Tectonic Applied Geology, Faculty of Geology and Geoenvironment, National and Kapodistrian University of Athens, University Campus, ZC. 15784 Zografou, Athens, Greece; soukis@geol.uoa.gr

[3] Department of Historical Geology-Palaeontology, Faculty of Geology and Geoenvironment, National and Kapodistrian University of Athens, University Campus, ZC. 15784 Zografou, Athens, Greece; ekosker@geol.uoa.gr

* Correspondence: gbathrellos@geol.uoa.gr; Tel.: +30-210-727-4882

Received: 11 August 2018; Accepted: 7 September 2018; Published: 11 September 2018

Abstract: Historic data and old topographic maps include information on historical floods and paleo-floods. This paper aims at identifying the flood hazard by using historic data in the drainage basin of Pinios (Peneus) River, in Thessaly, central Greece. For this purpose, a catalogue of historical flood events that occurred between 1979 and 2010 and old topographic maps of 1881 were used. Moreover, geomorphic parameters such as elevation, slope, aspect and slope curvature were taken into account. The data were combined with the Geographical Information System to analyze the temporal and spatial distribution of flood events. The results show that a total number of 146 flood events were recorded in the study area. The number of flood events reaches its maximum value in the year 1994, while October contains the most flood events. The flood occurrences increased during the period 1990–2010. The flooded area reaches its maximum value in the year 1987, and November is the month with the most records. The type of damages with the most records is for rural land use. Regarding the class of damages, no human casualties were recorded during the studied period. The annual and monthly distribution of the very high category reaches the maximum values, respectively, in the year 2005 and in June. The analysis of the spatial distribution of the floods proves that most of the occurrences are recorded in the southern part of the study area. There is a certain amount of clustering of flood events in the areas of former marshes and lakes along with the lowest and flattest parts of the study area. These areas are located in the central, southern, south-eastern and coastal part of the study area and create favorable conditions for flooding. The proposed method estimates the localization of sites prone to flood, and it may be used for flood hazard assessment mapping and for flood risk management.

Keywords: historic flood data; old topographic maps; GIS; temporal and spatial distribution of flood events; marshy areas and lakes; flood hazard assessment

1. Introduction

Natural hazards are physical events that can cause significant damages to the natural and human environment. Endogenic or exogenic processes such as active tectonics and climate changes are capable of changing landforms and triggering natural hazards, which in some cases control human activities [1–11].

Floods are physical phenomena active in geological time and the result of excess runoff. When rivers overtop their banks, the excess water goes to the floodplain. Floodplains represent favorable sites for man to settle because they are fertile, level, easy to excavate and near water. These features have contributed to increased development and urbanization of floodplains, thereby increasing the chances of the flood occurrences that cause disasters. Despite the construction of flood control works such as dams, levees and channeling rivers, the flood damage has increased [12–16].

Floods occur frequently in some parts of the world. While for some areas, yearly flooding is necessary to sustain crops, for other areas, flooding spells disaster. Especially, in urban areas, floods are considered among the most dangerous natural hazards due to the increasing number of events. Their consequences are not only environmental, but social and economic, as well, since they may cause damages to urban areas and agricultural lands and may even result in the loss of lives [17]. Worldwide, flood events have caused the largest amount of deaths and property damage during the last decades [18]. In Europe, the annual average flood damage in the last two decades was about €4 billion per year [19]. In the Mediterranean countries, floods tend to be greater in magnitude compared to the inner continental countries, and they frequently cause catastrophic damages [20].

Flood occurrence can be estimated in space and time through a sound basis of knowledge acquired by the scientific use of a large number of historical documents [21]. It is possible to gather useful information about historical floods that occurred during a period ranging from 1–2 centuries by using historical data [22]. Payrastre et al. [21] used discharges of historical floods that occurred during a period ranging from 1–2 centuries in four watersheds of the French Mediterranean area. They estimated the flood peak discharges and the associated low and high bound of its possible values for the main floods of each studied area. Historic data are useful for floodplain management, flood insurance rating, emergency planning and flood risk management. A careful search for data and their suitable arrangement may result in a powerful tool for flood disaster prevention and land use planning [23–25].

One essential step in any flood hazard analysis and estimation is the flood hazard recognition. This statement includes the record of possible geomorphologic evidence of flood activity, historic accounts and records [26]. Hydrologic models using time series flood data can be applied to assess flood peaks, depths, volumes and to map flood hazard areas [27–31]. However, these methods require data that are often unavailable [32]. Alternatively, researchers have used historical documents and historic flood data to estimate flood hazard [22,33]. Tropeano et al. [22] analyzed historical documents for a statistical elaboration of the frequency of landslide and flood events in Northern Italy in the last five centuries. Diakakis et al. [33] used an extensive catalogue of flooding phenomena during the last 130 years to examine flood events in Greece. According to their statistical and spatial analysis, urban areas tend to present higher flood recurrence rates than mountainous and rural ones and an increasing trend in reported flood event numbers during the last few decades. Moreover, a diachronic appraisal of the landscape evolution throughout the study of old topographic maps may include information on ungauged historical floods and paleofloods [34,35]. Skilodimou et al. [35] used old topographic maps to investigate the causes of flooding generation in the southwestern coast of Attica in Greece. They proved that the former wetlands and the lagoon of the study area have been dried up and covered by buildings, causing flood occurrences.

This paper aims to provide recognition of flood hazard by using historic data in the drainage basin of Pinios (Peneus) River, in Thessaly, central Greece. For this purpose, a catalogue of historical flood events that have caused damages was studied. These data, along with old topographic maps and geomorphic parameters such as elevation, slope, aspect and slope curvature were evaluated to record the temporal and spatial distribution of flood events of the study area. The analysis, processing and the evaluation of the old maps, geomorphic parameters and flood events were performed in a GIS environment.

2. Study Area

The drainage basin of the study area is located in Thessaly, central Greece (Figure 1a), and it covers an area of about 11,200 km^2. The elevations of the study area vary from 0–2678 m a.s.l. (Figure 1b). The hydrologic basin is drained by Pinios River, which has an approximate length of 205 km. It is the third longest river in Greece and crosses a large part of the eastern part of central Greece discharging into the Aegean Sea, where it forms a delta.

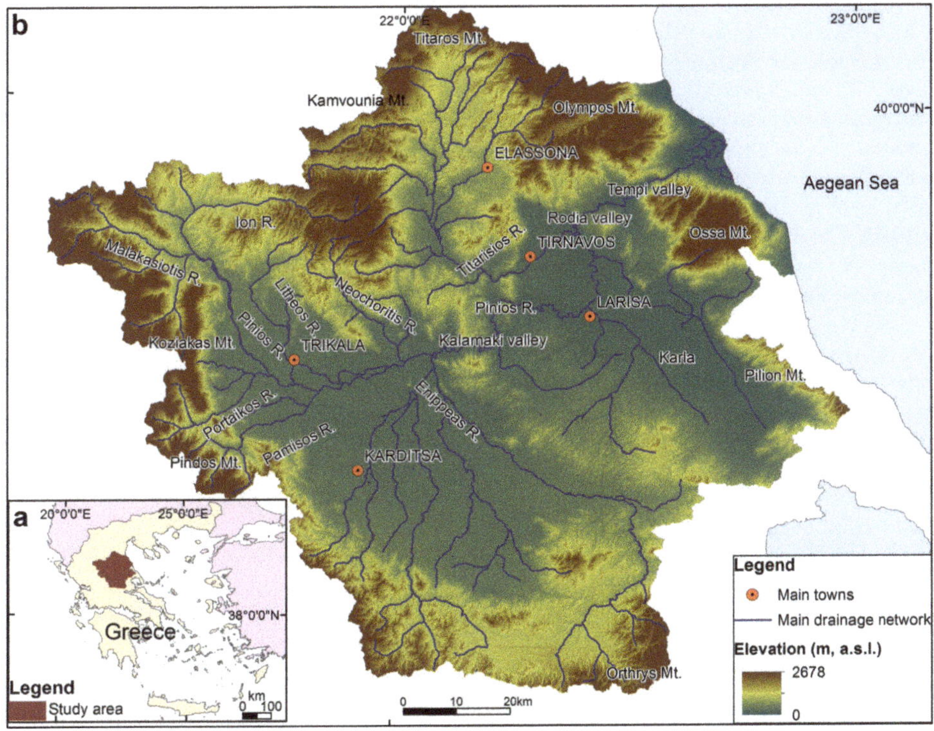

Figure 1. (**a**) Location map of the study area; (**b**) the elevations of the study area, the drainage network, the road network and the main settlements (modified from [36]).

Its course starts at the northwestern part of the Thessaly plain, from the confluence of the Ion and Malakasiotis rivers. It is surrounded by mountainous areas, which enclose its drainage basin and form its watershed (Figure 1b). To the north are the Titaros Mt. (1837 m) and the Kamvounia Mt. (1615 m); to the northeast are the Olympos Mt. (2917 m) and the Ossa Mt. (1978 m); to the east is the Pilio Mt. (1548 m); to the south is the Orthrys Mt. (1726 m); and finally, to the west are the Pindos Mt. (2204 m) and the Koziakas Mt. (1901 m).

The major tributaries of Pinios River are the Malakasiotis, Portaikos, Pamisos and Enippeas rivers to the west and south and the Ion, Lithaios, Neochoritis and Titarisios rivers to the north, which all drain large, heterogeneous areas, through extensive hydrographic networks (Figure 1b).

The plain area crossed by Pinios River is divided by the presence of a low-lying hill area into two parts (Figure 1b): (a) a western part (Trikala-Karditsa), whose altitude varies between 80 and 200 m, and (b) an eastern part (Larisa), whose altitude varies between 45 and 100 m. The mountainous and hilly area crossed by Pinios River has a quite rugged surface with altitudes exceeding 200 m. The most important gorges are the ones of Kalamaki (internal low-lying hill area), Rodia and Tempi (between Olympos and Ossa). Through these areas, the water flow velocity is high, while erosion is intense. All the above, in conjunction with the fertility of the plain areas, necessitated the construction of drainage and irrigation projects.

Fifty five percent (55%) of the study is mainly rural, semi-urban area. Irrigated agricultural land comprises 56% of the total cultivated area in the basin. The intense and extensive cultivation has led to a remarkable water demand increase, which is usually covered by the over-exploitation of groundwater resources [37].

The climate in the western and central side of the drainage basin of Pinios River is continental with cold winters and hot summers with a large temperature difference. The coastal area of the basin has a typical Mediterranean climate. Summers in the study area are usually very hot and dry, and in July and August, temperatures can reach up to 40 °C. The mean annual precipitation over the basin is about 779 mm. Rainfall is distributed unevenly in space, and it varies from about 360 mm at the eastern part of the basin to more than 1850 mm at the western mountain peaks. Generally, the rainiest months are November and December, while rainfall is rare from June–August. The mountain areas receive significant amounts of snow during the winter months. The mean annual total flow of the river is 3500×10^6 m^3. The hydrologic regime of the river is affected by both winter rainfalls and spring snowmelt. The river flow regime at the main stem is perennial. The mean average flow near its delta ranges from more than 150 m^3/s in February and March to about 10 m^3/s in August and September. The water of Pinios River is used primarily for irrigation [37,38].

The Pinios River crosses alpine formations of various units and post-alpine formations. More specifically, the alpine formations at its western part belong to the Koziakas unit, at the central part to the Sub-Pelagonian and Pelagonian units and at its eastern part to the Pelagonian and Olympos-Ossa units. At the western sub-basin, a great part of its length passes through molassic formations of the Meso-Hellenic trough (Eocene-Pliocene) and Quaternary formations, while at the eastern part, it crosses mostly Neogene and Quaternary formations. The permeable rocks and sediments of the study area are divided into two major units, the karstic one consisting of carbonate formations (limestones, marbles and dolomites) and the granule materials, which include Neocene and Quaternary sediments. Semi-permeable formations are comprised of loose to semi-coherent Quaternary and Neocene deposits along with thin-bedded limestones and ophiolites. Impermeable formations consist of schists and fine-grained Quaternary and Neocene deposits [39–41].

The plain of Thessaly has suffered from flooding since ancient times, when several structures had been constructed in order to control the Pinios River 2500 years ago. Today, despite the construction of flood protection works, floods remain an important problem of the area [42].

3. Materials and Methods

3.1. Data

In the present study, the historical flood database of Greece was used [43]. It refers to floods that occurred during the period 1979–2010. This database was required by the Directive 2007/60/EC concerning floods in Europe. In Article 4 of the Directive 2007/60/EC, the construction of a historic flood data database is assigned to every state-member as a part of the preliminary assessment of flood risk, as well as finding the sensitive to flooding areas in each country, for the database will be used for the construction of flood risk maps [44]. The database contains information such as the location of flood (county and village or locality), geographical details, dates of occurrence, characteristics of the flood (e.g., flash flood, snow melt flood, etc.), the sources of flooding (e.g., pluvial, fluvial, groundwater, etc.), the causes of flooding, the mechanism of the flood, the extent of flooded area (in km^2), the maximum distance of the flood (in km), the type of damages, the total cost of the damage caused by the flood event, the number of human casualties and the class of the total damage caused by the flood (very high, high, medium or low).

Four old topographic maps of 1881 with the scale 1:200,000 were used in the current work. The used sheets are: Ioannina-Metsovon-Grevena-Kozani-Servia [45], Larissa-Elasson-Katerini [46],

Arta-Trikala-Karditsa [47] and Volos-Farsalos-Lamia [48]. Figure 2 shows a mosaic of the four old topographic maps used.

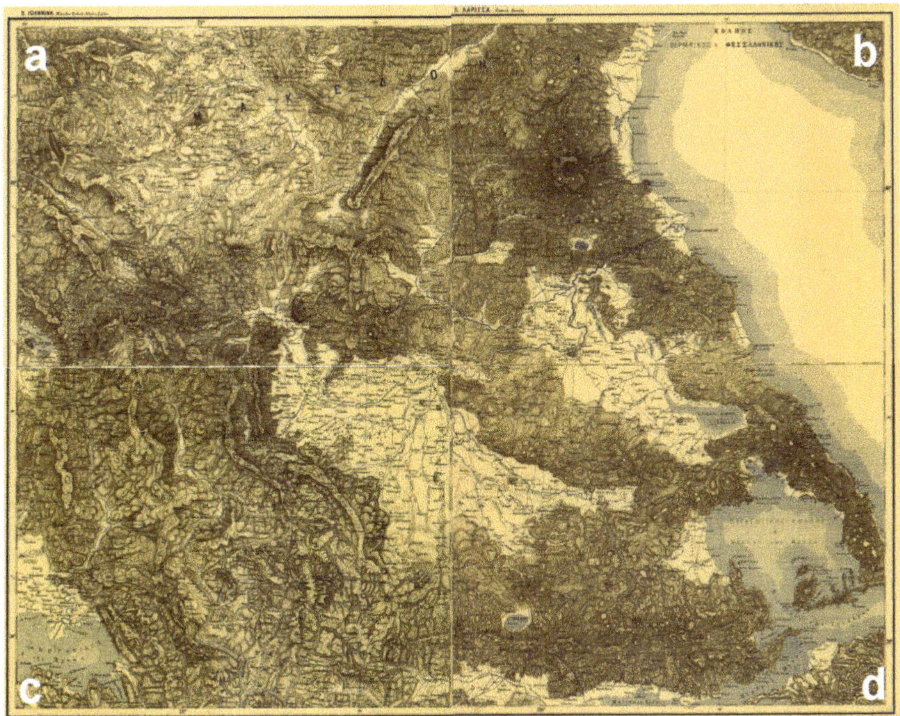

Figure 2. Old topographic maps of 1881, sheets: (**a**) Ioannina-Metsovon-Grevena-Kozani-Servia; (**b**) Larissa-Elasson-Katerini; (**c**) Arta-Trikala-Karditsa; (**d**) Volos-Farsalos-Lamia.

Finally, thematic maps were used. These maps are the digital elevation model (DEM) of the study area, which is presented in Figure 1 and the geological map of the study area. They were acquired from previous work [36]. The DEM map was used to produce the elevation, slope, aspect and slope curvature maps. The geological map of the study area was used to identify the semi-permeable and impermeable formations of the study area.

3.2. Statistical Analyses

Several works have used historic flood data to analyze flood frequency, flood peak discharges and estimate flood hazard maps [21,27,29,30]. This kind of flood frequency analysis requires hydro-climatic data, such as flow and rainfall. In many cases, this is difficult to do, because detailed data are unavailable. On the other hand, researchers have analyzed the temporal and spatial distribution of floods based on the count of events [22,33]. The scope of the present work is to recognize flood hazard and examine the temporal and spatial distribution of flood events in the study area, rather than to create a flood hazard map. Additionally, high resolution data such as discharge or rainfall during flood events were not readily available. For these reasons, the statistical analysis depends on the location of floods, dates of occurrences, the flooded area, the type of damages and the total damage caused by the floods.

For the description and assessment of historical floods, information was sourced from various national and regional authorities, scientific reports and newspaper articles. In general, the accuracy of the data used was considered acceptable as multiple checks between different sources were carried out to ensure consistency of the data. The location of events shows some uncertainty. According to [49], when there was no reference to a particular community, but the geographical determination was different (i.e., reference to river or torrent), location was determined on the basis of the other descriptive information. Thus, in some cases, as an event location is given the center of the municipal department. The historic database includes information about the causes of flooding, the flood mechanism and characteristics along with the maximum distance of the flood. The data of these records were not systematic and accurate, and they were not taken into account in the present work.

As a first step, a temporal statistical analysis was performed to identify the number of events and flooded area distribution during this period. The statistical significance of these variables was tested by *p*-values. Additionally, to identify the seasonal distribution of flood events, these parameters were analyzed for each month.

Flood events that occurred in the study area had adverse consequences and caused damages to economic activity. The type of damage was categorized into three categories: economic (positions of infrastructures, industrial and commercial centers), rural land use (positions of farmland with significant economic value production) and property [49]. The distributions of the number of each type of damage from 1979–2010 along with their monthly distribution were examined.

The total damages caused by the floods were categorized into five classes taking into account the number of human casualties, amount of monetary compensation (state compensation for damages to agriculture and for damages to settlements) or the size of flooded area [50]. The classes of the total damage and the limits of the above-mentioned parameters are presented in Table 1.

Table 1. Categories of the total damages, amount of monetary compensation and flooded area of the historical flood events.

Classes of Total Damages	Human Casualties	Compensation (Euro)	Flooded Area (Hectares)
Low		<50,000	<200
Medium		50,000–200,000	200–500
High		200,000–500,000	500–1000
Very High	≥1	>500,000	>1000

The annual and monthly distributions of the two above-mentioned parameters were studied during the period 1979–2010.

3.3. Spatial Analyses

Furthermore, a spatial analysis of the flood occurrences was carried out. The study area was divided into seven individual drainage basins. The spatial distribution of flood events in each drainage basin was examined. A spatial database was created, and ArcGIS 10.3 software was used to process the collected data.

Old maps may describe or simply locate areas that have flooded in the past. In many cases, former wetlands and lagoons that have been dried up cause flood events. Thus, the study of old topographic maps is very helpful to examine the causes of flood genesis [22,34,35]. The comparative observation of the old topographic maps of 1881 and the recent topographic maps of the study area led to a mapping of the previous wetlands. Marshy areas and lakes were digitized from the older edition survey maps, and they compared with the spatial distribution of current flood events. Finally, bibliographic data [36] were used to examine the association of the paleo-environment of Pinios River with the spatial frequency of floods.

The geomorphological setting of an area affects flood occurrences. Lowland morphology, gentle slopes and the flat areas create favorable conditions for flooding [51]. For this reason, the elevation, slope, aspect and slope curvature of the study area were examined. The selection of these geomorphological parameters and the determination of the class numbers, as well as their boundary values was based on the literature [51–54]. The elevation map was classified into two categories: (i) <100 m and (ii) >100 m. Similarly, the slope was divided into two categories: (i) <5° and (ii) >5°. The aspect map shows the slope direction and identifies the flat areas that have values = −1°. Thus, the aspect map was categorized into two classes: (i) = −1° and (ii) >−1°. In the case of curvature, negative curvatures represent concave, zero curvature represents flat and positive curvatures represent convex, respectively. Figure 3 shows the elevation, slope, aspect and the curvature maps and their categories. The areas with elevation < 100 m, slope < 5°, aspect = −1° and curvature = 0 were combined to identify the lowest and flattest areas of the study area. This was accomplished by the intersection tool of ArcGIS 10.3.

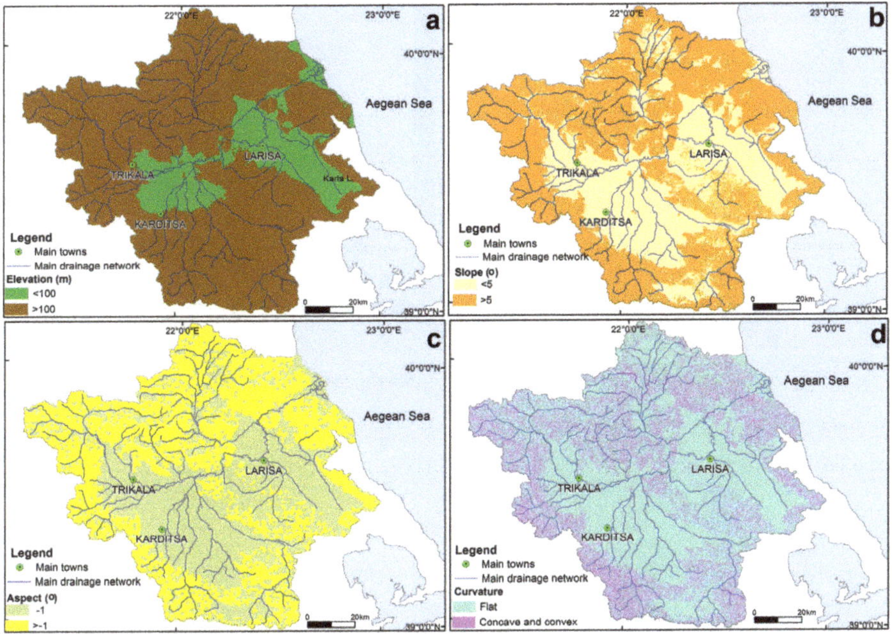

Figure 3. Thematic maps of the four geomorphological parameters: (**a**) elevation; (**b**) slope; (**c**) aspect; (**d**) curvature.

4. Results

4.1. Temporal Distribution of Flood Events

A total number of 146 flood events were recorded in the drainage basin of Pinios River during the period 1979–2010. Figure 4 shows the temporal distribution of flood occurrences during this period.

The number of flood events reached its maximum value (38) in 1994. The value of flood events was high (36) in the year 1987 and was relatively high (19) in 2002. In 1994, flood events were two-times higher than in 2002. In the 1980s, a total number of 40 flood events were recorded, while in the 1990s and 2000s, 55 and 47 were observed, respectively. Thus, there was a trend of increasing flood occurrences over the last two decades. The statistical analysis of the number of events showed that $p = 0.009$. Thus, the data used have values $p < 0.05$ and are statistically significant.

Figure 5 presents the seasonal distribution of flood events. The statistical analysis proved that October contained the most flood events (42). High values of flood occurrences were observed in March (35), in June (11) and in November (10). In October, the number of flood events was four-times higher than in June and November.

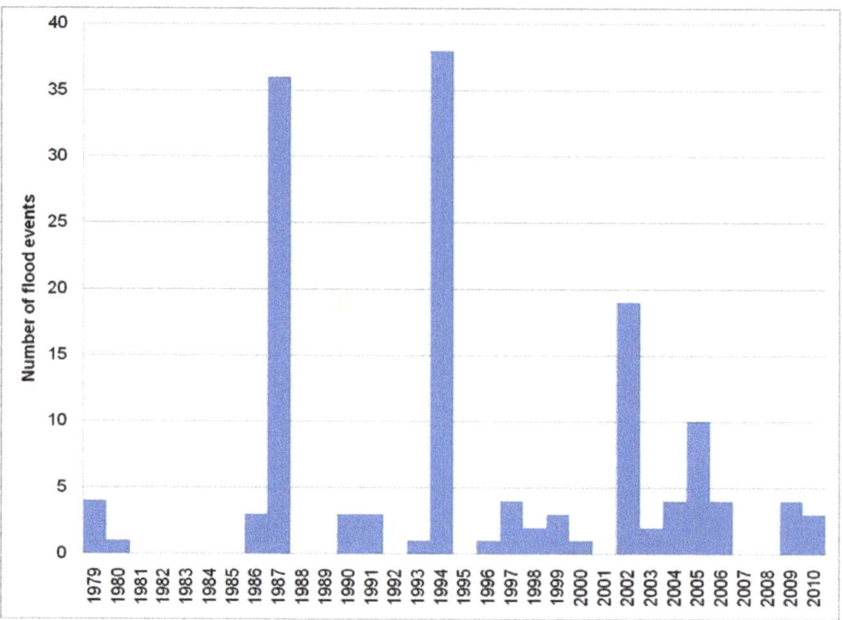

Figure 4. The annual distribution of flood events from 1979–2010.

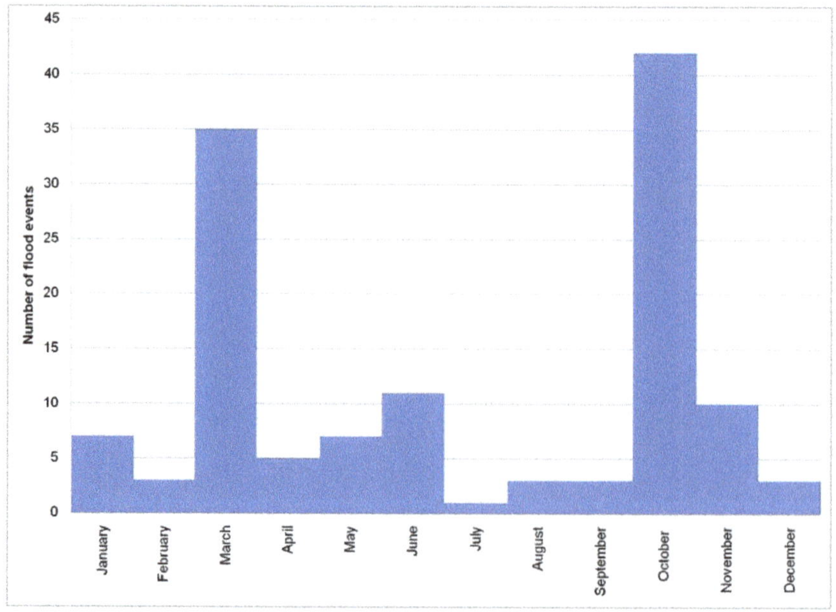

Figure 5. The monthly distribution of flood events during 1979–2010.

On the other hand, substantial differences appeared in the temporal distribution of the flooded area (Figure 6). It reached its maximum value (87.4 km^2) in 1987. Additionally, relatively high values were calculated in the years 1994 (69.5 km^2) and 1997 (64.8 km^2). The p-value was found to be 0.01, and thus, the data used were statistically significant. Regarding the seasonal distribution of flooded area (Figure 7), it reached its maximum value (136.5 km^2) in November. It presented relatively high values in March (83.2 km^2) and October (73.1 km^2).

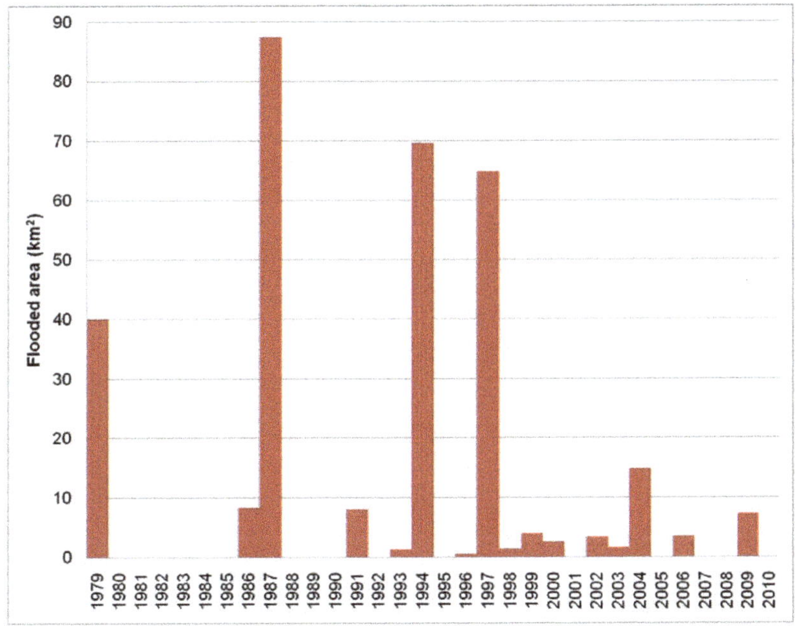

Figure 6. The monthly distribution of the flooded area during 1979–2010.

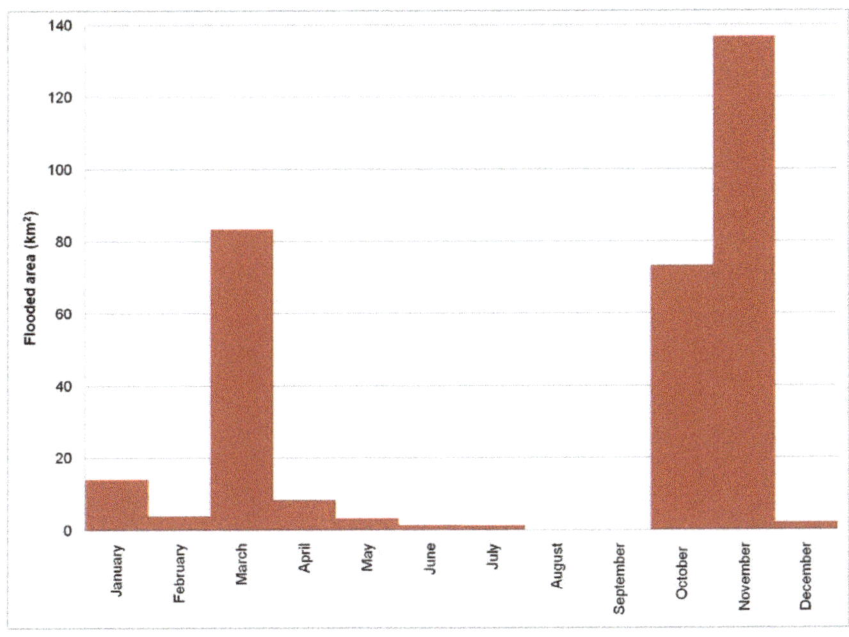

Figure 7. The monthly distribution of the flooded area during 1979–2010.

4.2. Temporal Distribution of Damages

Concerning the type of damages, their temporal and seasonal distributions are presented, respectively, in Figures 8 and 9. It is important to note that no human casualties were observed during the studied period. The values of economic damages were low during the studied period and reached the maximum value (3) in the years 1987, 1990 and 2010 (Figure 8). Similarly, their monthly distributions had relatively low values. The maximum value (4) of economic damages was recorded in March (Figure 9).

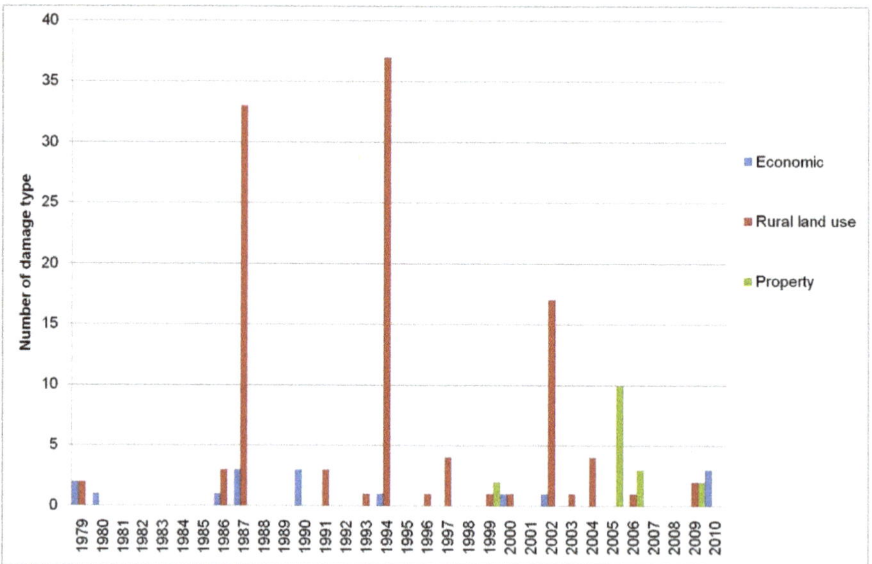

Figure 8. The annual distribution of type of damages in the study area during 1979–2010.

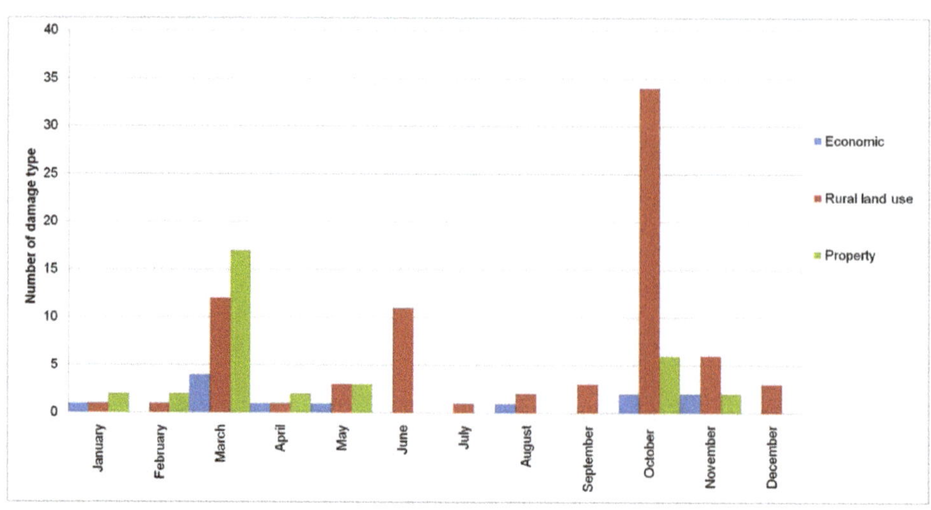

Figure 9. The monthly distribution of the type of damage.

On the contrary, the values of rural land use were the highest among all the types of damages (Figure 8). It reached its maximum value (37) in 1994, while it had relatively high values in the years 1987 (33) and 2002 (17). October was the month with the most records (34) in this type of damage (Figure 9). Additionally, high values were observed in March (12) and June (11). The maximum value (10) of the property damages was recorded in 2005 (Figure 8). March included the most number of property damages (17), while relatively high values (Figure 9) were recorded in October (six).

Figures 10 and 11 show the temporal and monthly distribution of categories of damages in the study area. The maximum value (10) of the very high category was observed in 2005, whilst its relatively high values were recorded in the years 1979 (three) and 1987 (three). Moreover, this category reached its maximum value (10) in June; a relatively high value (four) was calculated in November. The high class of damages reached its maximum value (six) in 1987 and a relatively high value (three) in the year 1994. March contained most records (five) of this class, and a high value was computed in October. The medium class reached its maximum value (24) in 1994, and a high value (12) was observed in 1987. The month with the most records (28) of this class was October; while a high value (13) was calculated in March. The annual and monthly maximum values of the low damages were observed in the year 1987 (14) and in March (13). Finally, the maximum value of the unknown class was recorded in the year 2002 (17).

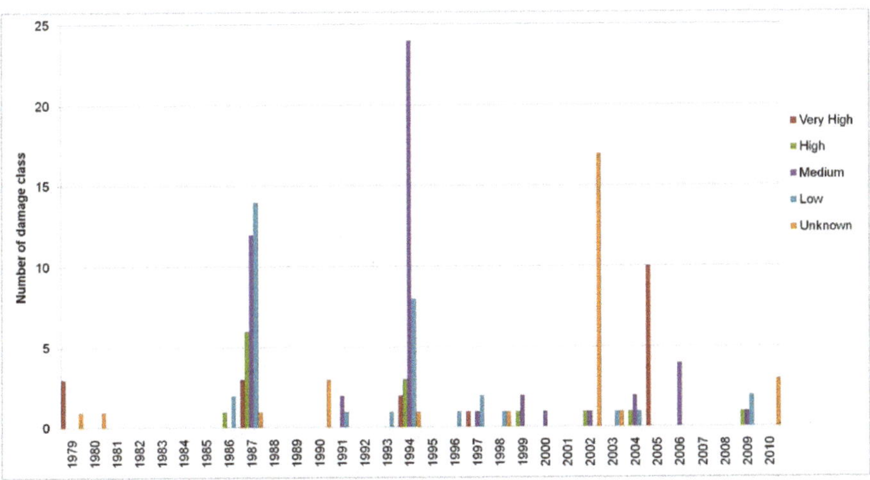

Figure 10. The annual distribution of categories of damages in the study area during 1979–2010.

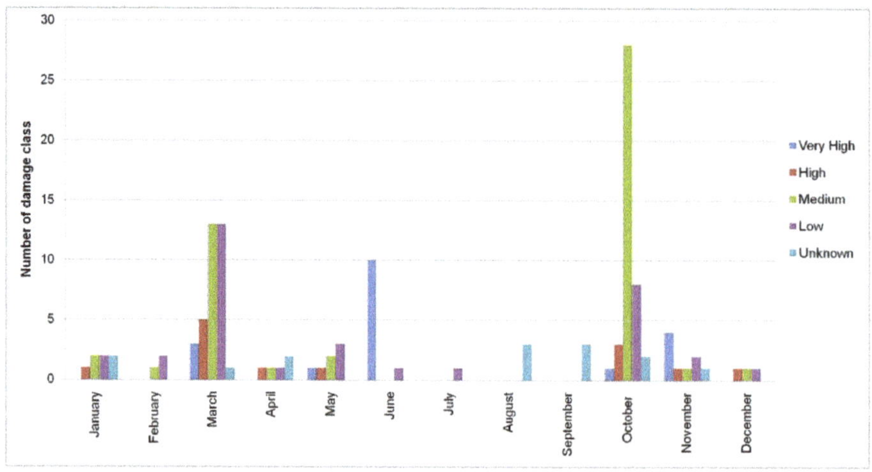

Figure 11. The monthly distribution of categories of damages during 1979–2010.

4.3. Spatial Distribution of Flood Events

The drainage basin of Pinios River were divided into seven sub-basins, which were: (a) the upper reaches of Pinios River; (b) Trikala-Neochoritis River; (c) Karditsa-Enippeas River; (d) Larisa-Karla; (e) Tempi Valley; (f) Titarisios River; and (g) the lower reaches of Pinios River. The spatial distribution of flood events in each drainage basin was examined. Table 2 shows the spatial distribution of flood events expressed as the number of events per drainage basin during the studied period. The most of floods were located in the drainage basin of Karditsa-Enippeas River.

Table 2. Spatial distribution of flood occurrences in each drainage sub-basin of the study area during the period 1979–2010.

Sub-Basin	Number of Floods
Upper reaches of Pinios River	4
Trikala-Neochoritis River	21
Karditsa-Enippeas River	63
Larisa-Karla	33
Tempi Valley	6
Titarisios River	7
Lower reaches of Pinios River	12

The mapping of the marshy area and lakes of the study area was based on the topographic map of 1881. They covered an area of about 161 km^2. Moreover, according to [36], two paleo-lakes existed in the study area during the Quaternary. The former marshy areas and lakes along with the boundaries of two paleo-lakes were compared with the historic flood events (Figure 12).

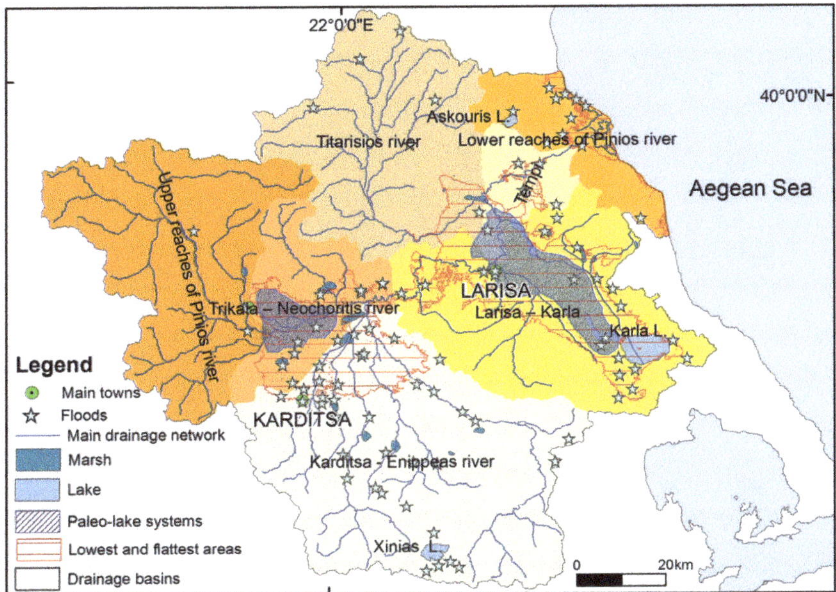

Figure 12. Marshy areas, lakes, paleo-lakes and lowest and flattest areas vs. flood occurrences in each drainage sub-basin of the study area during the period 1979–2010.

In the drainage basins of Trikala-Neochoritis River and Larisa-Karla, many flood events (12 events) were located in marshy areas, Lake Karla and paleo-lakes. In the drainage basin of Karditsa-Enippeas River, many floods (eight events) were located in or near marshy areas and Lake Xinias. In the drainage basins of Titarisios River and the lower reaches of Pinios River, few floods (two events) were located in or near marshy areas and Lake Askouris. Finally, in the drainage basins of the upper reaches of Pinios River and Tempi, no floods were associated with marshy areas.

The intersection of areas with an elevation <100 m, slope <5°, aspect = −1°and curvature = 0 is shown in Figure 12. The lowest and flattest areas were mainly located in the drainage basins of Trikala-Neochoritis River, Larisa-Karla and the lower reaches of Pinios River. In the drainage basins of Trikala-Neochoritis River and Larisa-Karla, they were mainly located in the areas where the former marshy areas and lakes were located. Regarding the spatial distribution of flood events, many flood events (61 events) were located in these areas.

5. Discussion and Conclusions

In the present study, the historical flood database of Greece and the old topographic maps of 1881 were used. These data were combined in a GIS environment to analyze the temporal and spatial distribution of flood events. The study area was the drainage basin of Pinios River, in Thessaly, central Greece.

The database contains floods that occurred between 1979 and 2010. A total number of 146 flood events were recorded in the drainage basin of Pinios River during this period.

The number of flood events reaches its maximum value (38) in the year 1994 (Figure 4). This may be related to flood events that occurred on 21–22 October 1994. No information about the causes of flooding, the flood mechanism and characteristics along with the maximum distance of the flood were included in the historic database. According to [55], this flood was caused by a severe storm. This intense storm caused flooding along the Pinios River and its tributaries, as well as inundation of agricultural and residential areas.

The statistical analysis showed that the data used are statistically significant. The number of flood occurrences shows a rising trend during the period 1990–2010. The study of the monthly distribution of the events demonstrates that October contains the most flood events, while March is the second month with many flood records (Figure 5).

Regarding the flooded area, it reaches its maximum value (87.4 km^2) in the year 1987 (Figure 6). This is attributed to flood event that occurred on 24–27 March 1987. The flood of March 1987 was caused by intense rainfall that produced direct runoff, as well as snowmelt. The water level of the Pinios River measured during the flood event exceeded 6.3 m in Kalamaki Valley and 8.0 m at Tempi Valley when the normal water level was less than 1 m in both locations (Figure 1b). Owing to the narrow riverbed at both locations (Kalamaki and Tempi), significant parts of the Thessaly plain upstream of each location were inundated [56].

Although October is the most flood-prone month, substantial differences appear in the monthly distribution of the flooded area. It reaches its maximum value (136.5 km^2) in November (Figure 7). This difference may be related to several reasons such as the difference in storm characteristics.

In the study area, the economic damages have low values. The type of damage with the most records is the rural land use. The maximum value is recorded in the year 1994, while the month with the maximum value is October (Figures 8 and 9). According to [55], during the flood of 21–22 October 1994, more than 70 houses in about 20 communities were totally destroyed by the flood, more than 200 suffered severe damage and another 90 minor damage, whereas 80 km^2 of agricultural land (cotton fields) were flooded. Concerning the property damages, the maximum value is observed in the year 2005, while March contains the most events (Figures 8 and 9). A flood event occurred in the study area on 16 June 2005, which caused serious damage to infrastructures and properties.

Regarding the class of damage, no human casualties were recorded in the study area between 1979 and 2010. According to [57], the most destructive events regarding human victims occurred in the city of Trikala in 1907. This flood caused at least 200 fatalities. Thus, the classification of the total damages is based on the amount of monetary compensation and flooded area of the historical flood events. The annual maximum value of the very high class is observed in the year 2005, while its monthly maximum value is recorded in June (Figures 10 and 11). This fact is related to the flood event of 16 June 2005. The amount of monetary compensation was 1,027,146 euros and was given for damages to infrastructures and properties [43]. The maximum value of the high class is recorded in the year 1987, and the maximum value of its monthly distribution is observed in March. This may be related to the flood event of 24–27 March 1987. Finally, the medium class has its maximum values in the year 1994 and in October, and this is related to the flood event of 21–22 October 1994.

The study area was divided into seven sub-basins to examine the spatial distribution of flood events (Table 2). The most occurrences are recorded in the southern part of the study area and specifically in the drainage basin of Karditsa-Enippeas River. Moreover, increased clustering of flood events is observed in the drainage basin of Larisa-Karla.

Old topographic maps of 1881 and bibliographic data were used to examine the relation of the paleo-environment of Pinios River with the spatial distribution of floods. The old topographic maps show earlier marshy areas and lakes, which nowadays, have dried up. According to [36], paleo-lake systems were located in the drainage basin of Pinios River during the Quaternary. One paleo-lake was located in the western part of the study area. Another, lower altitude paleo-lake, which was not connected to the previous one, was in the eastern part. During the Quaternary, the weathering of the carbonate rocks led to the connection of independent paleo-environments, and the Pinios River emerged. There is a certain amount of clustering of flood events in the areas of former marshes, lakes and paleo-lakes, which are located in the drainage basins of Trikala-Neochoritis River, Larisa-Karla and Karditsa-Enippeas (Figure 12).

In the study area, drainage and irrigation works along with levees were built 90 years ago, to protect it from flooding. However, incidents of annual flood events are reported in specific areas, and still, floods remain a big problem in the region [58]. Figure 12 presents the sites with earlier marshes and lakes, which are located in the lowest and flattest parts of the study area. The locations with earlier marshes and lakes are characterized by gentle slopes with low elevation and form the bottom of the drainage basin of the drainage network. These areas are developed mainly over semi-permeable or impermeable formations. The underlying geology and soil type affect the quality and the rate of infiltration. More specifically, the sediments within these wetlands may consist of materials such as clays and fine sands. These materials create a saturated layer on the surface during intense rainfall. Therefore, the specific runoff may be unable to further infiltrate the surface water into the subsurface. Consequently, these specific locations are sites where surface water runoff naturally collects and can lead to floods. Although, the wetlands and lakes have dried up, the morphology of the relief in these areas has not changed, causing flooding. According to studies carried out in other similar regions [35,52], the drainage basins that were 15% covered by wetlands had a flooding supply of 60–65% more than those not bearing wetlands. Consequently, drying up of the marshy areas and lakes of the region has strongly favored the flood events in the drainage basin of Pinios River.

The low and flat areas are located in the central, south-eastern and coastal part of the study area (Figure 12), and they contain many flood events. Usually, when a flood occurs, the lowest and flattest areas will be flooded first. Thus, these locations are areas prone to flood. This lowland morphology of the plain and the narrow passages along the river course such as Kalamaki, Rodia and Tempi gorges are the main reasons for the flooding [58]. According to [55], other reasons favoring the flood genesis due to human activities are some bridges with inadequate heights that are across the river and the construction by the farmers of "handy" barriers in the river channel for storage of irrigation water.

Endogenic processes such as active tectonics of the study area probably will not affect the flooding hazard in the near future. On the other hand, exogenic processes such as climate changes are capable of influencing future flood occurrences. Possible climate changes will cause sea level rise. Different scenarios were produced by the Intergovernmental Panel on Climate Change (IPCC), which predicts sea level rise from 0.3 to about 1.0 m until 2100 [59]. The possible sea level rise will increase the flood risk in the coastal area of the study area.

The historic flood data and old topographic maps provide valuable information for land use planning at a regional scale, leading to the determination of the safe and non-safe areas for urban activities [60]. The proposed methodology estimates the localization of sites prone to flood, and it may be utilized for flood hazard assessment mapping and for flood risk management. In areas prone to flooding, the appropriate land use planning along with the selection of the proper constructions are essential to prevent and mitigate the consequences of flood hazard occurrence. Consequently, proper land use for specific areas (i.e., parks, residential areas, etc.) may be determined. Additionally, construction of flood control works such as dams and levees, ponds, lakes and lagoons may be selected. Thus, planners, engineers and policy makers may use the applied approach for new and existing land use planning projects and for floodplain management.

Author Contributions: G.D.B conceived of the research. H.D.S. and G.D.B. designed the research and the data analysis. K.S. prepared and analyzed the data. G.D.B., H.D.S., K.S. and E.K completed the field work. H.D.S. and E.K. created the figures. G.D.B and H.D.S. wrote the paper.

Funding: This research received no external funding.

Conflicts of Interest: The authors declare no conflict of interest.

References

1. Cerdà, A. Effect of Climate on Surface Flow along a Climatological Gradient in Israel: A Field Rainfall Simulation Approach. *J. Arid Environ.* **1998**, *38*, 145–159. [CrossRef]
2. Bathrellos, G.D.; Skilodimou, H.D.; Maroukian, H. The spatial distribution of Middle and Late Pleistocene cirques in Greece. *Geogr. Ann. A* **2014**, *96*, 323–338. [CrossRef]
3. Skilodimou, H.D.; Bathrellos, G.D.; Maroukian, H.; Gaki-Papanastassiou, K. Late Quaternary evolution of the lower reaches of Ziliana stream in south Mt. Olympus (Greece). *Geogr. Fis. Din. Quat.* **2014**, *37*, 43–50. [CrossRef]

4. Bathrellos, G.D.; Skilodimou, H.D.; Maroukian, H.; Gaki-Papanastassiou, K.; Kouli, K.; Tsourou, T.; Tsaparas, N. Pleistocene glacial and lacustrine activity in the southern part of Mount Olympus (central Greece). *Area* **2017**, *49*, 137–147. [CrossRef]

5. Kamberis, E.; Bathrellos, G.; Kokinou, E.; Skilodimou, H. Correlation between the structural pattern and the development of the hydrographic network in a portion of the Western Thessaly basin (Greece). *Cent. Eur. J. Geosci.* **2012**, *4*, 416–424. [CrossRef]

6. Kokinou, E.; Skilodimou, H.D.; Bathrellos, G.D.; Antonarakou, A.; Kamberis, E. Morphotectonic analysis, structural evolution/pattern of a contractional ridge: Giouchtas Mt., Central Crete, Greece. *J. Earth Syst. Sci.* **2015**, *124*, 587–602. [CrossRef]

7. Chousianitis, K.; Del Gaudio, V.; Sabatakakis, N.; Kavoura, K.; Drakatos, G.; Bathrellos, G.D.; Skilodimou, H.D. Assessment of Earthquake-Induced Landslide Hazard in Greece: From Arias Intensity to Spatial Distribution of Slope Resistance Demand. *Bull. Seismol. Soc. Am.* **2016**, *106*, 174–188. [CrossRef]

8. Bathrellos, G.D.; Skilodimou, H.D.; Maroukian, H. The significance of tectonism in the glaciations of Greece. *Geol. Soc. Spéc. Publ.* **2017**, *433*, 237–250. [CrossRef]

9. Papadopoulou-Vrynioti, K.; Bathrellos, G.D.; Skilodimou, H.D.; Kaviris, G.; Makropoulos, K. Karst collapse susceptibility mapping considering peak ground acceleration in a rapidly growing urban area. *Eng. Geol.* **2013**, *158*, 77–88. [CrossRef]

10. Tsolaki-Fiaka, S.; Bathrellos, G.D.; Skilodimou, H.D. Multi-criteria decision analysis for abandoned quarry restoration in Evros Region (NE Greece). *Land* **2018**, *7*, 43. [CrossRef]

11. Skilodimou, H.D.; Bathrellos, G.D.; Koskeridou, E.; Soukis, K.; Rozos, D. Physical and anthropogenic factors related to landslide activity in the Northern Peloponnese, Greece. *Land* **2018**, *7*, 85. [CrossRef]

12. Morisawa, M. *Geomorphology Laboratory Manual*; Willey & Sons: Hoboken, NJ, USA, 1976; 253p.

13. Slaymaker, O. *Geomorphic Hazards*; John Wiley & Sons: Hoboken, NJ, USA, 1997; 204p.

14. Bathrellos, G.D. An overview in Urban Geology and Urban Geomorphology. *Bull. Geol. Soc. Greece* **2007**, *40*, 1354–1364. [CrossRef]

15. Bathrellos, G.; Skilodimou, H. Geomorphic Hazards and Disasters. *Bull. Geol. Soc. Greece* **2006**, *39*, 96–103. (In Greek)

16. Goudie, S.A. *Encyclopedia of Geomorphology*; Goudie, S.A., Ed.; Taylor & Francis Group: New York, NY, USA, 2006; pp. 378–379.

17. Merz, B.; Kreibich, H.; Schwarze, R.; Thieken, A. Assessment of economic flood damage. *Nat. Hazards Earth Syst. Sci.* **2010**, *10*, 1679–1724. [CrossRef]

18. CEOS. *The Use of Earth Observing Satellites for Hazard Support: Assessments and Scenarios*; Final Report of the CEOS Disaster Management Support Group (DMSG); CEOS: Rome, Italy, 2003.

19. European Environment Agency (EEA). *Mapping the Impacts of Natural Hazards and Technological Accidents in Europe: An Overview of the Last Decade*; Office for Official Publications of the European Communities: Luxembourg; European Environment Agency (EEA): Copenhagen, Denmark, 2010.

20. Gaume, E.; Bain, V.; Bernardara, P.; Newinger, O.; Barbuc, M. A compilation of data on European flash floods. *J. Hydrol.* **2009**, *367*, 70–78. [CrossRef]

21. Payrastre, O.; Gaume, E.; Andrieu, H. Use of historical data to assess the occurrence of floods in small watersheds in the French Mediterranean area. In *Advances in Geosciences*; European Geosciences Union: Munich, Germany, 2005; Volume 2, pp. 313–320.

22. Tropeano, D.; Turconi, L. Using historical documents for landslide, debris flow and stream flood prevention. Applications in Northern Italy. *Nat. Hazards* **2004**, *31*, 663–679. [CrossRef]

23. Van Alphen, J.; Martini, F.; Loat, R.; Slomp, R.; Passchier, R. Flood risk mapping in Europe, experiences and best practices. *J. Flood Risk Manag.* **2009**, *2*, 285–292. [CrossRef]

24. Rijal, S.; Rimal, B.; Sloan, S. Flood Hazard Mapping of a Rapidly Urbanizing City in the Foothills (Birendranagar, Surkhet) of Nepal. *Land* **2018**, *7*, 60. [CrossRef]

25. Samanta, S.; Koloa, C.; Kumar Pal, D.; Palsamanta, B. Flood Risk Analysis in Lower Part of Markham River Based on Multi-Criteria Decision Approach (MCDA). *Hydrology* **2016**, *3*, 29. [CrossRef]

26. Jakob, M. Debris flow hazard analysis. In *Debris-Flow Hazards and Related Phenomena*; Jakob, M., Hungr, O., Eds.; Springer: Berlin, Germany, 2005; p. 794.

27. Langbein, W.B. Annual floods and the partial-duration flood series. *Trans. Am. Geophys. Union* **1949**, *30*, 879–881. [CrossRef]

28. Onyutha, C. On rigorous drought assessment using daily time scale: Non-stationary frequency analyses, revisited concepts, and a new method to yield non-parametric indices. *Hydrology* **2017**, *4*, 48. [CrossRef]

29. Lang, M.; Ouarda, T.B.M.J.; Bobée, B. Towards operational guidelines for over-threshold modeling. *J. Hydrol.* **1999**, *225*, 103–117. [CrossRef]

30. Willems, P. A time series tool to support the multi-criteria performance evaluation of rainfall-runoff models. *Environ. Model. Softw.* **2009**, *24*, 311–321. [CrossRef]

31. Smith, R.L. Threshold Methods for Sample Extremes. In *Statistical Extremes and Applications*; de Oliveira, J.T., Ed.; NATO ASI Series; Springer: Dordrecht, The Netherlands, 1985; Volume 131, pp. 623–638.

32. De Moel, H.; van Alphen, J.; Aerts, J.C.J.H. Flood maps in Europe—Methods, availability and use. *Nat. Hazards Earth Syst. Sci.* **2009**, *9*, 289–301. [CrossRef]

33. Diakakis, M.; Mavroulis, S.; Deligiannakis, G. Floods in Greece, a statistical and spatial approach. *Nat. Hazards* **2012**, *62*, 485–500. [CrossRef]

34. Skilodimou, H.; Stefouli, M.; Bathrellos, G. Spatio-temporal analysis of the coastline of Faliro Bay, Attica, Greece. *Estud. Geol.-Madrid.* **2002**, *58*, 87–93. [CrossRef]

35. Skilodimou, H.; Livaditis, G.; Bathrellos, G.; Verikiou-Papaspiridakou, E. Investigating the flooding events of the urban regions of Glyfada and Voula, Attica, Greece: A contribution to Urban Geomorphology. *Geogr. Ann. A* **2003**, *85*, 197–204. [CrossRef]

36. Migiros, G.; Bathrellos, G.; Skilodimou, H.; Karamousalis, T. Pinios (Peneus) River (Central Greece): Hydrological-geomorphological elements and changes during the quaternary. *Cent. Eur. J. Geosci.* **2011**, *3*, 215–228. [CrossRef]

37. Mylopoulos, N.; Kolokytha, E.; Loukas, A.; Mylopoulos, Y. Agricultural and water resources development in Thessaly, Greece in the framework of new European Union policies. *Int. J. River Basin Manag.* **2009**, *7*, 73–89. [CrossRef]

38. Goumas, K. The irrigation in Thessaly plain: Effects to underground and surface waters. In Proceedings of the Greek Hydrotechnical Association Conference, Larissa, Greece, 2 February 2006; pp. 39–53. (In Greek)

39. Katsikatsos, G.; Migiros, G.; Vidakis, M. La structure géologique de la région de la Théssalie orientale (Grèce). *Ann. Soc. Géol. Nord* **1982**, *CI*, 177–188.

40. Migiros, G. The lithostratigraphic-tectonic structure of Orthris (Central Greece). *Bull. Geol. Soc. Greece* **1990**, *26*, 107–120. (In Greek)

41. Ferriere, J.; Reynaud, J.; Pavlopoulos, A.; Bonneau, M.; Migiros, G.; Proust, J.N.; Gardin, S. Geological evolution and geodynamic controls of Tertiary intramontane piggyback Meso-Hellenic Basin, Greece. *Bull. Soc. Geol. Fr.* **2004**, *175*, 361–381. [CrossRef]

42. Mimikou, M.; Koutsoyiannis, D. Extreme Floods in Greece: The Case of 1994. In Proceedings of the U.S.-Italy Research Workshop on the Hydrometeorology, Impacts and Management of Extreme Floods, Perugia, Italy, 13–17 November 1995.

43. Ministry of Environment and Energy. Special Secretariat for Water. Floods, Historic Floods. 2017. Available online: http://www.ypeka.gr/Default.aspx?tabid=252&locale=el-GR&language=en-US (accessed on 17 June 2018).

44. European Commission. *Directive 2007/60/EC of the European Parliament and of the Council on the Assessment and Management of Flood Risks*; European Environment Agency: Copenhagen, Denmark, 2007.

45. Hrisohoou, M. *Sheet "Ioannina-Metsovon-Grevena-Kozani-Servia"*; Topographic Map, Scale 1:200,000; Colman George: Athens, Greece, 1881.

46. Hrisohoou, M. *Sheet "Larissa-Elasson-Katerini"*; Topographic Map, Scale 1:200,000; Colman George: Athens, Greece, 1881.

47. Hrisohoou, M. *Sheet "Arta-Trikala-Karditsa"*; Topographic Map, Scale 1:200,000; Colman George: Athens, Greece, 1881.

48. Hrisohoou, M. *Sheet "Volos-Farsalos-Lamia"*; Topographic Map, Scale 1:200,000; Colman George: Athens, Greece, 1881.

49. Ministry of Environment, Energy and Climate Change. *Preliminary Flood Risk Assessment in Greece*; Ministry of Environment and Energy: Athens, Greece, 2012. (In Greek)

50. Ministry of Environment and Energy. *Special Secretariat for Water, Management Plan for the Drainage Basins of Eastern Peloponnese*; Ministry of Environment and Energy: Athens, Greece, 2017. (In Greek)

51. Bathrellos, G.D.; Skilodimou, H.D.; Chousianitis, K.; Youssef, A.M.; Pradhan, B. Suitability estimation for urban development using multi-hazard assessment map. *Sci. Total Environ.* **2017**, *575*, 119–134. [CrossRef] [PubMed]
52. Bathrellos, G.D.; Karymbalis, E.; Skilodimou, H.D.; Gaki-Papanastassiou, K.; Baltas, E.A. Urban flood hazard assessment in the basin of Athens Metropolitan city, Greece. *Environ. Earth Sci.* **2016**, *75*, 319. [CrossRef]
53. Rozos, D.; Bathrellos, G.D.; Skilodimou, H.D. Comparison of the implementation of Rock Engineering System (RES) and Analytic Hierarchy Process (AHP) methods, based on landslide susceptibility maps, compiled in GIS environment. A case study from the Eastern Achaia County of Peloponnesus, Greece. *Environ. Earth Sci.* **2011**, *63*, 49–63. [CrossRef]
54. Pradhan, B. Flood susceptible mapping and risk area delineation using logistic regression, GIS and remote sensing. *J Spat. Hydrol.* **2010**, *9*, 1–18.
55. Koutsoyiannis, D.; Mimikou, M. *Management and Prevention of Crisis Situations: Floods, Droughts and Institutional Aspects*; 3rd EURAQUA Technical Review; Country Paper for Greece: Rome, Italy, 1996; pp. 63–77.
56. Koukis, G.; Koutsoyiannis, D.; Embleton, C.; Embleton-Hamann, C. (Eds.) *Geomorphological Hazards of Europe*; Elsevier: New York, NY, USA, 1997; Volume 5, pp. 215–242.
57. Bathrellos, G. Geological, Geomorphological and Geographic Study of Urban Areas in Trikala Prefecture—Western Thessaly. Ph.D. Thesis, National & Kapodistrian University of Athens, Athens, Greece, 2005; p. 561, (In Greek with Extended English Abstract).
58. Bathrellos, G.D.; Gaki-Papanastassiou, K.; Skilodimou, H.D.; Papanastassiou, D.; Chousianitis, K.G. Potential suitability for urban planning and industry development by using natural hazard maps and geological—Geomorphological parameters. *Environ. Earth Sci.* **2012**, *66*, 537–548. [CrossRef]
59. Intergovernmental Panel on Climate Change (IPCC). Climate Change 2014 Synthesis Report Summary for Policymakers, August 2018. Available online: https://www.ipcc.ch/pdf/assessment-report/ar5/syr/AR5_SYR_FINAL_SPM.pdf (accessed on 17 June 2018).
60. Bathrellos, G.D.; Gaki-Papanastassiou, K.; Skilodimou, H.D.; Skianis, G.A.; Chousianitis, K.G. Assessment of rural community and agricultural development using geomorphological-geological factors and GIS in the Trikala prefecture (Central Greece). *Stoch. Environ. Res. Risk Assess.* **2013**, *27*, 573–588. [CrossRef]

 *land*

Article

Flood Hazard Mapping of a Rapidly Urbanizing City in the Foothills (Birendranagar, Surkhet) of Nepal

Sushila Rijal [1,*], **Bhagawat Rimal** [2,3] **and Sean Sloan** [4]

[1]   Faculty of Humanities and Social Science, Mahendra Ratna Multiple Campus, Ilam 57300, Nepal
[2]   The State Key Laboratory of Remote Sensing Science, Institute of Remote Sensing and Digital Earth, Chinese Academy of Sciences (CAS), Beijing 100101, China; bhagawat@radi.ac.cn
[3]   College of Applied Sciences (CAS)-Nepal, Thapathali, Tribhuvan University, Kathmandu 44613, Nepal
[4]   College of Science and Engineering, Center for Tropical Environmental and Sustainability Science, James Cook University, Cairns QLD 4870, Australia; sean.sloan@jcu.edu.au
*   Correspondence: sushilarijal@ymail.com

Received: 26 March 2018; Accepted: 3 May 2018; Published: 5 May 2018

Abstract: Flooding in the rapidly urbanizing city of Birendranagar, Nepal has been intensifying, culminating in massive loss of life and property during July and August 2014. No previous studies have monitored underlying land-cover dynamics and flood hazards for the area. This study described spatiotemporal urbanization dynamics and associated land-use/land-cover (LULC) changes of the city using Landsat imagery classifications for five periods between 1989 and 2016 (1989–1996, 1996–2001, 2001–2011, 2011–2016). Areas with high flood-hazard risk were also identified on the basis of field surveys, literature, and the Landsat analysis. The major LULC changes observed were the rapid expansion of urban cover and the gradual decline of cultivated lands. The urban area expanded nearly by 700%, from 85 ha in 1989 to 656 ha in 2016, with an average annual growth rate of 23.99%. Cultivated land declined simultaneously by 12%, from 7005 ha to 6205 ha. The loss of forest cover also contributed significantly to increased flood hazard. Steep topography, excessive land utilization, fragile physiographic structure, and intense monsoonal precipitation aggravate hazards locally. As in Nepal generally, the sustainable development of the Birendranagar area has been jeopardized by a disregard for integrated flood-hazard mapping, accounting for historical land-cover changes. This study provides essential input information for improved urban-area planning in this regard.

Keywords: urbanization; flood; remote sensing/GIS; Birendranagar; Nepal

---

1. Introduction

Hazards, defined as natural or human-induced activities that elevate the probability of material, social, or natural loss [1], are typified by the nexus of uncontrolled urbanization in contexts susceptible to natural flood, landslides, and earthquakes in Nepal. Here, urbanization is understood as processes leading toward increased population density, socioeconomic activities, and expanded built up areas and associated infrastructures [2]. Natural hazards and urbanization can interact to amplify land-use change, such as that negatively affecting agricultural areas [3,4]. In hazardous areas, including newly formed urban areas, land management and planning that would enhance resilience depend therefore on an understanding of those land-use/land-cover change (LULC) patterns that accentuate latent hazards [5].

Floods are the most common and devastating natural hazards [6] on a global scale and have been increasingly frequent and devastating since the mid-20th century [4]. Of all flood events recorded between 1950 and 2011, most have occurred during recent decades. Some 2% occurred during the 1950s, rising rapidly each decade thereafter to 3.9%, 6.6%, 13.2%, 21.9%, and 52.2% for the 1960s,

1970s, 1980s, 1990s, and 2000s, respectively [7]. Flood events also accounted more than two-third of all hydrological disasters during 2012–2014 [8] and 90.9% during 2015–2016 [9]. Flood hazards are global occurrences, yet associated economic damage and fatalities concentrate disproportionately among the continents [7]. Due to favorable geomorphological, metrological, and anthropogenic factors [10], more than 60% of the total economic and human losses during the period 1950 to 2011 were concentrated in Asia [7]. South Asia accounts for 33% of all Asian floods, 50% of associated fatalities, and 38% of the effected regions. As a proportion of South Asian totals, Nepal accounts for 7.2% of fatalities, 7.4% of the total victims, and 3.1% of economic losses, thereby ranking the country after India, Bangladesh, Pakistan, and Afghanistan [11]—8th globally in terms of flood fatalities [12] and 30th globally in terms of flood-hazard risk [13]. Of the five million Nepalese effected by natural hazards between 1971 and 2007, 68% were flood-related [14]. Between 1982 and 2014, nearly 9000 Nepalese lost their lives in flood and landslide events [15,16]. The Koshi flood event during 2008 affected 65,000 people and 700 ha of fertile land in the eastern lowland Tarai region [17]. In August 2017, floods affected 35 districts and destroyed 190,000 homes, fully or partially, with total economic losses estimated at $584.7 million [18]. While flooding in Nepal is triggered by monsoonal rainfall, steep and erosive topography, and wide catchments [4,19], its frequency and intensity increased largely due to increasing anthropogenic factors, namely, improper land-use, poorly-planned urbanization, deforestation, and settlements along river banks [20].

The study area, Birendranagar city of Nepal's Surkhet district, is rapidly urbanizing due to high rates of migration. Locals have migrated to the city in the quest of better quality of life and opportunities in the formal and informal markets [21], resulting in a 720% urban population growth rate for the period 1982–2015 (Table 1). It is a major socioeconomic hub and administrative center and an important gateway to Karnali zone, as well as a migrant-receiving area, mainly from the Dailekh district and the Karnali zone. Urban development has been haphazard due to deficient urban plans/policies and weak implementation. New and expanding urban settlements are characterized as spatially dispersed or discontiguous, and frequently arise over prime farm lands surrounding historic Birendranagar city. These urban areas, collectively defining Birendranagar city, are frequently devastated by annual flood events affecting transportation networks, and other infrastructure (e.g., bridges, markets), livelihoods, fatalities (numbering 38 in 2014 [16]), and soil erosion and transport to neighboring districts (27,092.94 Mt) of soil annually [22]. Despite their frequency, or perhaps because of their frequency, such events struggle to sustain the attention of policymakers [4]. Resilience is a vital tool with which to reduce the vulnerability [23]; however, the integration of flood-hazard risk management in regional plans and policies is hindered due to the lack of routine-based researches. Urban planning and flood hazard management to address these issues is inadequate, in part due to the lack of reliable, updated spatial databases.

Table 1. Population of Birendranagar city.

Year	1981	1991	2001	2011	2015
Population	13,859	22,973	31,381	93,718	100,458

Source: Central Bureau of Statistics (CBS), 2014.

To this end, this study uses remote sensing and GIS techniques to observe LULC changes and identify flood-risk hazards underlying the urbanization of Birendranagar city. This research will remain as an important benchmark for Nepalese planners/policymakers and land-change researchers, since its insights and outputs may serve as essential inputs for sustainable land-use plans and strategies for flood-hazard mitigation.

2. Method

2.1. Study Area

Nepal, like the focal region of this study, is increasingly urban. Nationally, the urban population grew from 2.9% of the general population in 1952/54 to 17.1% in 2011 [24] to more than 50% by 2017 [25]. Urban centers similarly increased in number from 10 to 58 over this period, and to 292 following after the local level reconstruction in 2017. Urbanization has been driven largely by internal migration, in turn sustained by regionally unequal development and economic opportunities [21,26]. Nepal experienced decade long political armed conflict during 1996–2006 [27,28], which displaced or otherwise spurred the migration of rural dwellers to growing cities in search of security. As the conflict subsided by 2006, several development activities were advanced throughout the country [29], again concentrating largely in select urban areas. Coincidentally, a peri-urban land market boom resulted in rapid, disorganized settlement expansion at the expense of arable lands [21,29].

The study area, Birendranagar city is the capital city of Karnali Province (Figure 1). After the reclassification of local administrative units as mandated by the New Constitution of Nepal, 2015, the former Village Development Committees (VDCs) were consolidated as Birendranagar city. This city spans 24,582 ha area and is divided amongst 16 wards housing 100,458 residents as of 2015 (Table 1) [24].

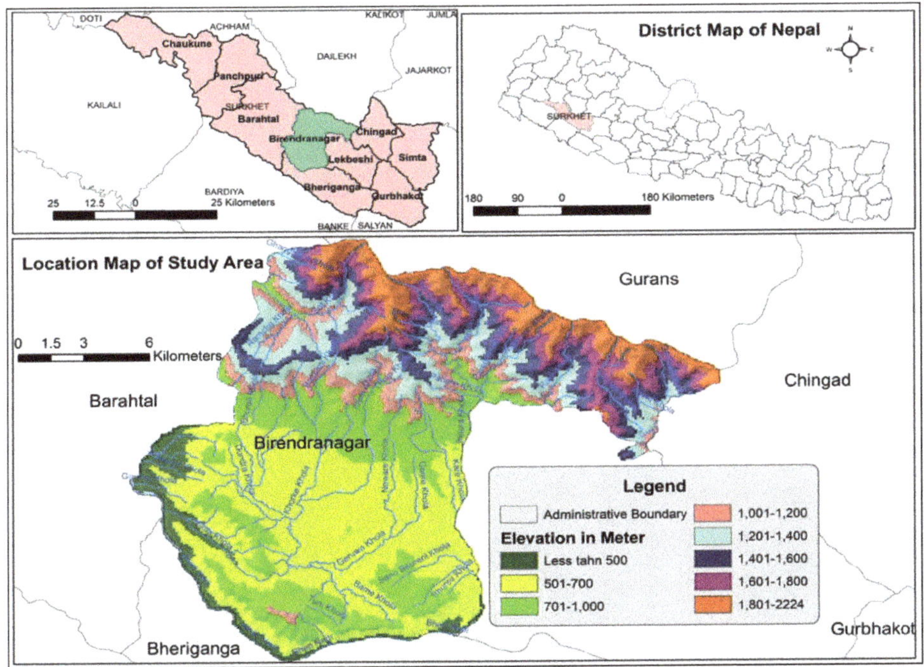

Figure 1. Location Map of the study area.

The Birendranagar city spans Nepal's sub-Himalayan and lesser-Himalayan zones, which are characterized by a warm-moist temperate, hot-dry sub-tropical, warm-dry sub-tropical, and cool-moist temperate climates, with annual temperatures ranging from 10 °C to 30 °C [22]. Elevation ranges from 380 to 2224 MASL. Birendranagar city receives intensive monsoonal precipitation, mainly between June and September, causing flooding and soil erosion, including along river banks. The study region is composed of fluvio-lacustrine sediments (sand, silt, clay, cobbles, and pebbles) deposited from the northern and southern parts of Siwalik Hills. Hard rocks in the region are comprised of sedimentary, meta-sedimentary, and metamorphic rock.

2.2. Extraction of LULC Change

Remote Sensing (RS) and Geographic information System (GIS) are successfully applied in various fields. They are effectively used for the LULC analyses [4,30–34], and flood area mapping. Multi-temporal time series of Landsat images [35] is capable of exploring accurate [36] LULC dynamics in specific time and space [21].

A city level land-use/land-cover (LULC) change analysis was realized by classifying time-series Landsat satellite imagery for 1989, 1996, 2001, 2006, 2011, and 2016. Atmospherically-corrected and maximal cloud-free Landsat images were sourced from United States Geological Survey (USGS) data portal (https://earthexplorer.usgs.gov). All scenes were verified for geometric accuracy, and data were projected on to UTM 44N (WGS 1984). For land-cover classification, six images with path/row 144/40 were selected for analysis. A "region of interest" (ROI) boundary representing municipality level study area was delineated for the classification. ENVI software was used to stack, subset, and classify the Landsat images via a maximum-likelihood (ML) algorithm [21,37,38]. A supervised approach with the ML-classifier algorithm was applied for the extraction of LULC. The land-cover classification system as recommended by Anderson et al. 1976 [39] was taken into consideration to classify the LULC and mainly seven LULC classes: urban/built-up, cultivated land, forest, shrub, sand, water, and others were identified (Table 2). Google Earth images (http://earth.google.com) and topographical maps from Survey Department, Government of Nepal (scale 1: 50,000) [40] were used to verify the LULC results. LULC transitions between each observation year were assessed using land change modeler (LCM) of TerrSet software developed by Clark Lab (www.clarklabs.org). Socioeconomic information on the drivers of urbanization were obtained from Central Bureau of Statistics [24] and use of Surkhet district profile [22].

Table 2. Land-cover classification scheme.

Land-Cover Types	Description
Urban (built up)	Urban and rural settlements, commercial areas, industrial areas, construction areas, traffic, airports, public service areas (e.g., schools, colleges, hospitals)
Cultivated land	Wet and dry crop lands, orchards
Forest	Evergreen broad leaf forest, deciduous forest, scattered forest, degraded forest
Shrub	Mix of trees (<5 m tall) and other natural covers
Sand	Sand area, other open field area, river bank
Water	River, lake/pond, canal, reservoir
Others	Cliffs/small landslide, bare rocks,

2.3. Identification of Flood Hazards

Areas of relatively acute flood hazard were further identified through field verification and informal surveys of local people during a field visit in 2014. High-hazard areas were geo-located using a GPS, and post-hazard observations were realized using high-resolution images of Google Earth. Contextual information was derived from the disaster reports of the National Planning Commission (NPC) [41] and the Ministry of Home Affairs (MoHA) [13,18], with additional information were garnered from the District Development Committee (DDC) [22]. Precipitation data for Birendranagar station were sourced from Department of Hydrology and Meteorology [42].

2.4. Urban/Built-Up Area Expansion Rate

Total urban area was estimated each observation year to measure the rate of urban expansion [21,37]. The rate of urban expansion describes the average annual growth of urban area during a given period, thus

$$\mathrm{BER} = (B_2 - B_1)/(T_2 - T_1) * 100 \tag{1}$$

in which BER implies the urban expansion rate (*ha/year*), and B_1, B_2 refers to the urban area (*ha*) between times T_1, T_2 in years.

2.5. Classiifcation Accuracy Assessment

The verification of each Landsat classifications was realized with reference to high-resolution satellite imagery in Google Earth (http://earth.google.com), 1:50,000 scale topographical maps from the Nepalese Survey Department [40], and GPS points collected during the field visit, with the mixture of these reference data varying by classification year. Over the all-time periods of 1989–2016, some 350 sample points were interpreted across the city. Points were distributed via stratified random sampling method, with at least 50 points for each land-use/cover class. The overall accuracy, user's accuracies, and producer's accuracies were obtained for each observation year [21].

3. Result and Discussion

3.1. Land-Use Land-Cover Change between 1989 and 2016

The overall accuracies obtained for respective years were 83%, 82%, 85%, 85%, 84%, and 86% in the city. Over the 27 years of observation, the predominant LULC changes were (a) the rapid increase in urban cover after 2001, and more gradual increase in shrub lands; (b) the simultaneous losses of cultivated lands, as well as the steady but lesser decline in forest cover (Table 3, Figures 2 and 3).

During 1989–2016, urban area increased 571 ha, from 85 ha to 656 ha, with an average annual growth rate of 23.99% (Figure 2). Virtually all urban growth occurred after 2001; urban area was steady until 2001, but more than doubled between 2001 and 2006, from 113 ha to 289 ha. The decrease in cultivated lands paralleled this growth of urban areas in both timing and areas (Figure 2). Cultivated lands have declined 800 ha since 1989, from 7005 ha to 6205 ha. The increases in sandy areas and water after 2011 are mainly associated with the major flood events during 2013 and 2014.

Cultivated land was the major source of the newly expanded urban area. During 1989–1996, with the annual urban growth rate of 0.67%, urban area increased slightly by 4 ha, of which 75% (3 ha) was previously cultivated land. Subsequently, during the period 1996–2001, 24 ha new urban area was established (5.4% growth rate), 98% of which was cultivated previously (Figure 3). Between 2001 and 2006, all 176 ha of new urban cover occurred over cultivated lands (31% growth rate). The period 2006–2011 experienced unprecedented 295 ha of urban growth (20% growth rate, with 97% sourced from cultivated). This conversion rate continued for the last period, 2011–2016, despite its urban growth rate dropping to only 72 ha (2.5%).

Similar to the replacement of cultivated lands by expanding urban areas, forests have been steadily replaced by expanding shrub lands as economic development proceeded, aggravating flood hazard. The inflow of migrants to Birendranagar city and encroachments upon forest areas accelerated after the eradication program of malaria regionally in 1958. Forests were felled to supply resources for the development of Indian railway line, and subsequent development activities widely provoked deforestation and forest degeneration [43]. Also, devastating practices of deforestation under the Rana regime (1846–1951) were continued under the subsequent Panchayat system (1960–1990). More recently, the protection of forest cover to safeguard environmental integrity and ecological functions such as hydrological flow and flood protection has been prioritized in vulnerable regions like Birendranagar. The national government has launched various community-based forest management plans and

President Chure-Tarai conservation program to maintain current forest cover. These efforts likely contributed to the cessation of forest loss after 2006 in Birendranagar following earlier losses (Figure 2).

Table 3. LULC change of Birendranagar city during 1989–2016 (Area in ha).

LULC	1989	1996	2001	2006	2011	2016	1989–2016
Urban/built up	85	89	113	289	584	656	571
Cultivated	7005	6958	6935	6794	6371	6205	−800
Forest	13,703	13,482	13,406	13,341	13,494	13,342	−361
Shrub	3281	3566	3640	3644	3608	3798	517
Sand	276	258	253	272	274	324	48
Water	140	140	129	129	157	157	17
Others	91	89	105	111	95	99	8
Total	24,582	24,582	24,582	24,582	24,582	24,582	

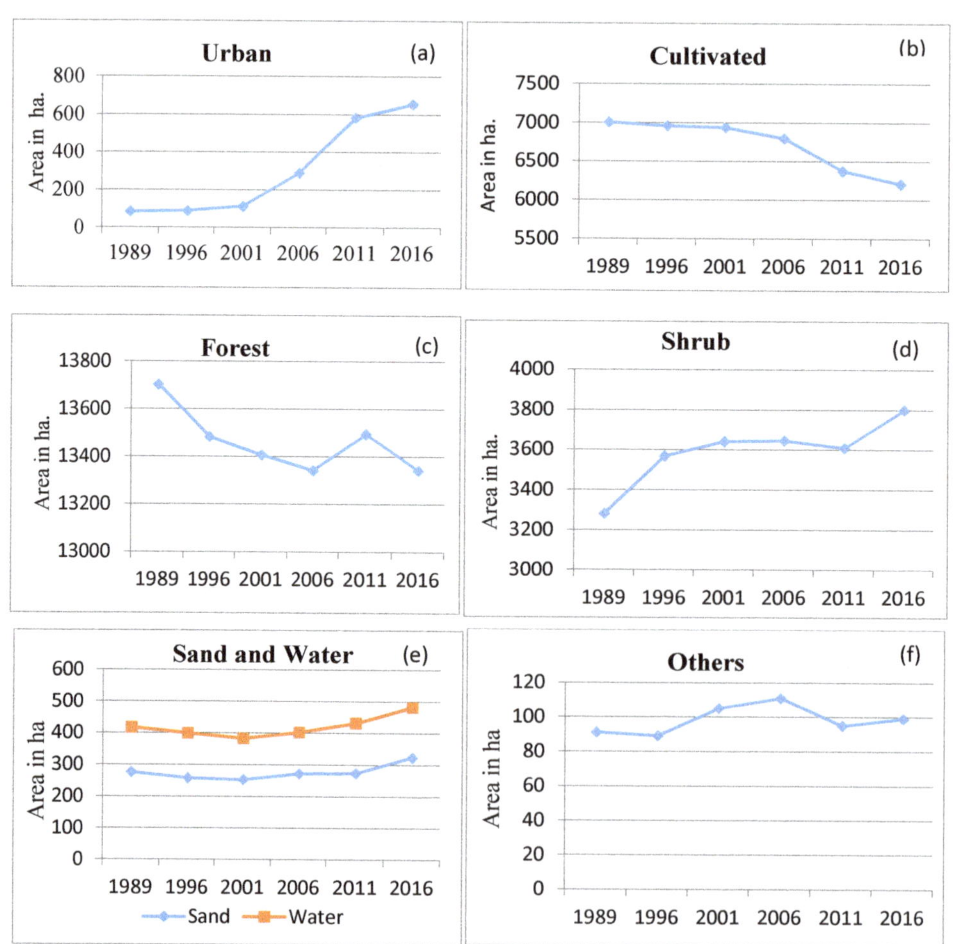

Figure 2. LULC change trend during 1989–2016. (**a**) Urban/built-up, (**b**) cultivated Land, (**c**) forest, (**d**) shrub (**e**) sand and water body, and (**f**) others.

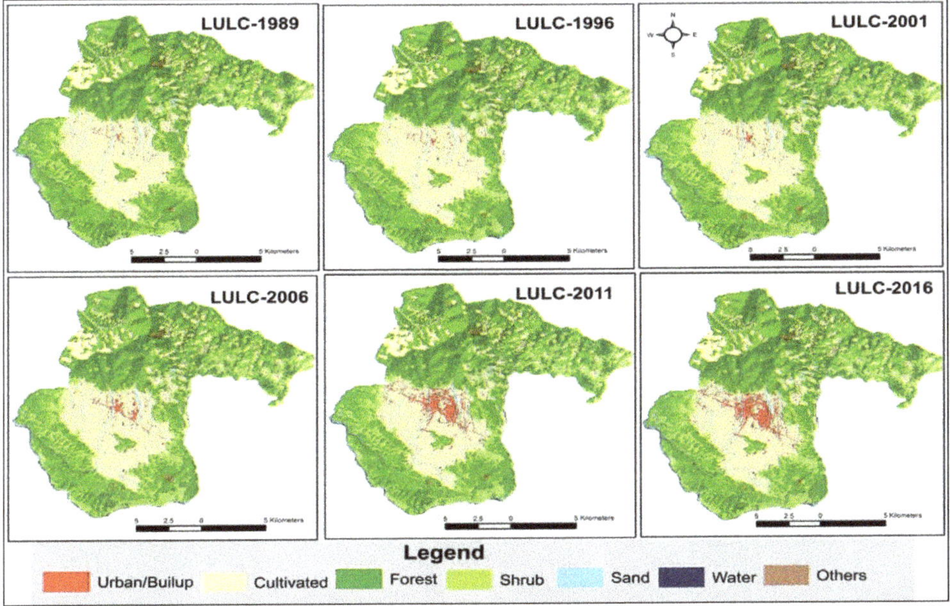

Figure 3. LULC change Maps of 1989 to 2016.

Nepal's New Constitution of 2015 mandated the reconstruction of all administrative areas and their reclassification within the federal administrative hierarchy. Birendranagar was declared the capital of Karnali Province. Hence, its recent history of urbanization is expected to continue as new governmental investments in economic and infrastructure development attract additional migrants. A ring road around the city, currently under construction, is expected to enhance the industrial, commercial, and development activities and thus to provoke urbanization and LULC change widely in the near future [44].

3.2. Flood Area Analysis

The Siwalik (Churiya) range and the southern slopes of the Mahabharat range are highly prone to geo-disasters [14,45] due to their fragile geology, steep slopes, and intense monsoonal precipitation during June through September (Figure 4), causing regular landslides and debris flow along creeks and steep slopes [20]. The highest average monthly rainfall was 488.77 mm and 428.64 mm for July and August, respectively, between 2000 and 2015 [42]. The southern belt of Churiya region and mid-hills are especially subject to intense rainfall from monsoonal low-pressure systems entering from the Bay of Bengal. The topographical slope of the study area ranges from below 4 degrees in the valley plain to 48 degrees in the Mahabharat range (Figure 5). The rocky soils cannot absorb intense rainfalls, resulting in major overland runs-off carrying soil and debris that have caused significant economic and human losses as urban expansion and deforestation have proceeded [41]. New settlements and urban expansion along river banks have disturbed river channels and drainage system. Deforestation and sand/gravel mining on the Siwalik range and upstream river beds have aggravated landslide hazards during the dry seasons. The major erosive and destructive forces of swollen rivers, which locals claim have increased in recent decades, dissipate only once steep riverine channels give way to gentler slopes, frustrating potential geo-engineering solutions to flood disasters.

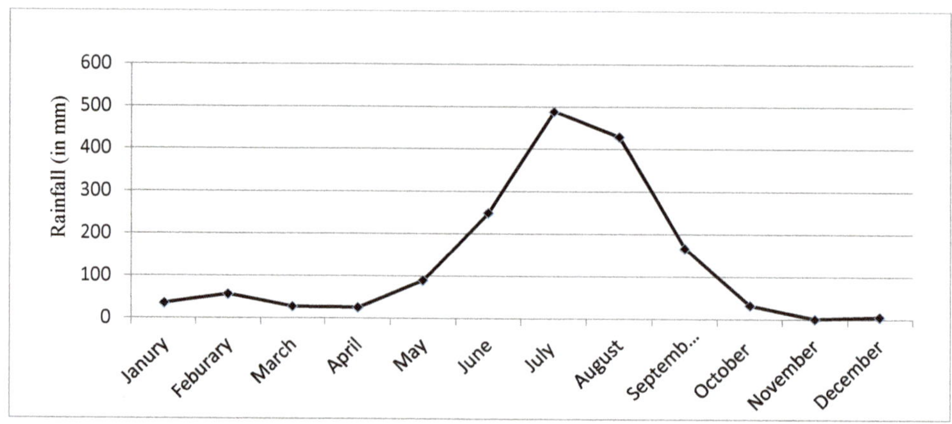

Figure 4. Trend of monthly mean rainfall of Birendranagar station during 2000–2015.

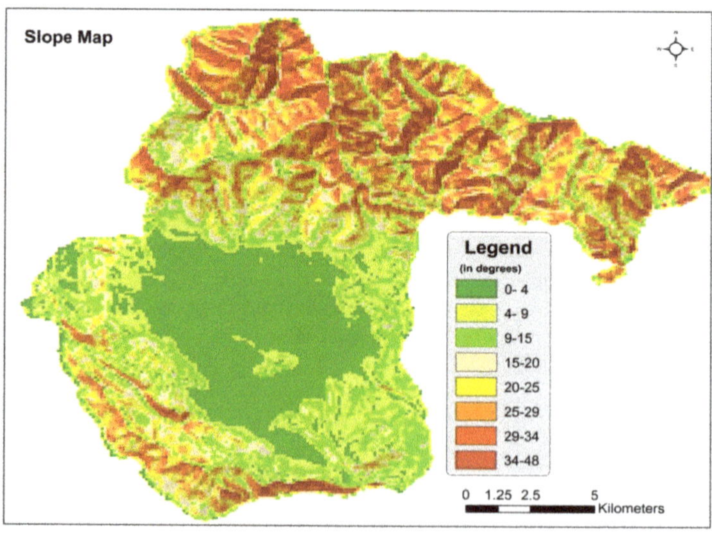

Figure 5. Slope Map of the Study Area.

Surkhet district is affected annually by monsoonal flood events, the impacts of which have been increasingly devastating in recent decades. During August, 2014, more than 12,385 people from 2327 families were displaced [46], 1581 houses were washed away, 301 houses were damaged, 15 government schools were destroyed, and 31 more were damaged, resulting in the economic loss of NRs 6 billion. Additionally, 411 irrigation projects, 99 drinking water schemes, 11 child development centers, and 663 ha of forest were swept away [43]. Latikoili, Ghatgaun, Satakhani, Chinchhu, Lekhparajul, Hariharpur, Babiyachaur, Tatapani, Taranga, Dharapani, and Kunathari settlements were the most affected areas [46]. Birendranagar city was particularly highly affected by the flooding [44] on Khorke River (Figure 6) and Itram River (Figure 7). This 2014 flood not only impacted the settlements and cultivated area, but also claimed human lives.

Figure 6. River bank erosion and sediment transportation by Khokre River (photo taken by author, 2014).

Figure 7. Affected infrastructures from the floods in the Itram River (photo taken by author, 2014).

Our survey of flooding events highlighted that the floods along the Neware river, Gagre river, Geruwa river, Tuni river, Dwari river, and Dundra river also effected the nearby settlements and farm lands. This study has identified several settlements and cultivated lands along these river banks that are at high risk of flood hazards, which are presented in Figure 8.

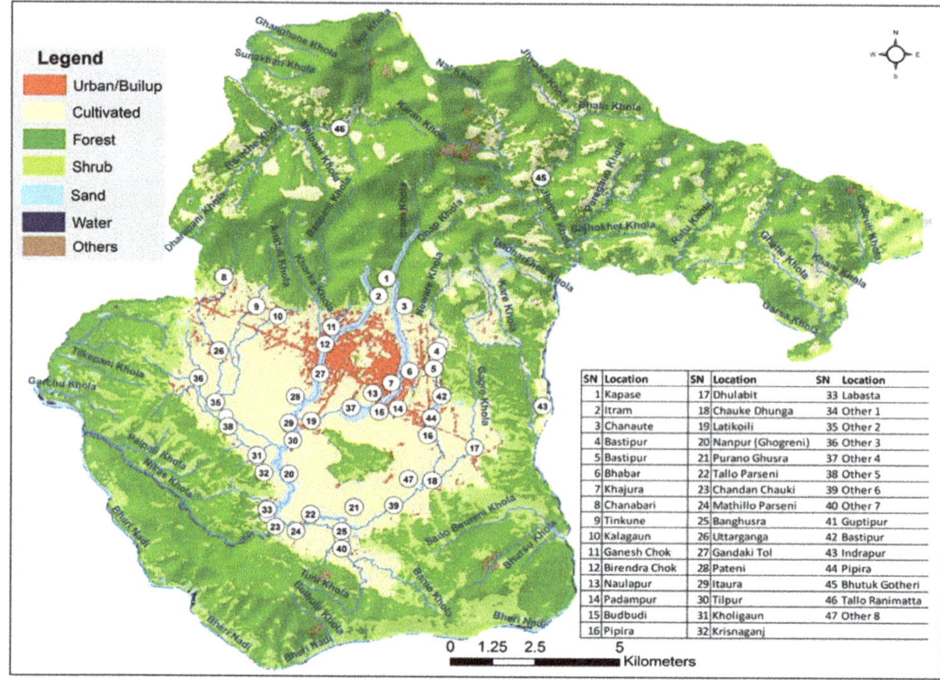

SN	Location	SN	Location	SN	Location
1	Kapase	17	Dhulabit	33	Labasta
2	Itram	18	Chauke Dhunga	34	Other 1
3	Chanaute	19	Latikoili	35	Other 2
4	Bastipur	20	Nanpur (Ghogreni)	36	Other 3
5	Bastipur	21	Purano Ghusra	37	Other 4
6	Bhabar	22	Tallo Parseni	38	Other 5
7	Khajura	23	Chandan Chauki	39	Other 6
8	Chanabari	24	Mathillo Parseni	40	Other 7
9	Tinkune	25	Banghusra	41	Guptipur
10	Kalagaun	26	Uttarganga	42	Bastipur
11	Ganesh Chok	27	Gandaki Tol	43	Indrapur
12	Birendra Chok	28	Pateni	44	Pipira
13	Naulapur	29	Itaura	45	Bhutuk Gotheri
14	Padampur	30	Tilpur	46	Tallo Ranimatta
15	Budbudi	31	Kholigaun	47	Other 8
16	Pipira	32	Krisnaganj		

Figure 8. Flood hazard map.

4. Conclusions

The study describes the LULC changes and urbanization process surrounding Birendranagar city during 1989–2016 using Landsat imagery and extensive consultations with local residents and officials. It highlights the rapid expansion of urban/built-up area from 2001. Urban areas have expanded most aggressively over fertile farm lands and become agglomerated within the flood-prone valley, particularly along Surkhet-Jumla-Karnali Highway and cross sectional roads. Forest conservation programs have helped slow deforestation during the last two decades. While wider reforestation measures are probably required to counter soil erosion and downstream transport during monsoons, this alone is probably insufficient to reduce flood hazard to reasonable levels. New settlements and cultivated lands established along river banks have especially high monsoonal flood hazards. Therefore, careful sustainable urban planning, migration control, and redirection measures, as well as flood hazard management and monitoring programs, are also required. Flood hazard management and monitoring strategies are not yet explicitly incorporated into local urban development plans, let alone regional land-use plans [47]. Indeed, much recent urban development has been informal or otherwise not subject to specific plans and zoning. Addressing the challenges of flood hazard mitigation in Nepal is thus fundamentally a challenge of instituting good governance practices based on solid empirical foundations, whereby development is subject to local zoning plans and laws.

Author Contributions: S.R. and B.R. designed the project and wrote the manuscript. B.R. contributed to remote sensing and GIS part including field data acquisition and verification. S.S. contributed to writing and English editing. All authors revised and edited the final manuscript.

Acknowledgments: The authors thank all the scientists and local participants who have participated in the establishment of database.

Conflicts of Interest: The authors declare no conflict of interest.

References

1. UNISDR. *Living with Risk: A global Review of Disasterreduction Initiatives*; United Nation, Inter Agency Secretariat of the International Strategy for Disaster Reduction: Geneva, Switzerland, 2004; Volume 1.
2. Zhao, M.; Cheng, W.; Zhou, C.; Li, M.; Huang, K.; Wang, N. Assessing Spatiotemporal Characteristics of Urbanization Dynamics in Southeast Asia Using Time Series of DMSP/OLS Nighttime Light Data. *Remote Sens.* **2018**, *10*, 47. [CrossRef]
3. Eskandari, M.; Omidvar, B.; Modiri, M.; Nekooie, M.; Alesheikh, A. Geospatial Analysis of Earthquake Damage Probability of Water Pipelines Due to Multi-Hazard Failure. *ISPRS Int. J. Geo-Inf.* **2017**, *6*, 169. [CrossRef]
4. Rimal, B.; Zhang, L.; Keshtkar, H.; Sun, X.; Rijal, S. Quantifying the Spatiotemporal Pattern of Urban Expansion and Hazard and Risk Area Identification in the Kaski District of Nepal. *Land* **2018**, *7*, 37. [CrossRef]
5. Halimi, M.; Sedighifar, Z.; Mohammadi, C. Analyzing spatiotemporal land use/cover dynamic using remote sensing imagery and GIS techniques case: Kan basin of Iran. *GeoJournal* **2017**, 1–11. [CrossRef]
6. Sanyal, J.; Lu, X.X. Application of Remote Sensing in Flood Management with Special Reference to Monsoon Asia: A Review. *Nat. Hazards* **2004**, *33*, 283–301. [CrossRef]
7. Dewan, A.M. *Geospatial Techniques in Assessing Hazards, Risk and Vulnerability*; Springer: Dordrecht, The Netherlands; Heidelberg, Germany; New York, NY, USA; London, UK, 2013.
8. Guha-Sapir, D.; Hoyois, P.; Below, R. *Annual Disaster Statistical Review 2014: The Numbers and Trends*; Centre for Research on the Epidemiology of Disasters (CRED): Brussels, Belgium, 2015.
9. Guha-Sapir, D.; Hoyois, P.; Below, R. *Annual Disaster Statistical Review 2015: The Numbers and Trends*; Centre for Research on the Epidemiology of Disasters (CRED): Brussels, Belgium, 2016.
10. Karymbalis, E.; Katsafados, P.; Chalkias, C.; Gaki-Papanastassiou, K. An integrated study for the evaluation of natural and anthropogenic causes of flooding in small catchments based on geomorphological and meteorological data and modeling techniques: The case of the Xerias torrent (Corinth, Greece). *Z. Geomorphol.* **2012**, *56*, 45–67. [CrossRef]
11. Shrestha, M. Impacts of Floods in South Asia. *J. South Asia Disaster Stud.* **2008**, *1*, 85–106.
12. Upreti, B. Impact of natural disaster on development of Nepal. In *Bhandary NP. Disasters and Development, Investing in Sustainable Development in Nepal*; Subedi, J.K., Ed.; Ehime University, Japanand Bajra Publication: Kathmandu, Nepal, 2010.
13. Ministry of Home Affairs (MoHA). *Nepal Disaster Report 2011*; Ministry of Home Affairs (MoHA) and Disaster Preparedance Network (DPNeT): Kathmandu, Nepal, 2011.
14. United Nations Development Programme (UNDP). Global Assessment of Risk. In *Nepal Countary Report*; UNDP Pulchowk: Kathmandu, Nepal, 2009.
15. Department of Water Induced Disaster Prevention (DWIDP). *Annual Disaster Review 2012. Government of Nepal*; Ministry of Irrigation Department of Water Induced Disater Management, Ed.; DWIDP: Kathmandu, Nepal, 2013.
16. Ministry of Home Affairs (MoHA). *Nepal Disaster Report 2015*; Disaster Preparedance Network (DPNeT); Ministry of Home Affairs (MoHA): Kathmandu, Nepal, 2015.
17. Kafle, K.R.; Khanal, S.N.; Dahal, R.K. Consequences of Koshi flood 2008 in terms of sedimentation characteristics and agricultural practices. *Geoenviron. Disasters* **2017**, *4*. [CrossRef]
18. MoHA. Nepal Disaster Report. Available online: http;;//drrportal.gov.np (accessed on 19 November 2017).
19. Dahal, R.K.; Bhandary, N.P. Geo-Disaster and Its Mitigation in Nepal. In *Progress of Geo-Disaster Mitigation Technology in Asia*; Springer: Berlin/Heidelberg, Germany, 2013; pp. 123–156.
20. International Centre for Integrated Mountain Development/United Nation Educational Scientific and Cultural Organization (ICIMOD/UNESCO). *Preparing for Flood Disaster Mapping and Assessing Hazard in the Ratu Watershed, Nepal*; Nepal, K., United Nations Educational Scientific and Cultural Organization (UNESCO), Eds.; ICIMOD: New Delhi, India, 2007.
21. Rimal, B.; Zhang, L.; Stork, N.; Sloan, S.; Rijal, S. Urban Expansion Occurred at the Expense of Agricultural Lands in the Tarai Region of Nepal from 1989 to 2016. *Sustainability* **2018**, *10*, 1341. [CrossRef]
22. District Development Committee (DDC). *District Profile, Surkhet*; Government of Nepal, Ed.; DDC: Surkhet, Nepal, 2010.

23. Morrison, A.; Westbrook, C.J.; Noble, B.F. A review of the flood risk management governance and resilience literature. *J. Flood Risk Manag.* **2017**. [CrossRef]
24. Central Bureau of Statistics (CBS). *Population Monograph of Nepal*; National Planning Commission Secretariat; CBS: Kathmandu, Nepal, 2014.
25. Ministry of Federal Affairs and Local Development (MoFALD). *Local Level Reconstruction Report*; MoFALD: Kathmandu, Nepal, 2017.
26. Thapa, R.B.; Murayama, Y. Drivers of urban growth in the Kathmandu valley, Nepal: Examining the efficacy of the analytic hierarchy process. *Appl. Geogr.* **2010**, *30*, 70–83. [CrossRef]
27. Muzzini, E.; Aparicio, G. *Urban Growth and Spatial Transition in Nepal*; The World Bank: Washington, DC, USA, 2013.
28. Do, Q.T.; Iyer, L. Geography, Poverty and Conflict in Nepal. In *Working Paper -07-065*; Harvard Business School: Boston, MA, USA, 2009.
29. Rimal, B.; Zhang, L.; Fu, D.; Kunwar, R.; Zhai, Y. Monitoring Urban Growth and the Nepal Earthquake 2015 for Sustainability of Kathmandu Valley, Nepal. *Land* **2017**, *6*, 42. [CrossRef]
30. Dewan, A.M.; Yamaguchi, Y. Using remote sensing and GIS to detect and monitor land use and land cover change in Dhaka Metropolitan of Bangladesh during 1960–2005. *Environ. Monit. Assess.* **2009**, *150*, 237–249. [CrossRef] [PubMed]
31. Thapa, R.B.; Murayama, Y. Examining Spatiotemporal Urbanization Patterns in Kathmandu Valley, Nepal: Remote Sensing and Spatial Metrics Approaches. *Remote Sens.* **2009**, *1*, 534–556. [CrossRef]
32. Rimal, B.; Zhang, L.; Keshtkar, H.; Haack, B.; Rijal, S.; Zhang, P. Land Use/Land Cover Dynamics and Modeling of Urban Land Expansion by the Integration of Cellular Automata and Markov Chain. *ISPRS Int. J. Geo-Inf.* **2018**, *7*, 154. [CrossRef]
33. Keshtkar, H.; Voigt, W. Potential impacts of climate and landscape fragmentation changes on plant distributions: Coupling multi-temporal satellite imagery with GIS-based cellular automata model. *Ecol. Inform.* **2016**, *32*, 145–155. [CrossRef]
34. Keshtkar, H.; Voigt, W. A spatiotemporal analysis of landscape change using an integrated Markov chain and cellular automata models. *Model. Earth Syst. Environ.* **2015**, *2*, 10. [CrossRef]
35. Byun, Y.; Han, Y.; Chae, T. Image Fusion-Based Change Detection for Flood Extent Extraction Using Bi-Temporal Very High-Resolution Satellite Images. *Remote Sens.* **2015**, *7*, 10347–10363. [CrossRef]
36. Sloan, S.; Pelletier, J. How accurately may we project tropical forest-cover change? A validation of a forward-looking baseline for REDD. *Glob. Environ. Chang.* **2012**, *22*, 440–453. [CrossRef]
37. Rimal, B.; Zhang, L.; Keshtkar, H.; Wang, N.; Lin, Y. Monitoring and Modeling of Spatiotemporal Urban Expansion and Land-Use/Land-Cover Change Using Integrated Markov Chain Cellular Automata Model. *ISPRS Int. J. Geo-Inf.* **2017**, *6*, 288. [CrossRef]
38. Rimal, B.; Baral, H.; Stork, N.; Paudyal, K.; Rijal, S. Growing City and Rapid Land Use Transition: Assessing Multiple Hazards and Risks in the Pokhara Valley, Nepal. *Land* **2015**, *4*, 957–978. [CrossRef]
39. Anderson, J.R. *A Land Use and Land Cover Classification System for Use with Remote Sensor Data*; United States Government Printing Office: Washington, DC, USA, 1976.
40. Government of Nepal (GN). *Topographical Map*; Ministry of Land Resources and Management Survey Department; Topographic Survey Branch: Kathmandu, Nepal, 1998.
41. National Planning Commission (NPC). *Post Flood Recovery Needs Assessment*; NPC; Government of Nepal; Singha Durbar: Kathmandu, Nepal, 2017.
42. The Department of Hydrology and Meteorology (DHM). *Metrological Data of Birendranagar Station, Surkhet*; Government of Nepal, Ed.; DHM: Kathmandu, Nepal, 2016.
43. UN-Habitat. *District Land Use Plan of Surkhet*; Catalytic Support on Land Issues, Participatory Land Use Planning Project; UN-Habitat: Pulchowk, Nepal, 2015.
44. UN-Habitat. *Village Development Committee Land Use Plan of Latikoili*; Catalytic Support on Land Issues, Participatory Land Use Planning Project; UN-Habitat: Pulchowk, Nepal, 2015.
45. Ghimire, M. Historical Land Covers Change in the Chure-Tarai Landscape in the Last Six Decades: Drivers and Environmental Consequences. In *Land Cover Change and Its Eco-Environmental Responses in Nepal*; Li, A., Deng, W., Zhao, W., Eds.; Springer: Singapore, 2017; pp. 109–147.

46. Informal Sector Service Center (INSEC). *Flood Victims of Surkhet District Undergoing Pains due to State Apathy*; INSEC: Kathmandu, Nepal, 2017.
47. Baan, P.J.; Klijn, F. Flood risk perception and implications for flood risk management in the Netherlands. *Int. J. River Basin Manag.* **2004**, *2*, 113–122. [CrossRef]

Article

Physical and Anthropogenic Factors Related to Landslide Activity in the Northern Peloponnese, Greece

Hariklia D. Skilodimou [1], George D. Bathrellos [1,*], Efterpi Koskeridou [2], Konstantinos Soukis [3] and Dimitrios Rozos [4]

[1] Department of Geography and Climatology, Faculty of Geology and Geoenvironment,
    National and Kapodistrian University of Athens, University Campus, Zografou, ZC 15784, Athens, Greece;
    hskilodimou@geol.uoa.gr
[2] Department of Historical Geology-Palaeontology, Faculty of Geology and Geoenvironment,
    National and Kapodistrian University of Athens, University Campus, Zografou, ZC 15784, Athens, Greece;
    ekosker@geol.uoa.gr
[3] Department of Dynamic Tectonic Applied Geology, Faculty of Geology and Geoenvironment,
    National and Kapodistrian University of Athens, University Campus, Zografou, ZC 15784, Athens, Greece;
    soukis@geol.uoa.gr
[4] Department of Geological Sciences, School of Mining and Metallurgical Engineering,
    National Technical University of Athens, 9 Heroon Polytechniou str, Zografou, ZC 15780, Athens, Greece;
    rozos@metal.ntua.gr
*   Correspondence: gbathrellos@geol.uoa.gr; Tel.: +30-2107274882

Received: 27 June 2018; Accepted: 15 July 2018; Published: 19 July 2018

Abstract: The geological, geomorphic conditions of a mountainous environment along with precipitation and human activities influence landslide occurrences. In many cases, their relation to landslide events is not well defined. The scope of the present study is to identify the influence of physical and anthropogenic factors in landslide activity. The study area is a mountainous part of the northern Peloponnesus in southern Greece. The existing landslides, lithology, slope angle, rainfall, two types of road network (highway-provincial roads and rural roads) along with land use of the study area are taken into consideration. Each physical and anthropogenic factor is further divided into sub-categories. Statistical analysis of landslide frequency and density, as well as frequency and density ratios, are applied and combined with a geographic information system (GIS) to evaluate the collected data and determine the relationship between physical and anthropogenic factors and landslide activity. The results prove that Plio-Pleistocene fine-grained sediments and flysch, relatively steep slopes (15°–30°) and a rise in the amount of rainfall increase landslide frequency and density. Additionally, Plio-Pleistocene fine-grained sediments and flysch, as well as schist chert formations, moderate (5°–15°) and relatively steep slopes (15°–30°), along with the amount of rainfall of >700 mm are strongly associated with landslide occurrences. The frequency and magnitude of landslides increase in close proximity to roads. Their maximum values are observed within the 50 m buffer zone. This corresponds to a 100 m wide zone along with any type of road corridors, increasing landslide occurrences. In addition, a buffer zone of 75 m or 150 m wide zone along highway and provincial roads, as well as a buffer zone of 100 m or 200 m wide zones along rural roads, are strongly correlated with landslide events. The extensive cultivated land of the study area is strongly related to landslide activity. By contrast, urban areas are poorly related to landslides, because most of them are located in the northern coastal part of the study area where landslides are limited. The results provide information on physical and anthropogenic factors characterizing landslide events in the study area. The applied methodology rapidly estimates areas prone to landslides and it may be utilized for landslide hazard assessment mapping as well as for new and existing land use planning projects.

Keywords: landslides; geographic information system (GIS); frequency ratio; density ratio; human activities; land use planning

1. Introduction

Landslides are physical phenomena, active in geological time. They have affected the natural environment and existing biota since even before the appearance of man on Earth. Nowadays, they are considered as one of most significant natural hazards worldwide. Frequently, their associated consequences have adverse long-term effects. When these consequences have a major impact on human life and activities, they become natural disasters [1].

Mass movements are identified as the movement of a mass of soil, rock, debris or earth down a slope [2]. They can include falls, topples, slides, spreads and flows. These phenomena are part of the process of hill slope erosion which is responsible for the introduction of sediment into streams, rivers, lakes, reservoirs and finally the oceans [3].

Landslides can be triggered by a variety of external factors, such as intense rainfall, earthquakes and rapid stream erosion [3–9]. Additionally, human activities such as deforestation of slopes, removal of slope support in road cuts, alteration of surface runoff paths, have become important triggers for landslide manifestation [10,11].

Every year, landslides kill people and cause huge property damage in mountainous areas of the world [11]. Kumar et al. [12] reported that in the Himalayas over 2000 landslides killed more than 5000 people during 2013. In the USA, landslide events cause an estimated US$1–2 billion in economic losses and about 25–50 fatalities annually [13]. In Italy, at least 263 people were killed by mass movements during the period of 1990–1999 [14].

Much of Greece consists of hilly and mountainous terrain subject to landslide manifestation. Therefore, landslide events are common phenomena and cause significant damage to road networks, sections of urban areas and cultivated land [15–17].

Several physical process factors and human activities influence landslide activity. Landslides in the future will most likely occur under geomorphic, geologic, and hydrologic conditions that have produced past and present landslides. Types of bedrock, slope steepness, and precipitation zones represent, respectively, geologic, geomorphic and hydrologic factors. Weak, incompetent rock is more likely to fail than strong, competent rock; general steeper slopes have a greater chance of landsliding, while rainfall is considered as an important factor in slope stability, almost as important as gravity [2,3,11,15]. In addition, the frequency of landslide events commonly increases, when man-made structures induce changes in mountainous environments [3]. Human settlements, cut-and-fill construction for roads, the construction of buildings and railroads, changes in land cover, and terracing for agriculture contribute to the conditions that lead to slope failure [18,19].

However, in most cases the influence of physical and anthropogenic factors in landslide manifestations is not well known. Determination of the actual landslide zone induced by geologic, geomorphic parameters and rainfall, along with human activities such as the road network, urban and cultivated areas is an important tool for engineers, planners, and environmental managers. This procedure is very useful to identify areas prone to landslide and assess landslide hazard, and it is also a necessary step for land use and government urban planning policies worldwide [20–23]. Moreover, the determination of the landslide zone of disturbance is vital for road engineers who must deal with the costly and sometimes life-threatening problems caused by road-induced landslides. Realistic assessment of the impact of proposed transportation construction must be quantifiable as road costs may be significantly increased by landsliding during construction [12,18,24].

During the last decades geographical information systems (GIS) and Earth observation (EO) data have become integral tools for the evaluation of natural hazard events. These current geospatial technologies are very useful for assessing future hazard occurrences and identifying the vulnerability of communities to hazards. In this sense, GIS is an excellent tool in the spatial analysis of multi-dimensional phenomena such as landslides [25–30].

The scope of the present study is to identify the influence of physical and anthropogenic factors on landslide occurrences. To accomplish this, existing landslides, lithology, slope angle and rainfall in conjunction with two types of road network and land use were taken into consideration. Statistical analysis and GIS are applied to process and evaluate the landslides and factors. Thus, the spatial distribution of landslide frequency, magnitude and the association of lithology, slope, rainfall, roads and land use with landslide occurrences in mountainous terrain were determined. The case study area was a mountainous part of the northern Peloponnesus in southern Greece.

2. Study Area

The study area is located in the northern part of the Peloponnesus, in southern Greece (Figure 1a). The region covers an area of about 194 km^2 and its altitude varies from 0 to 1400 m. The morphology of the area comprises southern mountainous land with very steep slopes reaching an altitude of 1400 m a.s.l.; intermediate semi-mountainous land with lower altitudes and steep slopes; and a northern coastal area of limited extent with low altitudes and gentle slopes. The drainage networks of the area flow with a main direction from SW to NE and discharge into the Gulf of Corinth (Figure 1b).

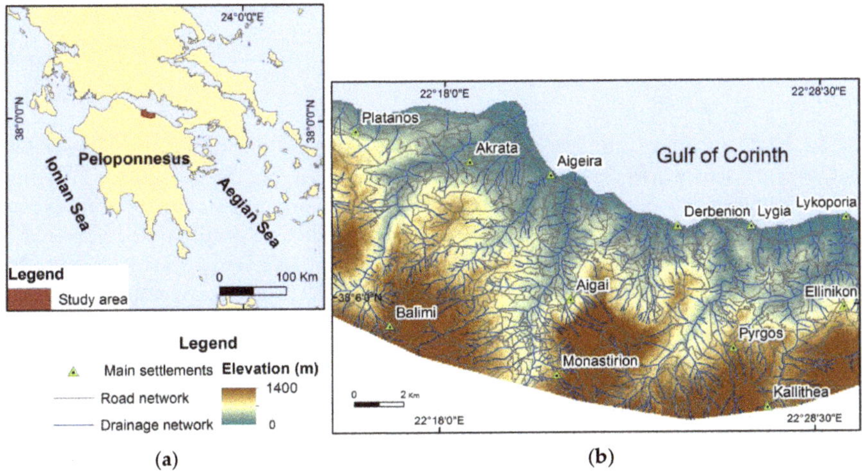

Figure 1. (**a**) Location map of the study area; (**b**) the elevations of the study area, the drainage network, the road network and the main settlements.

Climatologically, the study area is classified as Mediterranean, because of its coastal extension, without considerable temperature variations. The annual precipitation is about 800 mm in the mountainous part of the study area and about 550 mm at the lowlands.

The study area is located on the southern part of the Gulf of Corinth, and its landscape evolution is controlled by the neotectonic action of the graben, which forms the gulf. The presence of Pleistocene marine terraces in the northern part of the area proves that it has been affected until today by significant neotectonic uplift. This fact has been clearly expressed by the intense seismicity of the Gulf of Corinth [31].

The geological formations that crop out in the study area comprise both Neocene–Quaternary deposits and alpine formations. Alpine formations from two Hellenic geotectonic zones (Olonos–Pindos and Gavrovo–Tripolis) participate in the southern part of the study area. The study area is constructed by: (a) Holocene alluvial deposits, talus cones and scree; (b) Plio-Pleistocene sediments consisting of conglomerates such as clayey marls, marls, silty sands and sandstones; (c) schist-chert formations (Upper Cretaceous–Eocene); (d) Cretaceous limestones; and (e) schist-chert formations (Jurassic–Lower Cretaceous) [32].

Therefore, the study area mainly consists of Neocene deposits and, as a part of the Corinthian graben, it is characterized by intense neotectonic activity. Consequently, it is affected by many landslide events. It is an appropriate case study area because it comprises small urban areas, where there is still a potentiality of future urban growth and planning. Moreover the study area has often suffered from the consequences of several landslides, which usually cause serious damage in inhabited areas, road networks, and cultivated areas [33].

3. Materials and Methods

This study was carried out using the following:

- a topographic map (1:50,000 scale) from the Hellenic Military Geographical Service (HAGS);
- a satellite image via Google Earth, Landsat 7/Copernicus, with an acquisition date of July 2015;
- rainfall records from six stations, belonging to: (a) the Hellenic National Meteorological Service; (b) the Ministry for the Environment, Physical Planning and Public Works; and (c) the Ministry of Agriculture and Ministry of Development. These records referred to mean annual precipitation for the period of 1975–2010;
- field work data involving observations on landslide sites.

A spatial database was created, and ArcGIS 10.0 software was used to process the collected data.

3.1. Landslide Inventory Map

The landslide inventory map compilation has involved the following steps: (a) landslides were recorded from previous works [33]; (b) landslide locations were recognized on satellite images; (c) the manifestations of landslide were verified and mapped by field work. The main scarp of every recorded landslide during the field work was depicted in topographic maps at a proper scale and then digitized as a polygon layer. According to Yilmaz [34], the scarp sampling strategy gives better results than the point one. The landslides were used for the compilation of the landslide inventory map (Figure 2). A number of 270 sites of landslide manifestation were examined throughout the study area, with a varied size from 4130 m^2 to 91,000 m^2, having affected a total area of about 6.5 km^2.

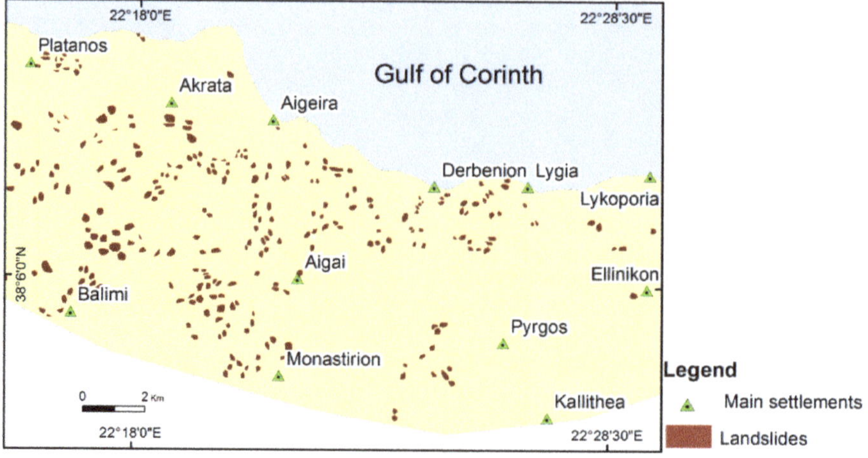

Figure 2. The landslide inventory map of the study area.

3.2. Physical and Anthropogenic Factors

The occurrence of landslides is largely a function of the interaction of several factors such as lithology, tectonic settings, geomorphological settings, earthquake, rainfall and human activities. These factors, which are directly or indirectly related with the occurrences of landslides, are commonly known as landslide-related factors [35–39]. In landslide susceptibility mapping studies, it is believed that the accuracy of the results increase when numerous landslide-related factors are included in the analytical process [40]. In many cases, this is usually difficult to do, because detailed data is hard to find. The aim of the present work is to determine the influence of physical and anthropogenic factors on landslide occurrences rather than the production of a landslide susceptibility map. Moreover, data for all the aforementioned factors was not available. For these reasons, analyses in this study depend on the physical factors which are lithology, slope angle and rainfall, while the anthropogenic factors are the road network and land use.

Each factor is separated into sub-classes. The determination of the classes' number as well as their boundary values were based on: (a) the literature review [3,11,15–21,33–39]; (b) the extended field observations in the framework of this study; and (c) personal experience from previous studies.

3.2.1. Lithology

Lithology is the main predisposing factor controlling landslide development. Different lithological units have different engineering geological behaviors and they are very important in providing data for landslide-related factors studies [3,36,39,40]. For the study area the geological formations were digitized and unified according to their engineering geological behavior [20,32,33,35], in relation to landslide manifestation. Thus, lithology includes five classes as follows: (a) fine, fine-coarse to coarse and loose to semi-coherent Quaternary formations; (b) cyclothematic formations (Plio-Pleistocene fine-grained sediments and Flysch sediments); (c) Plio-Pleistocene coarse-grained sediments; (d) thin-bedded schist chert formations; and (e) moderate to thick bedded limestones (Figure 3).

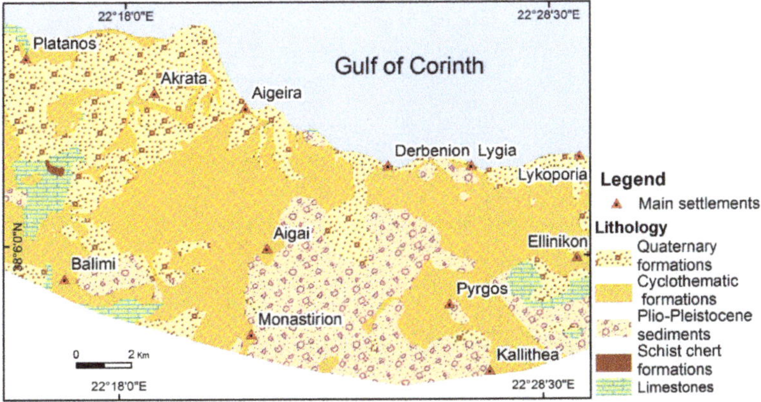

Figure 3. The lithology of the study area consisting of: Quaternary formations (fine, fine-coarse to coarse and loose to semi-coherent sediments), Cyclothematic formations (Plio-Pleistocene fine-grained sediments and Flysch sediments), Plio-Pleistocene coarse-grained sediments, thin bedded schist chert formations, and moderate to thick bedded limestones.

3.2.2. Slope Angle

The slope angle has an effect on slope stability, increasing the landslide hazard [35,39,40]. Contours with 20 m intervals and height points were digitized from topographic map (scale 1:50,000) and saved as line and point layer correspondingly. A digital elevation model (DEM) was derived from the digitized elevation data using the 3D Analyst extension of ArcGIS, and the slope layer was extracted from DEM. The slopes were classified into five classes: (a) <5°, (b) 5°–15°, (c) 15°–30°, (d) 30°–45°, and (e) >45° (Figure 4).

Figure 4. Map showing the spatial distribution of slopes of the study area.

3.2.3. Rainfall

Precipitation is a triggering factor for landslide occurrences [35,39]. The mean annual rainfall of the area varies between 550.7 mm and 789.8 mm. The intensity of rainfall was not analyzed due to lack of data. For the necessities of this study, the precipitation map was produced, using the data of the main meteorological stations in the area and applying the Inverse distance weighted (IDW) interpolation method. The distribution of representative rainfall depends upon the spatial distribution of the stations and elevation [41]. The stations used are uniformly distributed in the study area both hypsometrically and territorially giving accurate precipitation distribution.The precipitation map was separated into three classes, i.e.,: (a) <600 mm, (b) 600 mm–700 mm, and (c) >700 mm (Figure 5).

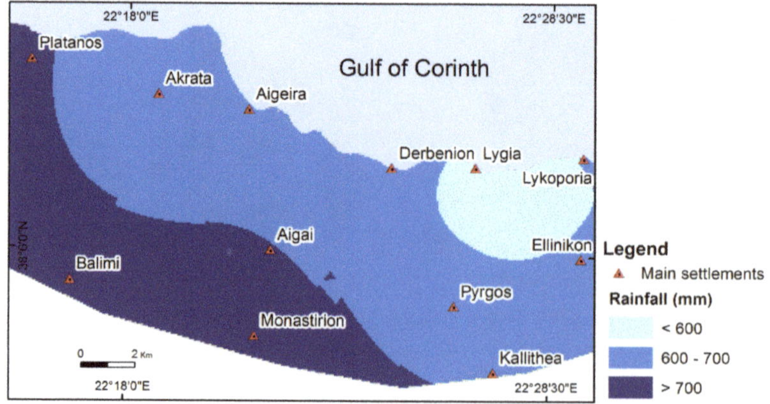

Figure 5. Map showing the spatial distribution of rainfall of the study area.

3.2.4. Road Network

The artificial and natural parts of the slopes around a road network are more sensitive in landslide manifestations [35]. The road network of the study area was digitized as a polyline layer using the topographic map. The complete road network included about: 24 km of highway, 322 km of provincial roads, 389 km rural roads, and totaled about 735 km in length (Figure 6).

The proximity to the road network is often related to an increase of landslide occurrences, so disturbance zones were created around the road network of the area. The creation of disturbance zones is referred to using the GIS term 'buffer' zone along a road network using GIS software. The classes of the buffer zones used in this analysis were: 10, 25, 50, 75, 100, 150, 200, 300 m and >300 m (6). The buffer lengths were measured on both sides of road network and represent a zone that is twice as wide as the value listed above. Therefore, a 100 m buffer describes a zone or swath 200 m wide along the road. Two sets of the aforementioned buffer zones were created along the different road types: one for highways and provincial roads and the other for rural roads.

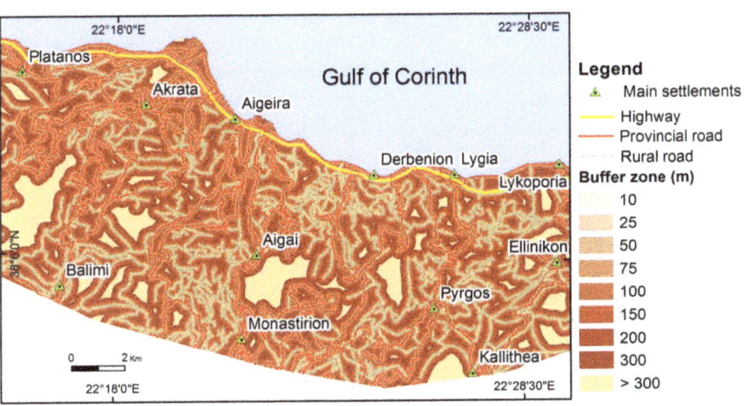

Figure 6. Buffer zones of increasing length along the highway, provincial, and rural roads of the study area.

3.2.5. Land Use

The land use of the study area was taken from the CORINE 2012 Land Cover (CLC) map, Copernicus Program [42]. The program contains land cover data for Europe including land cover class description at scale 1:100,000 published by the European Commission. The CORINE land use map was classified as follows: (a) urban area, (b) cultivated area, (c) forest, (d) shrubby area, and (e) bare area (Figure 7). The land use of the area was saved as polygon layer.

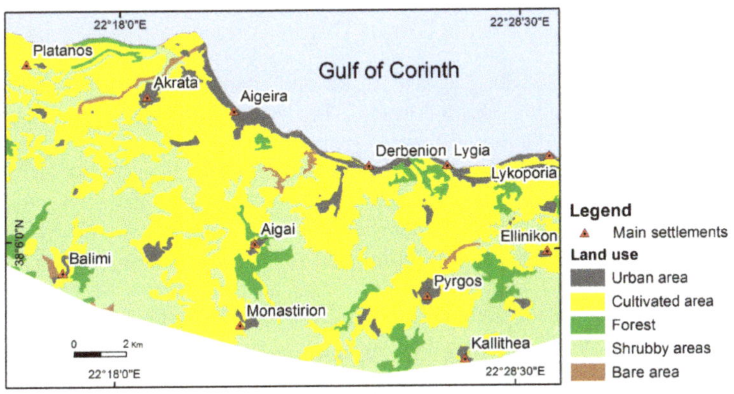

Figure 7. The land use of the study area.

To predict landslides, it is necessary to assume that landslide occurrence is determined by landslide-related factors, and that future landslides will occur under the same conditions as past landslides [3,36]. For this reason, as a first step, the relative frequency of landslides was calculated. The relative frequency is given from Equation (1):

$$FL = Ln/\Sigma Ln \tag{1}$$

where FL = the relative frequency of landslides, Ln = the number of landslide that located in each category of the landslide-related factors, and ΣLn = the total landslide events of the study area.

The landslides were transformed from polygon layer to point layer to calculate the landslide frequency. Thus, the number of landslide events, which were located in the classes of each factor was then determined by using the Spatial Analyst Tools of ArcGIS 10.0.

The next step was the estimation of the landslide frequency by applying a frequency ratio statistical analysis. This approach is based on the relationships between spatial distribution of landslides and each class of the involved factors. According to Lee and Pradhan [36], the frequency ratio is the ratio of the probability of a landslide event to a non-event for a given zone. The frequency ratio is estimated by the following Formula (2):

$$FR = FL/AC \tag{2}$$

where FR = the frequency ratio, FL = the relative frequency of landslides expressed in percentages, and AC = the area ratio for each category to the total area in a percentage form.

Furthermore, to determine the magnitude of a landslide area, the relative density of landslides was computed. The relative density of landslides is expressed by the Equation (3):

$$DL = La/\Sigma La \tag{3}$$

where DL = the relative density of landslides, La = the landslide area within each class, and ΣLa = the total landslide area.

Consequently, the area of landslides within the classes of each factor was computed by using spatial analyst capabilities. Similarly to landslide frequency, a density ratio statistical analysis was applied to specify the relationship between landslide area and the landslide-related factors. The density ratio is calculated according to the following mathematical operator (4):

$$DR = DL/AC \tag{4}$$

where DR = the density ratio, DL = the relative density of landslides expressed in percentages, and AC = the area ratio for each category to the total area in a percentage form.

4. Results

The statistical analysis described above allows discrimination of landslide frequency and magnitude in each category of the adopted factors. The physical and anthropogenic factors, their categories and the results of the statistical analysis are presented in Table 1.

Regarding the physical factors, Figure 8 shows the relative frequency (FL) and density (DL) of landslide distribution in each lithological formation of the study area. The maximum of FL (64%) and DL (61%) values are observed in the area underlain by cyclothematic formations consisting of Plio-Pleistocene fine-grained and flysch sediments. Additionally, high FL (24%) and DL (28%) values are presented in the Quaternary formations which consist of fine-grained to coarse-grained loose sediments. These values are almost two times lower in Quaternary formations than ones in cyclothematic formations. In Plio-Pleistocene coarse-grained sediments, the FL (10%) and DL (10%) values are six times lower than in cyclothematic formations. Schist chert formations have very low FL (1%) and DL (1%) values. The minimum of FR and DL values are observed in moderate to thick bedded limestones.

Table 1. The frequency (*FR*) and density (*DR*) ratio values of landslides into each category of physical and anthropogenic factors, A = area of each category in m^2, *AC* = the area ratio for each category to the total area in a percentage form, *Ln* = number of landslide in each category, *FL* = the relative frequency of landslides expressed in percentage, *La* = the landslide area within each class, and *DL* = the relative density of landslides expressed in percentage.

Lithology	A (m^2)	AC (%)	Ln	FL (%)	FR	La	DL (%)	DR
Quaternary formations	53,096,725	27	65	24	0.9	1,825,615	28	1.0
Cyclothematic formations	85,865,022	44	174	64	1.5	3,942,500	61	1.4
Plio-Pleistocene coarse-grained sediments	46,210,814	24	28	10	0.4	638,169	10	0.4
Schist chert formations	202,998	0.1	2	1	7.1	66,758	1	9.8
Limestones	8,754,203	5	1	0	0.1	10,947	0	0.0
Total	194,129,763	100	270	100	1.0	6,483,988	100	1.0
Slope angle (°)								
<5	14,298,664	7	3	1	0.2	125,782	2	0.3
5–15	71,833,677	37	124	46	1.2	2,827,163	44	1.2
15–30	90,318,029	47	136	50	1.1	3,369,873	52	1.1
30–45	16,176,882	8	6	2	0.3	154,197	2	0.3
>45	1,502,511	1	1	0	0.5	6974	0	0.1
Total	194,129,763	100	270	100	1.0	6,483,988	100	1.0
Rainfall (mm)								
<600	16,282,886	8	15	6	0.7	340,982	5	0.6
600–700	112,873,296	58	155	57	1.0	3,349,098	52	0.9
>700	64,973,582	33	100	37	1.1	2,793,907	43	1.3
Total	194,129,763	100	270	100	1.0	6,483,988	100	1.0
Highway and provincial roads network buffer zone (m)								
10	6,830,803	4	11	4	1.2	260,301	4	1.1
25	9,878,673	5	18	7	1.3	373,996	6	1.1
50	15,147,727	8	23	9	1.1	562,552	9	1.2
75	13,257,597	7	19	7	1.0	462,461	7	1.1
100	11,550,219	6	14	5	0.9	408,156	6	0.9
150	19,325,306	10	36	13	1.3	752,079	12	1.2
200	15,830,201	8	24	9	1.1	647,413	10	1.2
300	24,566,099	13	39	14	1.1	945,164	15	1.2
>300	77,743,138	40	86	32	0.8	2,071,866	32	0.8
Total	194,129,763	100	270	100	1.0	6,483,988	100	1.0
Rural roads network buffer zone (m)								
10	7,728,922	4	15	6	1.4	392,238	6	1.5
25	11,367,074	6	28	10	1.8	563,374	9	1.5
50	18,040,789	9	48	18	1.9	866,394	13	1.4
75	16,592,806	9	32	12	1.4	746,290	12	1.3
100	15,128,865	8	23	9	1.1	605,058	9	1.2
150	26,430,630	14	35	13	0.9	838,841	13	0.9
200	21,657,810	11	24	9	0.8	610,802	9	0.8
300	30,320,688	16	31	11	0.7	812,673	13	0.8
>300	46,862,180	24	34	13	0.5	1,048,317	16	0.7
Total	194,129,763	100	270	100	1.0	6,483,988	100	1.0
Land use								
Urban area	10,361,680	5	12	4	0.8	228,739	4	0.7
Cultivated area	89,597,161	46	188	70	1.5	4,158,855	64	1.4
Forest	11,307,988	6	11	4	0.7	249,236	4	0.7
Shrubby area	80,742,318	42	57	21	0.5	1,800,696	28	0.7
Bare area	2,120,616	1	2	1	0.7	46,461	1	0.7
Total	194,129,763	100	270	100	1.0	6,483,988	100	1.0

Apart from the *FL* and *DL*, the landslide frequency ratio (*FR*) and density ratio (*DR*) were calculated. According to Lee and Pradhan [36] the FR value of 1 is an average value. Thus, ratio values greater than 1 indicate a strong relationship between landslides and the given factor, and ratio values smaller than 1 indicate a poor relationship between landslides and the given factor. As demonstrated in Table 1, the schist chert formations have the maximum *FR* value (7.1), indicating a strong relationship between this lithological formation and landslide occurrences. Similarly, in cyclothematic formations the *FR* value is 1.5, showing a strong association with landslide events. For the remaining lithological formations of the study area, the *FR* values are found to be less than one, indicating a poor relation with landslides.

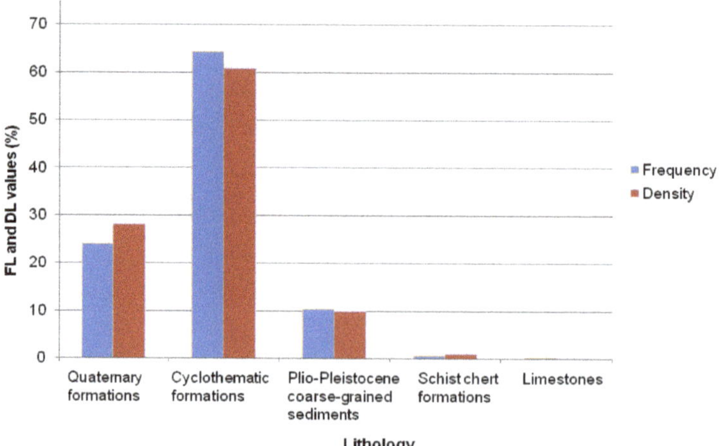

Figure 8. Relative frequency (*FR*) and density (*DL*) values within each category of lithology in the study area. The *FR* and *DL* values are expressed in percentages.

Alike the frequency ratio, the greater the density ratio values above one, the stronger the relationship between landslide occurrence and the given factor; the lower the ratio values below one, the lesser the relationship between landslide occurrence and the given factor. The maximum value of *DR* (9.8) is observed in the schist chert formations, while in the cyclothematic formations and Quaternary formations the *DR* values are 1.4 and 1.0 correspondingly (Table 1). This fact proves a strong relationship between the aforementioned formations and landslide manifestation.

Concerning the slope angle of the study area, Figure 9 shows the *FL* and *DL* distribution in each category of slope. The majority of landslide events (50%) and magnitudes (52%) are located in the class of slopes 15°–30°. Very high *FL* (46%) and *DL* (44%) values are observed in the category of medium slopes 5°–15°. The minimum *FR* and *DL* values are located in the class of very steep slopes >45° (Table 1).

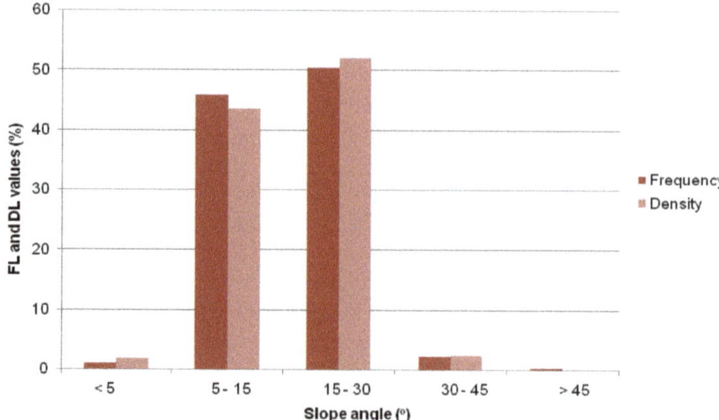

Figure 9. *FL* and *DL* values distribution in each class of slope angle. The *FR* and *DL* values are expressed in percentages.

The maximum *FR* and *DR* values are found to be 1.2 in the class of steep slopes 15°–30°. *FR* and *DR* values greater than one (1.1) have the class 5°–15°. Thus, these slopes are strongly related to landslide occurrences. The classes of slopes <5°, 30°–45°and >45° have *FR* and *DR* values <1, indicating a poor relationship between them and landslides (Table 1).

Figure 10 illustrates the *FL* and *DL* values distribution in each category of rainfall. The maximum values of *FL* (57%) and DL (52%) are attributed in the category of 600 mm–700 mm. Moreover, very high values of *FL* (37%) and *DL* (43%) are observed in the category >700 mm.

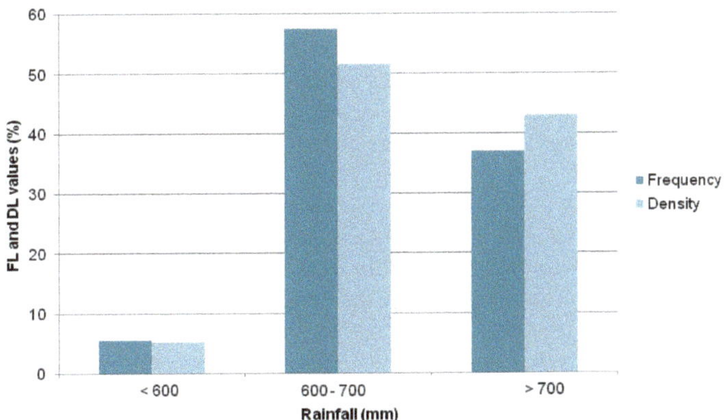

Figure 10. *FL* and *DL* values' distribution in each class of rainfall. The *FR* and *DL* values are expressed in percentages.

The maximum *FR* values 1.1 and 1.0 represent, respectively, the categories of rainfall >700 mm and 600–700 mm. These two categories are strongly related to landslide manifestation. On the contrary, the maximum *DR* value (1.3) is observed in the category >700 mm, indicating a strong relationship between this category and landslide events (Table 1).

The applied statistical analysis identifies the distribution of landslide frequency and magnitude in nine zones of increasing length measured around to the road network. The analysis was performed for two different types of road network. Figure 11 shows the distribution of *FL* and *DL* values in within each buffer zone of the highway and provincial roads. The relative frequency and density of landslides are expressed in percentages.

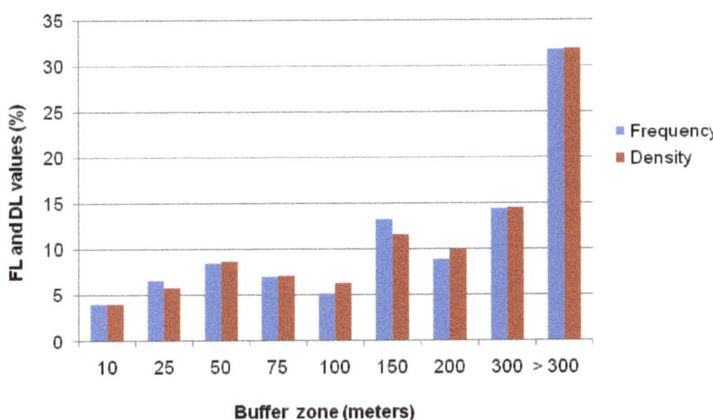

Figure 11. *FL* and *DL* values within each buffer zone extending 0 to 300 m from highway and provincial roads and >300 m in the study area. The *FL* and *DL* values are expressed in percentages.

The *FL* reaches its maximum value within the first 50 m of distance from any given highway and provincial road and decreases in distances beyond 50 m from roads. The *FL* value is greatest (9%) within 50 m buffer zone and is relatively high as far as 100 m from roads. The value of *FL* within buffer zone of 50 m is about two times higher than in distances: (i) approximately 10 m (4%) and (ii) between 75 and 100 m (5%) from roads. Moreover, the *FL* values become relatively high in distance beyond 150 m (13%) from roads. This increase is not related to roads but is likely due to other factors influencing landslide manifestation (Figure 11 and Table 1).

As in the case of *FL*, the *DL* reaches its maximum value (9%) within the buffer zone of 50 m and shows a gradational decline with buffer length (Figure 11). The *DL* value is relatively high in distances between 50 and 100 m from roads. The *DL* value in area of 10 m buffer zone is about two times lower (4%) than in 50 m buffer zone and for area of 100 m buffer zone is one and a half times lower (6%) than in 50 m buffer zone.

In the buffer zones of 10 and 25 m, the *FR* values are 1.2 and 1.3 respectively, indicating a strong relationship between these zones and the occurrence of landslides. For the distances of 50 m and 75 m from roads, the *FR* values were found to be 1.1 and 1.0, respectively, showing a strong association with landslide events. The *FR* values are <1 in distance beyond 100 m, proving a poor relation with landslide occurrences (Table 1).

The maximum *DR* value (1.3) is found to be at distance of 50 m from roads. High *DR* values (1.1) are observed at distances of 10, 25 and 75 m from roads. Thus, these zones are strongly correlated with landslide occurrences. The ratio values in areas beyond 100 m of roads are <1, showing a low probability of landslide occurrences (Table 1).

With regard to the rural roads, the maximum *FL* value (18%) is attributed to a 50 m buffer zone and is relatively high as far as 100 m from roads (Figure 12). The *FL* value within a buffer zone of 50 m is about three times higher than in distance approximately 10 m (6%) and two times higher than in distance between 75 and 100 m (9%) from roads. Similarly to *FL* values, the *DL* values have its maximum within 50 m buffer zone (13%). The *FL* and *DL* values are increased in distance beyond 150 m from rural roads (Figure 12).

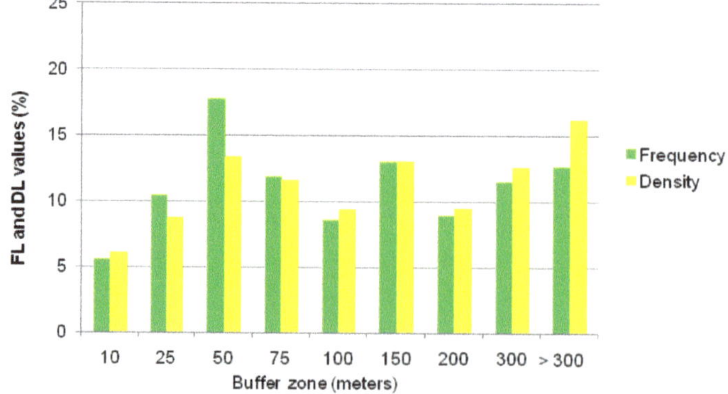

Figure 12. *FL* and *DL* values within each buffer zone extending 0 to 300 m from rural roads and >300 m in the study area. The *FL* and *DL* values are expressed in percentages.

The *FR* values reach their maximum (1.9) in 50 m buffer zone and its values are >1 in the area of the 10 m, 25 m, 75 m and 100 m buffer zones. Likewise, the *DR* values are >1 at distance 10 m, 25 m, 50 m, 75 m, and 100 m from rural roads. These observations indicate a strong relationship between landslides and the aforementioned distances (Table 1).

Concerning the land use of the study area, the calculation *FL* and *DL* provides an estimate of the influence of land use in landslide manifestation. The statistical analysis of *FL* shows that the vast majority of landslide events (70%) are located in cultivated areas, and a high value of *FL* within shrubby areas (21%). In urban and forest areas, it is seventeen and a half times lower (4% in each one) than in cultivated areas. The minimum value of *FL* is observed (1%) in a bare area (Figure 13 and Table 1).

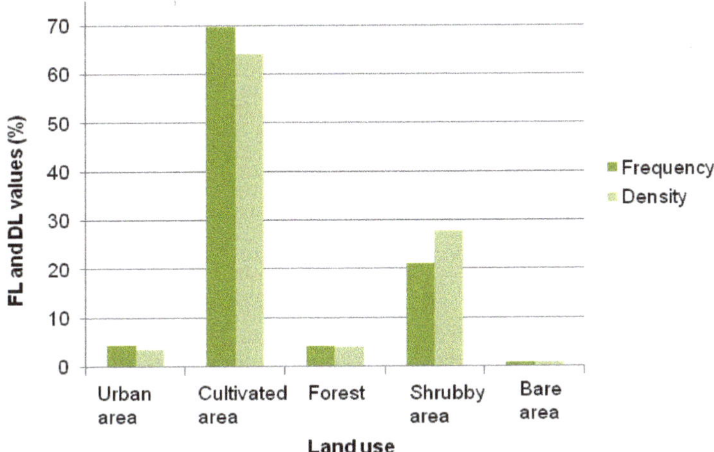

Figure 13. *FR* and *DR* values within each category of land use in the study area. The *FR* and *DR* values are expressed in percentages.

The *DL* values follow the distribution of *FL* values in each category of land use. More specifically, the maximum value of *DL* (64%) is attributed to cultivated areas. A relatively high density value (28%) is observed in shrubby areas, while the minimum density (1%) value is calculated in bare areas. The *DL* values in urban and forest areas (4% in each one) are low (Figure 13 and Table 1).

The *FR* was found to be 1.4 in cultivated areas, which proves a high probability of landslide manifestation. In the rest of the categories of land use, the ratio values are <1, indicating a strong relation with landslide manifestation (Table 1). The highest *DR* value is 1.4 and refers to the cultivated areas, showing a strong association between this land use and landslide occurrences. The remaining classes of land use have ratio values <1, indicating a low probability of landslide manifestation (Table 1).

5. Discussion

Hydrological parameters such as precipitation and human activities are capable of changing landscape characteristics. On the other hand, the geological, geomorphic conditions of a mountainous environment remain unchanged or vary little from a human perspective. These factors influence where landslides occur [3,19–24]. In many cases, the association of physical factors and human activities in landslide occurrences is not well defined. According to Lee and Pradhan [36], statistical approaches based on the observed relationship between each factor and the spatial distribution of landslides is very useful to reveal the correlation between landslide locations and factors.

In the present study physical and anthropogenic factors are analyzed and evaluated to determine their association with landslide occurrences. Lithology, slope angle, rainfall, two types of road network and land use are correlated with the existing landslides by using *FL*, *DL*, *FR* and *DR* statistical analysis and GIS. The case study area is a mountainous part of the northern Peloponnese in Greece. The results provide information on the physical the anthropogenic factors characterizing landslide events in the study area.

The statistical analysis proves that the lithological formations of the study area influence the occurrence of landslides. The events and magnitude of landslide areas reach their maximum in the area underlain by cyclothematic formations consisting of Plio-Pleistocene fine-grained sediments and flysch (Figure 8). Additionally the *FR* and *DR* values are >1 in areas where cyclothematic sediments and shist-chert formations outcrop (Table 1). The latter is possibly due to the limited extent of shist chert formations in the study area. Plio-Pleistocene sediments are fine-grained with a variety of lithological horizons. They consist of clays, marls, alternating sands of a varying degree of diagenesis and/or their mixed phases. Flysch are strongly folded sediments because of the tectonic action (nappes and upthrusts), which in many places results to the formation of thick weathering mantle. Schist chert formations consisting of alternations of cherts, siltstones, thin plated limestones and sandstones while volcanic tuffs rarely participate at places. A thick weathering mantle is formed, mainly in the cases of the surface occurrence of siltstones. The aforementioned soils, hard soil to soft rocky sediments, are weak and incompetent formations and are prone to landslides [4,32,33]. Consequently, the cyclothematic sediments and schist chert formations of the study area are strongly related to landslide occurrences.

Regarding the slope angle, the maximum *FL* and *DL* values are observed in the category of relatively steep slopes 15°–30°. High values of *FL* and *DL* are found to be in the category of moderate slopes 5°–15° (Figure 9). Additionally the *FR* and *DR* values are >1 in the classes of 5°–15° and 15°–30°, indicating a strong relationship between them and landslide events. *FR* and *DR* values <1 are observed in the remaining classes of gentle slopes (<5°–30°), steep slopes (30°–45°) and very steep slopes (>45°), showing a poor relationship between them and landslides (Table 1). According to Rozos et al. [33] this peculiar condition can be explained easily, as in nature slopes consisting of soil or hard soil to soft rocky formations (like those of the study area), and having high angles, fail almost immediately after their formation, resulting in lower slope angles. Finally the slopes with an inclination of around the angle of friction are those which fail after the action of triggering factors. On the other hand, rocky slopes are stable even in high angles suffering only from rock falls, wedge failures etc.

The *FL* and *DL* values increase with increasing in amount of precipitation (Figure 10). The maximum *FL* and *DL* values are attributed in the category of 600 mm–700 mm. FR and DR values >1 are observed in the category of rainfall >700 mm, showing a strong correlation with landslide events (Table 1). Therefore, the rainfall is an important factor in triggering manifestations of landslides [33,35].

Concerning the anthropogenic factors, nine buffer zones of increasing length measured around to two types of road network and five categories of land use are evaluated. The landslide events and magnitude of landslide areas are increased within a distance of approximately 0 to 50 m from any given road type (highway, provincial and rural road) and decreased in distances beyond 50 m from roads (Figures 11 and 12, Table 1). The relatively high values of *FL* and *DL*, which are measured in the 150 m buffer zone, are probably due to other landslide-related factors such as type of lithology or slope angle (Table 1). As already mentioned above, landslides of the study area tended to occur in locations with relatively steep slopes or consisting of cyclothematic sediments.

Consequently, the frequency and magnitude of landslides increase in close proximity to roads. According to Bathrellos et al. [35] this may be due to the fact that the road network sometimes destabilizes adjacent marginally balanced slopes, mainly by removing natural support for the upper part of the slope through undercutting the base of the slope during its construction and by adding extra weight on them. For the study area, landslide disturbance is associated with roads at distances of as much as 50 m from where they are constructed. Further away the frequency and magnitude of landslide occurrences gradually decreases. However, the 50 m buffer length, or 100 m swath, represents a sizeable area of the land surface. These results indicate that a 100 m-wide zone along road corridors increase the landslide activity.

When planning a new road, the alignments of the route should be carefully determined considering the probability of landslides during and after construction. The roads are likely to face unexpected problems of slope stability [12]. The quantification of landslide frequency, magnitude and distribution in mountainous terrain is an important consideration in calculating the cost of new roads construction and the maintenance of existing road networks [18]. Therefore, road engineers and planners in similar mountainous environments may utilize the buffer zone of 50 m during the construction of a new road or to protect the existing road network from future landslide occurrences.

The *FR* and *DR* values (Table 1) show that buffer zones with distances smaller than/or equal to 75 m from highway and provincial roads are strongly associated with landslide manifestation. For buffer zones with distances >75 m, the ratio is <1, indicating a poor relation with landslide occurrences. Conversely, the *FR* and *DR* values (Table 1) are >1 at distances lower than/or equal to 100 m from rural roads, showing a strong relationship between landslides and these distances. In the study area, a 150 m wide zone along highway and provincial road corridors or a 200 m wide zone along rural road corridors indicates a strong association with landslides. This difference may be related to the fact that the rural road network is long and dense in the study area. Thus, the closer the distance is to the road network, the greater is the relationship with landslide manifestation.

In the case of land use, the majority of frequencies and magnitudes of the landslide area are located in cultivated areas (Figure 13). The *FR* and *DR* values are >1 only in cultivated land areas, showing a strong relation to slope failures (Table 1). The variations of the vegetation in an area constitute an important parameter affecting the slope failures, as slope stability is very sensitive to changes in vegetation [33,39]. The soil cohesion is modified depending on the type of vegetation and, thus, cultivated or sparsely vegetated areas are more prone to landslide processes. Additionally, crops can increase the moisture in the soil and alter ground water conditions. In the study area, crops cover an area of about 90 km^2 and this represents 42% of the total area (Table 1). The extensive cultivated land combined with the altered ground water conditions is capable of causing landslide problems.

By contrast, the *FL* and *DL* values are limited in urban areas (Figure 13). Furthermore, the *FR* and *DR* values indicate a poor relationship between urban areas and landslide occurrences (Table 1). Most of the urban areas are located in the northern coastal part of the study area where there are few landslides.

6. Conclusions

In the present study, statistical analysis and GIS was applied to determine the relation of physical and anthropogenic parameters with landslide activity in a mountainous terrain.

Concerning the physical factors, the lithology of the study area controls landslide activity. The cyclothematic formations consisting of Plio-Pleistocene fine-grained sediments and flysch increase *FL* and *DL*. These sediments along with schist chert formations are strongly associated with landslide occurrences. Since *FL* and *DL* increase in relatively steep slopes, geomorphologic factors such as slope angle influence landslide occurrences. Moderate (5°–15°) and relatively steep slopes (15°–30°) are strongly associated with landslide events. Rainfall is an important factor in triggering the manifestation of landslides, as their frequency and density rise with an increase in the amount of rainfall. An amount of rainfall of >700 m is strongly related to landslide events.

In terms of the anthropogenic factors, the statistical analysis proves that the zone of landslide disturbance associated with roads is extensive. The frequency and magnitude of landslides increase in close proximity to roads. The maximum *FL* and *DL* values are observed within the 50 m buffer zone. This zone (100 m wide) along with any type of road corridors increase landslide occurrences. On the other hand, the buffer zone of 75 m or a 150 m wide zone along highway and provincial road corridors is strongly related to landslide manifestation. Since the rural road network is long and dense in the study area, the buffer zone of 100 m or a 200 m wide zone along rural roads is strongly connected with landslide events. The extensive cultivated land of the study area leads to an increase of *FL* and *DL*. This land use is strongly associated with landslides. Urban areas are poorly related to landslide activity, because most of the urban areas are located in the northern coastal part of the study area where landslides are limited.

The proposed methodology reveals a relatively simple and quick way of determining the association of physical and anthropogenic parameters with landslide activity in a mountainous terrain. Topographical, geological, hydrological, transportation and landslide location data can easily be found and their analysis and evaluation are simple and rapid. In regional studies, the applied procedure can be used for the localization of sites prone to landslides and for landslide hazard assessment mapping. Therefore, engineers, planners, decision-makers and environmental managers may utilize the proposed methodology in new and existing spatial planning projects. Additionally, it may be used by the local authorities to guide them in the adoption of policies and strategies aiming at landslide hazard mitigation.

Author Contributions: H.D.S. and D.R. conceived the research; H.D.S. and G.D.B. designed the research and the data analysis; K.S. prepared and analyzed the data; H.D.S., G.D.B., K.S., E.K. and D.R. completed the field work; H.D.S. and E.K. created the figures; H.D.S., G.D.B. and K.S. wrote the paper.

Funding: This research received no external funding.

Conflicts of Interest: The authors declare no conflict of interest.

References

1. Bathrellos, G.D.; Skilodimou, H.D.; Chousianitis, K.; Youssef, A.M.; Pradhan, B. Suitability estimation for urban development using multi-hazard assessment map. *Sci. Total Environ.* **2017**, *575*, 119–134. [CrossRef] [PubMed]

2. Cruden, D.M. A simple definition of a landslide. *Bull. Eng. Geol. Environ.* **1991**, *43*, 27–29. [CrossRef]

3. Varnes, D.J. Slope movement types and processes. In *Landslides, Analysis and Control*; Schuster, R.L., Krizek, R.J., Eds.; Transportation Research Board, Special Report 176; National Research Council: Washington, DC, USA, 1978; pp. 12–33.

4. Chousianitis, K.; Del Gaudio, V.; Sabatakakis, N.; Kavoura, K.; Drakatos, G.; Bathrellos, G.D.; Skilodimou, H.D. Assessment of Earthquake-Induced Landslide Hazard in Greece: From Arias Intensity to Spatial Distribution of Slope Resistance Demand. *Bull. Seismol. Soc. Am.* **2016**, *106*, 174–188. [CrossRef]

5. Rozos, D.; Skilodimou, H.D.; Loupasakis, C.; Bathrellos, G.D. Application of the revised universal soil loss equation model on landslide prevention. An example from N. Euboea (Evia) Island, Greece. *Environ. Earth Sci.* **2013**, *70*, 3255–3266. [CrossRef]

6. Bathrellos, G.D.; Skilodimou, H.D.; Maroukian, H.; Gaki-Papanastassiou, K.; Kouli, K.; Tsourou, T.; Tsaparas, N. Pleistocene glacial and lacustrine activity in the southern part of Mount Olympus (Central Greece). *Area* **2017**, *49*, 137–147. [CrossRef]

7. Skilodimou, H.D.; Bathrellos, G.D.; Maroukian, H.; Gaki-Papanastassiou, K. Late Quaternary evolution of the lower reaches of Ziliana stream in south Mt. Olympus (Greece). *Geogr. Fis. Din. Quat.* **2014**, *37*, 43–50. [CrossRef]

8. Kamberis, E.; Bathrellos, G.; Kokinou, E.; Skilodimou, H. Correlation between the structural pattern and the development of the hydrographic network in a portion of the Western Thessaly basin (Greece). *Cent. Eur. J. Geosci.* **2012**, *4*, 416–424. [CrossRef]

9. Kokinou, E.; Skilodimou, H.D.; Bathrellos, G.D.; Antonarakou, A.; Kamberis, E. Morphotectonic analysis, structural evolution/pattern of a contractional ridge: Giouchtas Mt., Central Crete, Greece. *J. Earth Syst. Sci.* **2015**, *124*, 587–602. [CrossRef]

10. Sidle, R.C.; Pearce, A.J.; O'Loughlin, C.L. *Hillslope Stability and Land Use*; American Geophysical Union Water Resources Monograph Series; AGU Publications: Washington, DC, USA, 1985; p. 140, ISBN 9781118665503.

11. Dai, F.C.; Lee, C.F.; Ngai, Y.Y. Landslide risk assessment and management: An overview. *Eng. Geol.* **2002**, *64*, 65–87. [CrossRef]

12. Kumar, K.; Jangpangi, L.; Gangopadhyay, S. Highway vs. landslides and their consequences in Himalaya. In *Landslide Science for a Safer Geoenvironment*; Sassa, K., Canuti, P., Yin, Y., Eds.; Springer: Berlin/Heidelberg, Germany, 2014; pp. 389–395.

13. Schuster, R.L.; Fleming, R.W. Economic losses and fatalities due to landslides. Environ. *Eng. Geosci.* **1986**, *xxiii*, 11–28. [CrossRef]

14. Guzzetti, F. Landslide fatalities and the evaluation of landslide risk in Italy. *Eng. Geol.* **2000**, *58*, 89–107. [CrossRef]

15. Sabatakakis, N.; Koukis, G.; Mourtas, D. Composite landslides induced by heavy rainfall in suburban areas. City of Patras and surrounding area, Western Greece. *Landslides* **2005**, *2*, 202–211. [CrossRef]

16. Bathrellos, G.D.; Gaki-Papanastassiou, K.; Skilodimou, H.D.; Papanastassiou, D.; Chousianitis, K.G. Potential suitability for urban planning and industry development by using natural hazard maps and geological-geomorphological parameters. *Environ. Earth Sci.* **2012**, *66*, 537–548. [CrossRef]

17. Bathrellos, G.D.; Gaki-Papanastassiou, K.; Skilodimou, H.D.; Skianis, G.A.; Chousianitis, K.G. Assessment of rural community and agricultural development using geomorphological-geological factors and GIS in the Trikala prefecture (Central Greece). *Stoch. Environ. Res. Risk A* **2013**, *27*, 573–588. [CrossRef]

18. Larsen, M.C.; Parks, J.E. How wide is a road? The association of roads and mass-wasting in a forested montane environment. *Earth Surf. Proc. Land.* **1997**, *22*, 835–848. [CrossRef]

19. Bozzano, F.; Cipriani, I.; Mazzanti, P.; Prestininzi, A. Displacement patterns of a landslide affected by human activities: Insights from ground-based InSAR monitoring. *Nat. Hazards* **2011**, *59*, 1377–1396. [CrossRef]

20. Bathrellos, G.D.; Kalivas, D.P.; Skilodimou, H.D. Landslide Susceptibility Assessment Mapping: A Case Study in Central Greece. In *Remote Sensing of Hydrometeorological Hazards*; Petropoulos, G.P., Islam, T., Eds.; CRC Press; Taylor & Francis Group: London, UK, 2017; pp. 493–512, ISBN-13 978-1498777582.

21. Papadopoulou-Vrynioti, K.; Bathrellos, G.D.; Skilodimou, H.D.; Kaviris, G.; Makropoulos, K. Karst collapse susceptibility mapping considering peak ground acceleration in a rapidly growing urban area. *Eng. Geol.* **2013**, *158*, 77–88. [CrossRef]

22. Papadopoulou-Vrynioti, K.; Alexakis, D.; Bathrellos, G.D.; Skilodimou, H.D.; Vryniotis, D.; Vasiliades, E. Environmental research and evaluation of agricultural soil of the Arta plain, western Hellas. *J. Geochem. Explor.* **2014**, *136*, 84–92. [CrossRef]

23. Papadopoulou-Vrynioti, K.; Alexakis, D.; Bathrellos, G.D.; Skilodimou, H.D.; Vryniotis, D.; Vasiliades, E.; Gamvroula, D. Distribution of trace elements in stream sediments of Arta plain (Western Hellas): The influence of geomorphological parameters. *J. Geochem. Explor.* **2013**, *134*, 17–26. [CrossRef]

24. Youssef, A.M.; Pradhan, B.; Al-Kathery, M.; Bathrellos, G.D.; Skilodimou, H.D. Assessment of rockfall hazard at Al-Noor Mountain, Makkah city (Saudi Arabia) using spatio-temporal remote sensing data and field investigation. *J. Afr. Earth Sci.* **2015**, *101*, 309–321. [CrossRef]

25. Lu, P.; Stumpf, A.; Kerle, N. Object-oriented change detection for landslide rapid mapping. *IEEE Geosci. Remote Sens. Lett.* **2011**, *8*, 701–705. [CrossRef]

26. Bathrellos, G.D.; Skilodimou, H.D.; Maroukian, H. The spatial distribution of Middle and Late Pleistocene Cirques in Greece. *Geogr. Ann.* **2014**, *96*, 323–338. [CrossRef]

27. Bathrellos, G.D.; Skilodimou, H.D.; Maroukian, H. The significance of tectonism in the glaciations of Greece. In *Quaternary Glaciation in the Mediterranean Mountains*; Hughes, P.D., Woodward, J.C., Eds.; Geological Society London Special Publications: London, UK, 2017; pp. 237–250. [CrossRef]

28. Youssef, A.M.; Pradhan, B.; Tarabees, E. Integrated evaluation of urban development suitability based on remote sensing and GIS techniques: Contribution from analytic hierarchy process. *Arab. J. Geosci.* **2011**, *4*, 463–473. [CrossRef]

29. Bathrellos, G.D.; Karymbalis, E.; Skilodimou, H.D.; Gaki-Papanastassiou, K.; Baltas, E.A. Urban flood hazard assessment in the basin of Athens Metropolitan city, Greece. *Environ. Earth Sci.* **2016**, *75*, 319. [CrossRef]

30. Van Westen, C.J. Remote sensing and GIS for natural hazards assessment and disaster risk management. In *Treatise on Geomorphology*; Schroder, J.F., Bishop, M.P., Eds.; Academic Press; Elsevier: New York, NY, USA, 2013; pp. 259–298.

31. Armijo, R.; Meyer, B.; King, G.; Rigo, A.; Papanastassiou, D. Quaternary evolution of the Corinth Rift and its implications for the late Cenozoic evolution of the Aegean. *Geophys. J. Int.* **1996**, *126*, 11–53. [CrossRef]

32. Koukis, G.; Rozos, D. Geotechnical conditions and landslide movements in the Greek territory in relation to the geological structure and geotectonic evolution. *Miner. Wealth* **1982**, *16*, 53–69.

33. Rozos, D.; Bathrellos, G.D.; Skilodimou, H.D. Comparison of the implementation of Rock Engineering System (RES) and Analytic Hierarchy Process (AHP) methods, based on landslide susceptibility maps, compiled in GIS environment. A case study from the Eastern Achaia County of Peloponnesus, Greece. *Environ. Earth Sci.* **2011**, *63*, 49–63. [CrossRef]

34. Yilmaz, I. The effect of the sampling strategies on the landslide susceptibility mapping by conditional probability and artificial neural networks. *Environ. Earth Sci.* **2010**, *60*, 505–519. [CrossRef]

35. Bathrellos, G.D.; Kalivas, D.P.; Skilodimou, H.D. Landslide susceptibility mapping models, applied to natural and urban planning, using G.I.S. *Estud. Geol.-Madrid* **2009**, *65*, 49–65. [CrossRef]

36. Lee, S.; Pradhan, B. Probabilistic landslide hazards and risk mapping on Penang Island, Malaysia. *J. Earth Syst. Sci.* **2006**, *115*, 661–672. [CrossRef]
37. Tsolaki-Fiaka, S.; Bathrellos, G.D.; Skilodimou, H.D. Multi-criteria decision analysis for abandoned quarry restoration in Evros Region (NE Greece). *Land* **2018**, *7*, 43. [CrossRef]
38. Migiros, G.; Bathrellos, G.; Skilodimou, H.; Karamousalis, T. Pinios (Peneus) River (Central Greece): Hydrological-geomorphological elements and changes during the quaternary. *Cent. Eur. J. Geosci.* **2011**, *3*, 215–228. [CrossRef]
39. Donati, L.; Turrini, M.C. An objective method to rank the importance of the factors predisposing to landslides with the GIS methodology: Application to an area of the Apennines (Valnerina; Perugia, Italy). *Eng. Geol.* **2002**, *63*, 277–289. [CrossRef]
40. Ayalew, L.; Yamagishi, H.; Ugawa, N. Landslide susceptibility mapping using GIS-based weighted linear combination, the case in Tsugawa area of Agano River, Niigata Prefecture, Japan. *Landslides* **2004**, *1*, 73–81. [CrossRef]
41. Cristiano, E.; Veldhuis, M.C.T.; Giesen, N.V.D. Spatial and temporal variability of rainfall and their effects on hydrological response in urban areas—A review. *Hydrol. Earth Syst. Sci.* **2017**, *21*, 3859–3878. [CrossRef]
42. Copernicus. Copernicus Land Monitoring Service. 2016. Available online: http://land.copernicus.eu (accessed on 31 May 2018).

MDPI

St. Alban-Anlage 66

4052 Basel

Switzerland

Tel. +41 61 683 77 34

Fax +41 61 302 89 18

www.mdpi.com

Land Editorial Office

E-mail: land@mdpi.com

www.mdpi.com/journal/land

9 783039 439256

34. Verderame, G.M.; Ricci, P.; Esposito, M.; Manfredi, G. *STIL v1.0–Software Per La Carat-Terizzazione Delle Proprietà Meccaniche Degli Acciai Da c.a. Tra il 1950 e il 2000*; ReLUIS: Naples, Italy, 2012.
35. Ibarra, L.F.; Medina, R.A.; Krawinkler, H. Hysteretic models that incorporate strength and stiffness deterioration. *Earthq. Eng. Struct. Dyn.* **2005**, *34*, 1489–1511. [CrossRef]
36. Haselton, C.B.; Liel, A.B.; Lange, S.T.; Deierlein, G.G. *Beam-Column Element Model Cal-Ibrated for Predicting Flexural Response Leading to Global Collapse of RC Frame Buildings*; Pacific Earthquake Engineering Research Center: Berkeley, CA, USA, 2008.
37. Braga, F.; Gigliotti, R.; Laterza, M.; D'Amato, M.; Kunnath, S.F. Modified Steel Bar Model Incorporating Bond-Slip for Seismic Assessment of Concrete Structures. *J. Struct. Eng.* **2012**, *138*, 1342–1350. [CrossRef]
38. Gesualdi, G.; Viggiani, L.R.S.; Cardone, D. Seismic performance of RC frame buildings accounting for the out-of-plane behavior of masonry infills. *Bull. Earthq. Eng.* **2020**, *18*, 1–39. [CrossRef]
39. Aslani, H.; Miranda, E. *Probabilistic Earthquake Loss Estimation and Loss Disaggregation in Buildings*; Report No., 157; The John A. Blume Earthquake Engineering Center Department of Civil and Environmental Engineering Stanford University: Palo Alto, CA, USA, June 2005.
40. Alath, S.; Kunnath, S.K. *Modelling Inelastic Shear Deformations in RC Beam-Column Joints*; American Society of Civil Engineers: Boulder, CO, USA, 1995.
41. De Risi, M.T.; Ricci, P.; Verderame, G.M.; Manfredi, G. Experimental assessment of un-reinforced exterior beam-column joints with deformed bars. *Eng. Struct.* **2016**, *112*, 215–232. [CrossRef]
42. Ricci, P.; Manfredi, V.; Noto, F.; Terrenzi, M.; De Risi, M.T.; Di Domenico, M.; Camata, G.; Franchin, P.; Masi, A.; Mollaioli, F.; et al. RINTC-e: Towards seismic risk assessment of existing residential reinforced concrete buildings in Italy. In Proceedings of the 7th International Conference on Computational Methods in Structural Dynamics and Earthquake Engineering, Crete, Greece, 24–26 June 2019. [CrossRef]
43. Sassun, K.; Sullivan, T.J.; Morandi, P.; Cardone, D. Characterising the in-plane seismic performance of infill masonry. *Bull. New Zealand Soc. Earthq. Eng.* **2016**, *49*, 100–117. [CrossRef]
44. Cardone, D.; Perrone, G.; Piesco, V. Developing collapse fragility curves for base-isolated buildings. *Earthq. Eng. Struct. Dyn.* **2019**, *48*, 78–102. [CrossRef]
45. Richard, W.N. *Seismic Evaluation and Retrofit of Concrete Buildings*; Report No. SSC 96-01; Applied Technology Council (ATC): Redwood City, CA, USA, November 1996.
46. *Eurocode 8: Design of Structures for Earthquake Resistance–Part 1: General Rules, Seismic Actions and Rules for Buildings, German Version*; EN 1998-1:2004; European Committee for Standardization: Brussels, Belgium, December 2010.
47. *In Bilancio di Previsione dello Stato per l'Anno Finanziario 2019 e Bilancio Pluriennale per il Triennio 2019–2021 (Legge di Bilancio 2019)*; Law 145/2018,.n.302 del; Gazzetta U_ciale della Repubblica Italiana: Roma, Italy, 2018. (In Italian)
48. *Recante Misure Urgenti in Materia di Salute, Sostegno al Lavoro e all'Economia, Nonche' di Politiche Sociali Connesse all'Emergenza Epidemiologica da COVID-19*; Law 77/2020, n.180 del; Gazzetta U_ciale della Repubblica Italiana: Roma, Italy, 2020. (In Italian)

 *buildings*

Review

Overview on the Nonlinear Static Procedures and Performance-Based Approach on Modern Unreinforced Masonry Buildings with Structural Irregularity

Abide Aşıkoğlu *, Graça Vasconcelos and Paulo B. Lourenço

ISISE, Department of Civil Engineering, University of Minho, 4800-058 Azurém, Guimarães, Portugal; graca@civil.uminho.pt (G.V.); pbl@civil.uminho.pt (P.B.L.)
* Correspondence: abideasikoglu@hotmail.com

Abstract: Performance-based design plays a significant role in the structural and earthquake engineering community to ensure both safety and economic feasibility. Its application to masonry building design/assessment is limited and requires straightforward rules considering the characteristics of masonry behavior. Nonlinear static procedures mainly cover regular frame system structures, and their application to both regular and irregular masonry buildings require further investigation. The present paper addresses two major issues: (i) the definition of irregularity in masonry buildings, and (ii) the applicability of classical nonlinear static procedures to irregular masonry buildings. It is observed that the irregularity definition is not comprehensive and has different descriptions among the seismic codes as well as among researchers, particularly in the case of masonry buildings. The lack of global language may result in the misuse of the procedures, while adjustments may be essential due to irregularity effects. Therefore, irregularity indices given by different codes and research studies are discussed. Furthermore, an overview of nonlinear static procedures implemented within the framework of the performance-based approach and improvements proposed for its application in masonry buildings is presented.

Keywords: seismic performance; deformation; unreinforced masonry; irregularity; performance-based design; nonlinear static procedures

Citation: Aşıkoğlu, A.; Vasconcelos, G.; Lourenço, P.B. Overview on the Nonlinear Static Procedures and Performance-Based Approach on Modern Unreinforced Masonry Buildings with Structural Irregularity. *Buildings* **2021**, *11*, 147. https://doi.org/10.3390/buildings11040147

Academic Editor: Alessandra Aprile

Received: 13 February 2021
Accepted: 23 March 2021
Published: 1 April 2021

Publisher's Note: MDPI stays neutral with regard to jurisdictional claims in published maps and institutional affiliations.

1. Introduction

Masonry construction is the oldest structural system, which, in fact, can be considered as the base for built heritage. Masonry has been used in different types of structures over centuries, mainly due to the easy accessibility of the material at its location. Yet, most of the traditional masonry structures were designed based on vertical loads only. Indeed, this led to the construction of massive walls to ensure both vertical and lateral stability. The earliest versions of building codes for masonry buildings covered empirical design rules based on this approach. However, designing a masonry structure in such a way leads to enormous dimensions that are not compatible neither with aesthetical and architectural contexts nor with economic sources. This is particularly relevant in the case of construction in regions with high seismic hazards. This is among the main reasons why masonry, as a structural material, has been replaced by other materials, such as reinforced concrete and steel. However, unreinforced masonry buildings, as isolated or in aggregates (with rigid or flexible diaphragm), are largely found in many countries in the world with both low and high seismicity [1–3], which justify the improvement of European and American seismic codes concerning masonry structures [4].

More recently, seismic design philosophies have been evolved to performance-based design (PBD) approaches, namely, in the case of masonry buildings. It aims at designing structures with acceptable damage levels under certain seismic intensity and, therefore, to avoid conservative design. Therefore, seismic performance levels, associated with

Buildings **2021**, *11*, 147. https://doi.org/10.3390/buildings11040147   https://www.mdpi.com/journal/buildings

a certain level of damage exhibited by the structures, which are commonly identified through deformations, must be defined [5]. The application of the performance-based design/assessment to masonry structures is not straightforward, but it has been successful in frame systems, such as reinforced concrete and steel constructions [5–9]. For instance, to achieve earthquake-resistant masonry buildings according to Eurocodes, general rules for masonry design [10] and seismic design [11] need to be integrated. The main issue here is that rules defined in the seismic design code, particularly the application of the performance-based approach, are not comprehensive for masonry structures. Furthermore, PBD and nonlinear static procedures (NSPs) were mainly developed for regular frame systems. It should be mentioned that regular configurations are not representative of real building stock because new buildings usually impose complex geometry due to architectural and functional concerns. Among the main issues with complex geometry is the presence of irregular structural configurations which show undesired torsional effects under seismic actions [12]. This causes additional difficulties in the application of the nonlinear static procedures in the case of masonry buildings.

NSPs aim at simulating the dynamic response of structures by simply using the pushover capacity of a multi-degree-of-freedom structure and its equivalent single-degree-of-freedom system. There are different modeling approaches available in the literature to perform nonlinear static analysis of masonry structures. An extensive literature review on the methodologies applied to the seismic assessment of masonry buildings is presented by D'Altri et al. (2019) [13]. According to [13], numerical strategies available for masonry structures can be categorized as four main groups based on the modeling approach: (i) block-based model,: this uses block elements aiming at considering the real masonry arrangement (units and mortar); (ii) continuum model,: a representative model is taken into account as a continuum deformable body with homogenous material behavior; (iii) macro-element model: the discretization of the model is carried out by panel elements, so-called macro-elements; and (iv) geometry-based model: the description of the model is based on the geometry of the structure only. Several studies have shown that results of the nonlinear static analysis are highly dependent on the numerical procedure adopted to simulate the structure [14–27]. Here, the performance-based applications are studied, regardless of the numerical modeling and simulations in the present paper.

In this context, the present paper intends to overview structural irregularity descriptions and indices derived for masonry structures and available NSP applicable to masonry in the literature. The main purpose is to point out the complexity of structural irregularities and their influence on the procedures adopted to achieve performance limits. To exemplify, two case studies are selected from the literature aiming at illustrating the application of different NSPs to evaluate the accuracy of the methods in structures with different irregularity levels. Finally, some improvements proposed by several researchers are addressed.

2. Description of Structural Irregularities: Is It Comprehensive Enough for Masonry Buildings?

It is widely recognized that geometrical configuration has an important role in the global behavior of structures. Past seismic events have demonstrated that buildings with structural irregularity suffer more damage than their regular counterparts [28,29]. Geometric irregularities can result in complex load patterns resulting in concentrated inelastic behavior at critical points, such as corners (Figure 1). A uniform load distribution among vertical resisting elements with similar stiffness is achieved by symmetrically designed plans once the center of mass, where the resultant force is imposed, coincides with the center of rigidity. Thus, regular structures are less likely to suffer significant torsional effects. Otherwise, the eccentricity results in undesired behavior, which is mainly controlled by a combination of lateral and torsional responses [30]. For instance, abrupt changes in the plan, such as the presence of setbacks, discontinuities in the in-plan stiffness of the floors due to openings or variable slab thickness, contribute to the torsional damage [31]. In this sense, most seismic codes cover designing rules mainly based on regular structures and impose certain penalties on structural irregularities. Yet, complex geometry usually results from space limitations and from architectural, economical and/or functional concerns. Irregularities in the geometry of structural systems are mostly inevitable, particularly in the case of masonry buildings. Even if such a building satisfies global regularity requirements, it may not be possible to achieve elevation once the masonry walls are composed of openings with different numbers, sizes, and alignments [28]. Furthermore, it is important to note that loadbearing masonry walls serve as both structural and architectural components which require a multidisciplinary approach and strict collaboration between engineer and architects during the design process. Additionally, masonry buildings are mainly found as aggregates which result in a high level of structural irregularities due, for example, to the misaligned of the floors between adjacent buildings [1–3].

Figure 1. Damage concentrated at the irregularities on masonry buildings [28].

Design codes classify the buildings as regular and irregular, and the irregularity is categorized into two types, in plan and in elevation. In general terms, the definition is based on the distribution of mass, stiffness and strength. However, the criteria given for the definition of irregularity are not sufficient to capture a variety of cases, such as irregularity due to progressive damage, or a combination of both plan, and vertical irregularities [32]. It should be noted that the definition of structural irregularity differs in different codes, as listed in Tables A1 and A2 in Appendix A. It appears that some of the codes describe irregularity mostly for framed-system structures, and very limited explanations are given for masonry buildings, as is the case of Turkish Earthquake Code (TEC) 2019 [33]. The ASCE standard presents more comprehensive and relatable definitions to masonry buildings.

Despite the rules given to define irregularity in the codes, it is noticed that the concept of irregularity has a wide definition spectrum among researchers. Considering masonry buildings, for instance, Abrams (1997) [34] studied a building with structural irregularity due to the different sizes and locations of openings in the walls. This results in varying stiffness and strength for two parallel shear walls, with various pier dimensions and aspect ratios, as seen in Figure 2a. Bairrão and Silva (2009) [35] classified their building as irregular in plan owing to the occurrence of a setback in one corner (Figure 2b). On the other hand, Giordano et al. (2008) [30] described the studied building as asymmetric in plan in both horizontal directions based on two features: (i) the position of the longitudinal inner wall that is not barycentric in the X direction; (ii) the lack of the second-last transverse wall in the Y direction (Figure 2c). The building shown in Figure 2d is also considered as irregular, as it has a setback in one corner, resulting in an asymmetric plan with irregularly distributed openings [36]. Another example of an irregular building (in plan) was studied by Lagomarsino et al. (2018) [37], with different diaphragm stiffness, window opening sizes and distribution of the openings on the outer walls (Figure 2e). Accordingly, it is seen that there is a need for introducing and improving irregularity indexes for masonry buildings into the current codes to attain straightforward and uniform definitions among the structural engineering community. It is important to underline that some definitions considered as a plan irregularity are attributed to structural irregularities in elevation or vice versa. The main reason for this is the influence of the stiffness of vertical elements on the location of the center of rigidity. Thus, it may be necessary to identify and include structural irregularities that might have a coupled influence in these cases.

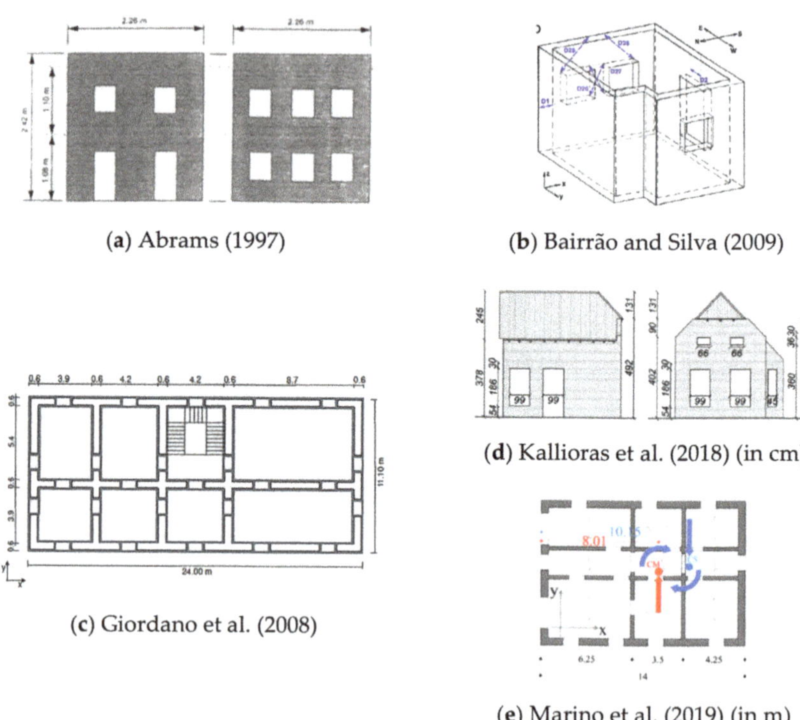

(**a**) Abrams (1997)

(**b**) Bairrão and Silva (2009)

(**d**) Kallioras et al. (2018) (in cm)

(**c**) Giordano et al. (2008)

(**e**) Marino et al. (2019) (in m)

Figure 2. Structural layouts described as irregular by different researchers, (**a**) Abrams (1997) [34], (**b**) Bairrão and Silva (2009) [35], (**c**) Giordano et al. (2008) [30], (**d**) Kallioras et al. (2018) [36], (**e**) Marino et al. (2019) [38].

To this end, Parisi and Augenti (2013) [28] and Berti et al. (2017) [39] developed studies to define irregularities of in-plane masonry walls in quantitative terms, and the details are given in Figures 3 and 4. They propose indices for geometry irregularity to obtain its typology and severity as listed in Table 1. The irregularities given by Parisi and Augenti (2013) [28] are classified into four categories: (i) horizontal irregularities, (ii) vertical irregularities, (iii) offset irregularities and (iv) variable opening numbers. It is noted that horizontal, vertical and offset irregularities are formulated based on openings in the same story level, while the last one considers the difference among openings between the stories. Hence, the regularity of a wall is indicated by the index $i = 0$, while an index within the range of $0 < i \leq 1$ represents different levels of irregularity. On the other hand, Berti et al. (2017) [39] characterize the most common irregularity types into six main groups based on the dimensions and alignments of openings, (i) horizontal and (ii) vertical alignments of openings, (iii) irregularity in opening width, (iv) irregularity in height, (v) global index as a combination of irregularities and (vi) the presence of non-rectangular openings.

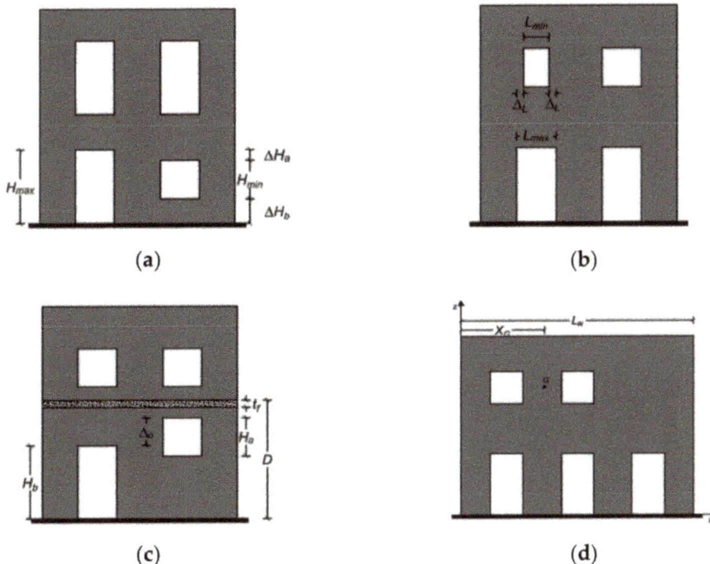

Figure 3. Classification of wall irregularities according to Parisi and Augenti (2013) [28]: (**a**) horizontal, (**b**) vertical, (**c**) offset and (**d**) variable opening number irregularity (figures adapted from [28]).

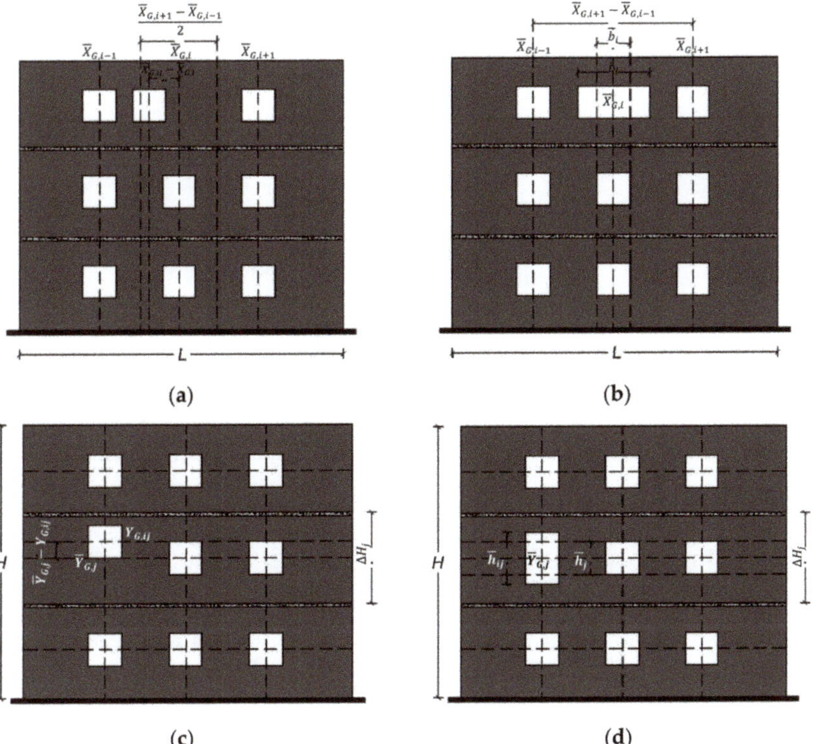

Figure 4. Types of irregularities given in Berti et al. (2017) [39]: (**a**) horizontal misalignment, (**b**) vertical misalignment, (**c**) irregularity in width and (**d**) irregularity in height (figures adapted from [39]).

Table 1. Irregularities described for masonry buildings by different authors (nomenclature for symbols is given in Appendix B).

Authors		Definition	Index
Authors	Parisi and Augenti (2013) [28]	**1—Horizontal Irregularity** The openings have different heights at the same story while equal lengths among the stories.	$i_H = \frac{\Delta_H}{2H_{med}} = \frac{H_{max}-H_{min}}{H_{max}+H_{min}} = \frac{\Delta H_a+\Delta H_b}{H_{max}+H_{min}}$
		2—Vertical Irregularity The openings have the same height at the same story while different lengths among the stories.	$i_V = \frac{\Delta_L}{2L_{med}} = \frac{L_{max}-L_{min}}{L_{max}+L_{min}} = \frac{\Delta L_r+\Delta L_l}{L_{max}+L_{min}}$
		3—Offset Irregularity The wall has horizontal and/or vertical offsets between openings of equal or different sizes.	$i_o = \frac{\Delta_o}{D-t_f-H_b}$
		4—Variable openings number irregularity The wall has a different number of openings per story.	$i_N = 1 - \frac{N_{min}}{N_{max}} \; ; i_D = \left\|1 - \frac{2x_G}{L_w}\right\|$
	Berti et al. (2017) [39]	**1—Horizontal misalignment** The centroid abscissa of an opening $X_{G,ij}$ differs from the vertical alignment of the i-th vertical opening array $X_{G,i}$.	$I_{X,ij} = \frac{2\left\|X_{G,ij}-\overline{X}_{G,i}\right\|}{\overline{X}_{G,i+1}-\overline{X}_{G,i-1}}$
		2—Vertical misalignment The centroid ordinate of an opening $Y_{G,ij}$ differs from the horizontal alignment of the j-th story $Y_{G,j}$	$I_{Y,ij} = \frac{\left\|Y_{G,ij}-\overline{Y}_{G,i}\right\|}{\Delta H_j}$
		3—Irregularity in width The opening width b_{ij} differs from the average one of the i-th vertical opening alignment b_i.	$I_{W,ij} = \frac{\left\|b_{ij}-\overline{b}_i\right\|}{\overline{X}_{G,i+1}-\overline{X}_{G,i-1}}$
		4—Irregularity in height The opening height h_{ij} differs from the average one in the j-th story h_j.	$I_{H,ij} = \frac{\left\|h_{ij}-\overline{h}_j\right\|}{\Delta H_j}$
		5—Global index Global irregularity measure in which the combination of irregularity indexes is obtained.	$I = \widetilde{I}\left(I_{X,ij},\, I_{Y,ij},\, I_{W,ij},\, I_{H,ij}\right)$

The differences found in the definition of structural irregularity in masonry buildings described previously reflects the difficulty in obtaining a uniform, straightforward and comprehensive description of it. Some attempts have been presented by different researchers, but it is considered that further research and descriptions are essential. The rules to define the irregularity of a masonry building should include a systematic approach to the different types of irregularity and provide indices to characterize the level of irregularity. In addition, it is considered that more details need to be identified when defining structural irregularity based on the geometry, stiffness and mass of vertical and horizontal structural elements to avoid any design/assessment errors. For instance, eccentricity of building is associated with the difference between the center of mass and the center of rigidity at the same floor level, which must be also characterized among different stories. In this case, the response of such a building is debatable, since the earthquake loads are attributed to the center of mass at each level.

3. Performance-Based Approach as a Design/Assessment Tool

The performance-based approach for the seismic assessment/design of a building is based on the comparison between the seismic demand and the seismic capacity in terms of deformation. Such a procedure is implemented in a simplified manner by considering certain assumptions to reduce computational and analysis efforts. The damage states can be defined qualitatively based on visual inspection as in the case of the macro-seismic post-earthquake assessment [40]. However, when seismic performance-based assessment is intended to be applied, it is necessary to define the performance level associated with a certain damage level that can develop in a building for a certain seismic hazard level in a quantitative manner. The performance level of a structure can be represented through a deformation level defined as a limit state, such as fully operational, operational, life safety and near collapse. According to [41], three design criteria are associated with the performance level and severity of an earthquake. Accordingly, the performance objective is chosen to ensure a low seismic risk for a given structure when submitted to a certain seismic action.

Although nonlinear dynamic analysis is widely recognized as the most accurate analysis method, nonlinear static procedures (NSP) have become the most practical method to assess and design structures [32] in which nonlinear static analysis play a central role. The main reason for this is that the seismic response of a structure highly depends on the dynamic motion and, therefore, a set of analyses with different levels of intensities is needed. Furthermore, nonlinear dynamic analysis requires the selection and scaling of seismic input and definition of hysteretic models [42]. On the other hand, nonlinear static analysis, namely, pushover analysis, provides fundamental information about the seismic performance of buildings by simply pushing the structure with an incremental lateral load until collapse. The pushover curve is an important tool to attain the seismic behavior and identify damage limit states. Essentially, NSP assumes that the seismic demand of a multi-degree-of-freedom (MDOF) system can be computed using an equivalent single-degree-of-freedom (SDOF) model of the same system. The procedure is composed of two main steps: (1) performing pushover analysis and (2) the application of a nonlinear static procedure, as depicted in Figure 5. Once the capacity of a building is attained by incremental static loading, it is converted from the MDOF to an equivalent SDOF system by using the response spectrum as a function of the site and seismic motion-specific parameters. Next, the target global displacement is calculated within the framework of a selected nonlinear static procedure. Afterward, the global building response is analyzed at the given displacement demand, which is calculated as the target displacement for the equivalent SDOF system. Accordingly, story drifts, internal forces and, therefore, component actions resulting due to target displacement are used to examine the seismic performance.

It should be stressed that there are several concerns about the application of NSPs. The maximum displacement demand is highly dependent on the empirical formulations, and, therefore, different results for the same nonlinear static response are observed. One important issue is associated with the difficulty to obtain clear stiffness and strength, which should reflect progressive damage [43]. In addition, the application of NSPs brings more uncertainty, as in some cases, the response of an MDOF system is not well represented by an SDOF counterpart, such as in the case of irregular buildings. Indeed, NSPs were developed for structures with seismic responses dominated by translational modes, i.e., regular buildings. Thus far, the N2 method [44], which is included in Eurocode 8; the Capacity Spectrum Method [45]; and the Displacement Coefficient Method [46] are the most preferred methods for regular structures [32]. Nevertheless, it is required to overcome the issue of torsional effects on the seismic response by using improved methodologies. Significant research has been carried out and has proposed methods that focus on (i) adaptive pushover algorithms and (ii) consideration of higher modes. Therefore, NSPs can be categorized into two groups, namely, classical and extended; see Figure 6. It is observed that classical procedures are associated with regular buildings, while extended versions intend to be applied in buildings in which the predominance of higher mode or torsional effects needs to be addressed.

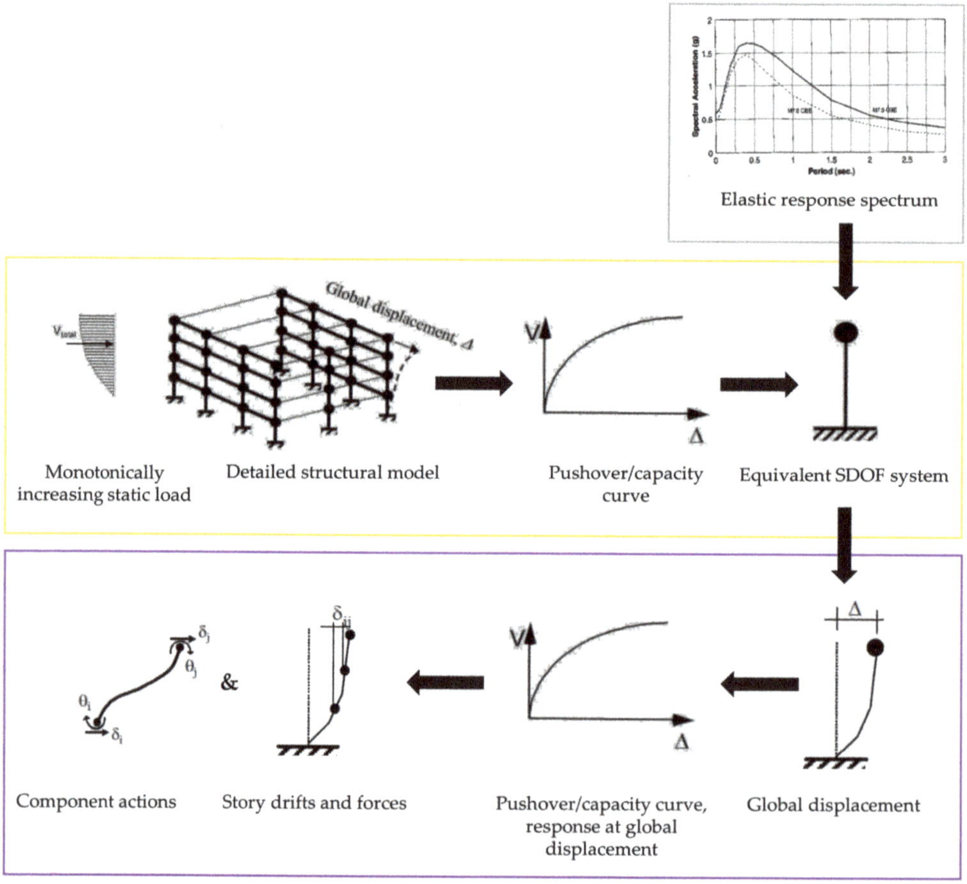

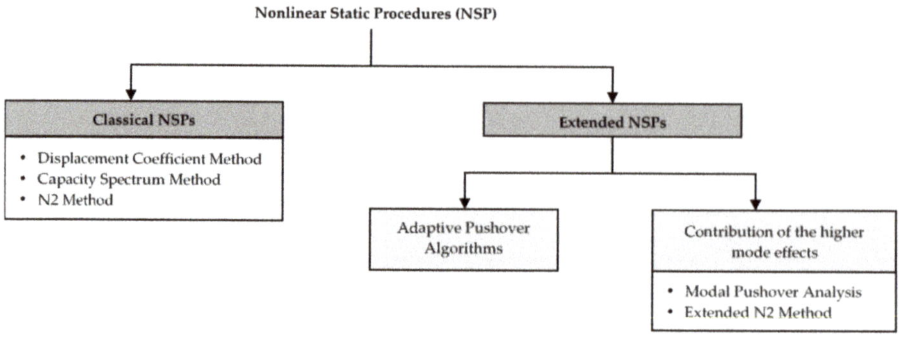

Figure 5. Schematic representation of nonlinear static procedures and performance-based assessment (adapted from [43]).

Figure 6. Classification of nonlinear static procedures.

3.1. Classical Nonlinear Static Procedures

The global deformation demand on a structure is computed based on an equivalent SDOF system, which is obtained from the pushover response of an MDOF system. In classical NSPs, the generation of the pushover curve is carried out by using a similar approach, as depicted in Figure 7. However, the technique used to evaluate the maximum displacement demand differs. For instance, the Coefficient Method takes into account the modification of base shear and roof displacement relation, while the capacity spectrum and N2 method use diagrams relating spectral acceleration and spectral displacement.

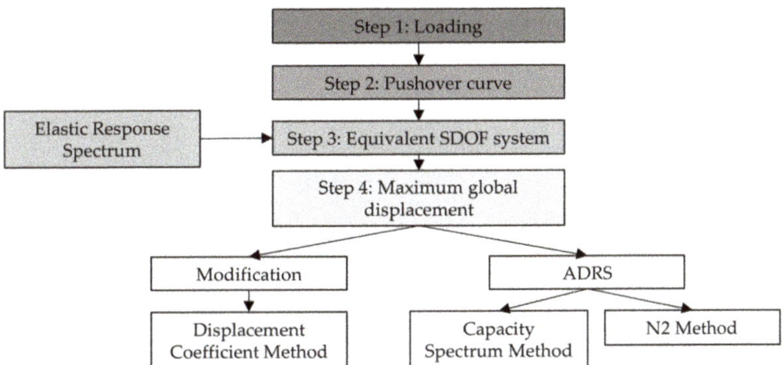

Figure 7. Main stages of classical nonlinear static procedures (NSPs).

3.1.1. Displacement Coefficient Method

The Displacement Coefficient Method (DCM) is considered a primary nonlinear static procedure that was introduced in FEMA 356 [46]. This method requires the modification of the linear elastic response of the equivalent SDOF system to predict the maximum global displacement, so-called target displacement. A set of coefficients are used to adjust the response. The procedure is illustrated in Figure 8 and summarized as follows.

1. A pushover curve, which is an idealization of force-deformation relation, is obtained through numerical analysis.
2. On the pushover curve, an effective period (T_{eff}) is calculated as a function of the initial period (T_i). In this way, stiffness loss observed during the transition from elastic to inelastic response is taken into account. Thus, an equivalent SDOF system is assumed to have the same elastic stiffness that corresponds to the effective period of the MDOF system obtained previously.
3. A maximum acceleration response of the SDOF system is obtained as a function of an effective period on an elastic response spectrum that is representative of the seismic ground motion.
4. The maximum global displacement demand is evaluated in terms of spectral displacement that is directly associated with the spectral acceleration through Equation (1).

$$S_d = C_0 C_1 C_2 C_3 \frac{T_{eff}^2}{4\pi^2} \cdot S_a\left(T_{eff}\right) \tag{1}$$

where:

- C_0 converts the SDOF spectral displacement to MDOF roof displacement (elastic); it can be considered as the first mode participation factor or an appropriate value given in Table 2.
- C_1 is the factor that relates the expected maximum inelastic displacement to elastic displacement.

- C_2 represents the effects of pinched hysteretic shape, stiffness degradation and strength deterioration. The values given in Table 2 are associated with different performance limit states.
- C_3 adjusts for second-order geometric nonlinearity *(P-Δ)* effects.
- $S_a(T_{eff})$ is the spectral acceleration at the effective period.

Table 2. Values suggested for coefficients in Equation (1) [46].

Coef.	Number of Stories	Shear Buildings [2]		Other Buildings
		Triangular Load Pattern	**Uniform Load Pattern**	**Any Load Pattern**
C_0 [1]	1	1.00	1.00	1.00
	2	1.20	1.15	1.20
	3	1.20	1.20	1.30
	5	1.30	1.20	1.40
	10+	1.30	1.20	1.50

$$C_1 = \begin{cases} 1.0 & for \quad T_{eff} \geq T_s \\ \dfrac{1.0 + \frac{(R-1)T_s}{T_{eff}}}{R} & for \quad T_{eff} < T_s \end{cases}$$
where T_s is the characteristic period of the response spectrum

However, it should be less than
$$C_1 = \begin{cases} 1.5 & for \quad T_{eff} < 0.1\ s \\ 1.0 & for \quad T_{eff} \geq T_s \end{cases}$$
and higher than 1.0.

Coef.	Structural Performance Level	$T < 0.1$ s [5]		$T > T_s$ s [5]	
		Framing Type 1 [3]	**Framing Type 2** [4]	**Framing Type 1** [3]	**Framing Type 2** [4]
C_2	Immediate Occupancy	1.0	1.0	1.0	1.0
	Life Safety	1.3	1.0	1.1	1.0
	Collapse Prevention	1.5	1.0	1.2	1.0

C_3
$$C_3 = 1.0 + \frac{|\alpha|(R-1)^{3/2}}{T_{eff}}$$
where α is the ratio of post-yield stiffness to effective elastic stiffness, and R is the strength ratio.

[1] Linear interpolation are used to calculate intermediate values. [2] Buildings in which, for all stories, inter-story drift decreases with increasing height. [3] Structure in which more than 30% of the story shear at any level is resisted by any combination of the following components, elements or frames: ordinary moment-resisting frames, concentrically braced frames, frames with partially restrained connections, tension-only braces, unreinforced masonry walls, shear-critical, piers and spandrels of reinforced concrete or masonry. [4] All frames not assigned to Framing Type 1. [5] Linear interpolation is used to calculate the intermediate values of T.

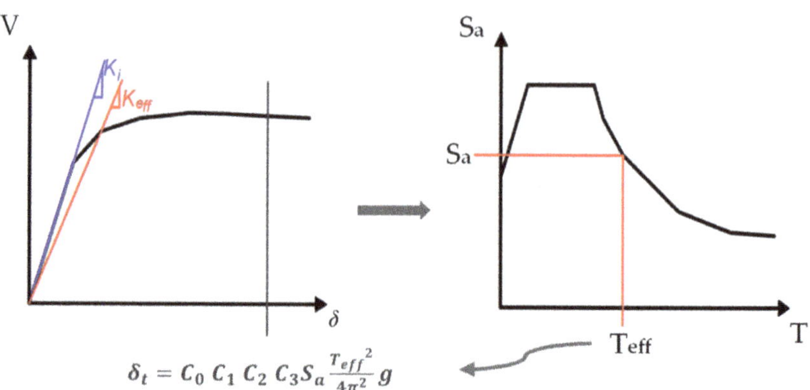

$$\delta_t = C_0\, C_1\, C_2\, C_3 S_a \frac{T_{eff}^2}{4\pi^2}\, g$$

Figure 8. Schematic representation of the Displacement Coefficient Method procedure [43].

Suggestions by [46] for the values of the coefficients are displayed in Table 2.

3.1.2. Capacity Spectrum Method of Equivalent Linearization

The Capacity Spectrum Method (CSM) was proposed by Freeman et al. (1975) [45] as a rapid assessment method. It represents the global force–displacement capacity graphically, which enables a comparison with the representative response spectrum. CSM assumes that the maximum inelastic deformation of a nonlinear SDOF system can be estimated from the maximum deformation of a linear elastic SDOF system which has a higher period and a higher damping ratio than the initial values of a nonlinear system. In brief, the procedure involves the following:

1. Definition of the structural response based on the force–deformation diagram, i.e., pushover curve.
2. The pushover curve is transformed into a capacity curve that is a function of spectral acceleration and spectral displacement of an SDOF system by using the modal properties of the structure. This format of the graph is termed as the acceleration–displacement response spectrum (ADRS).
3. The elastic response spectrum of representative seismic ground motion is converted into the ADRS format (Figure 9). This enables the drawing and comparison of both seismic capacity and demand curves on the same coordinate system.
4. As seen in Figure 9, the secant modulus is used to attain an equivalent inelastic period, and the inelastic displacement demand of the structure is estimated through the intersection of the capacity and overdamped demand curve.
5. In order to obtain the overdamped response spectrum, equivalent viscous damping is needed. Two different approaches can be used to estimate the value as follows:

   a. Analytical expression proposed in [47], according to Equation (2).

$$\beta_{eq} = \beta_{el} + \alpha \left(1 - \frac{1}{\mu^\rho} \right) \tag{2}$$

where:

- β_{el} is the elastic viscous damping, which is generally considered 5%.
- ρ is a factor, and values of 1.5 and 2.0 are suggested by [48] for buildings with box behavior and existing buildings without box behavior, respectively. This is dependent on the hysteretic behavior of the structure.
- α is the factor representing the asymptote of the hysteretic damping, and values of 25 and 20 are suggested by [48] for buildings with and without box behavior, respectively. This is also dependent on the hysteretic behavior of the structure.
- μ is the ductility.

   b. Cyclic pushover curve as a function of displacement (Figure 9).

$$\beta_{eq} = 0.05 + \left(\frac{1}{4\pi} \cdot \frac{E_D}{E_{S0}} \right) \tag{3}$$

3.1.3. N2 Method

The N2 method was proposed by Fajfar and Fischinger (1988) [44], and it was later improved by Fajfar (2000) [6]. It is a combination of two different mathematical procedures. The pushover analysis of an MDOF system is combined with the response spectrum of an equivalent SDOF system. This method is composed of steps such as the following:

1. Eigenvalue analysis to obtain modal properties of the MDOF system structure.
2. The representative seismic action is defined in the form of an elastic acceleration spectrum as a function of the natural period of the structure (T), converted to ADRS format (Figure 10).

3. The inelastic spectrum for constant ductility is determined by using the relations given in Equations (4) and (5).

$$S_a = \frac{S_{ae}}{R_\mu} \tag{4}$$

$$S_d = \frac{\mu}{R_\mu} S_{de} = \frac{\mu}{R_\mu} \frac{T^2}{4\pi^2} S_{ae} = \mu \frac{T^2}{4\pi^2} S_a \tag{5}$$

where μ is the ductility factor, and R_μ is the reduction factor due to hysteretic energy dissipation. It is important to mention that R_μ is different from the reduction factor R, which is used to modify the response of a building by taking into account both energy dissipation and overstrength.

4. The mode proportional pushover analysis is performed, and a capacity curve is obtained. The first mode shape of the vibration is assumed, and lateral loads are applied proportional to the 1st mode shape. Note that the displacement profile is assumed to be the initial first mode shape throughout the procedure.

5. The capacity curve of the MDOF system is then converted into a bilinear diagram, which represents the capacity of an equivalent SDOF system.

6. Seismic demand of the equivalent SDOF system is attained graphically from the demand versus capacity diagram given in ADRS format, as depicted in Figure 10. Alternatively, Equation (6) is used to compute the displacement demand.

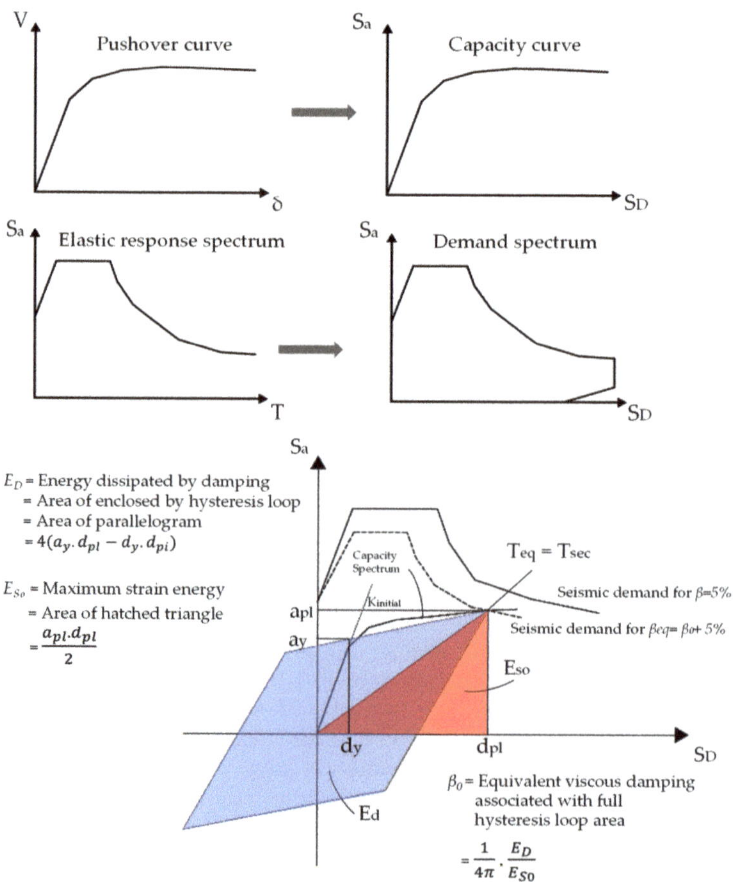

Figure 9. Schematic representation of the Capacity Spectrum Method [43].

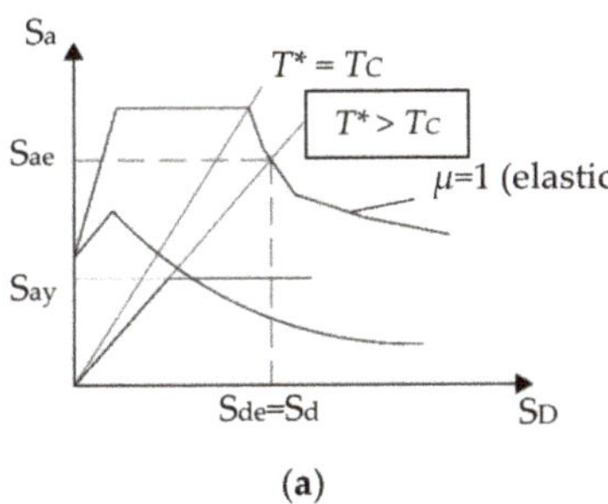
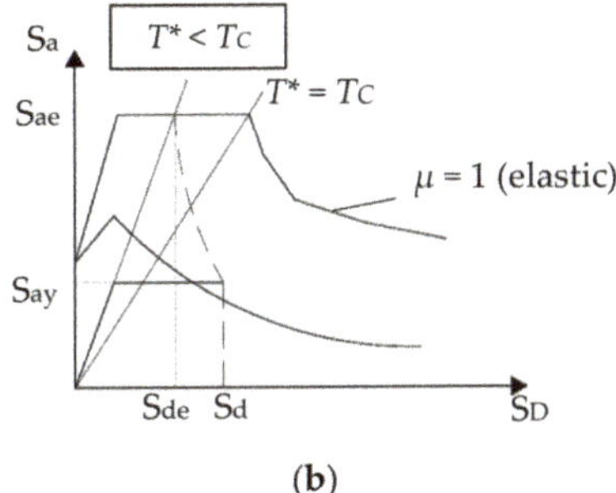

Figure 10. Graphical representation of N2 method and estimation of displacement demand for (**a**) low- and (**b**) high-period structures. Adapted from [6].

$$S_d = \frac{S_{de}}{R_\mu}\left[(R_\mu - 1)\left(\frac{T_C}{T}\right) + 1\right] \; if \; T^* < T_C \; S_d = S_{de} \; if \; T^* > T_C \tag{6}$$

3.2. Extended Nonlinear Static Procedures

3.2.1. Modal Pushover Analyses

Mode proportional pushover analyses are improved versions of classical pushover analysis in terms of higher modes, and there are several approaches. Mainly, this method is based on structural dynamics theory and considers the contribution of one or more than one mode of vibration. Multi-modal pushover analysis (MMP) is the first attempt to take into account more than one eigenmode that is proposed by Paret et al. (1996) [49]. The analysis of each mode is carried out individually, and then, the capacity of the structure concerning each modal shape is compared with seismic demand by using CSM. Modal pushover analysis (MPA) is developed by Chopra and Goel (2002) [50] for regular buildings and then improved for the building with asymmetric plans, which is called Modified Modal Pushover Analysis (MMPA) [51]. The former method assumes that the response of a building is governed by the first vibration mode being purely translational and, therefore, the analysis assumes a mode proportional loading pattern. On the other hand, in the latter approach, a set of pushover analyses considering several mode shapes is applied, and each mode is considered individually by imposing modal force distribution related to each mode. Then, quadratic combination rules, namely, Square-Root-of-Sum-of-Squares (SRSS) or Complete-Quadratic-Combination (CQC), are used to obtain the total demand of the inelastic structural system. It is important to notice that the use of quadratic combination rules is limited to the linear response. In the case of asymmetrically planned buildings, this drawback is overcome by an alternative approach performing pushover analysis based on the fundamental mode and consideration of several control points, since rotation on the structure results in different displacements.

3.2.2. Extended N2 Method

The extended N2 method is an improved version of the classic N2 method, and it takes into account higher mode effects on the nonlinear response in case the fundamental vibration mode of the building is not dominated by a single mode. The extended N2 method, proposed by Fajfar et al. (2005) [52] and further improved for both plan and elevation irregularities by Kreslin and Fajfar (2012) [8], is generally composed of pushover and response spectrum analyses in which the results are combined based on the SRSS rule. Pushover analysis mainly covers the fundamental mode shape in translation, and the contribution of the higher mode effects is accounted for by applying correction factors. The key assumption here is that higher mode effects are calculated within the linear range, and nonlinear displacement demands are updated with elastic displacement demand obtained by means of response spectrum analysis. In other words, the ratio of maximum inelastic and elastic displacements obtained from the pushover and response spectrum analysis, respectively, is considered as a correction factor. The procedure is summarized as follows:

1. Perform the basic N2 method and determine the displacement demand at the center of mass (CM) at the roof level. Neglect the higher mode effects at the roof level.
2. Perform the eigenvalue analysis and consider all the relevant modes. Use the SRSS rule to combine the results for both orthogonal directions. Next, obtain displacements and drifts at each level and normalize the results with respect to target displacement being equal to the roof displacement at CM.
3. Apply a set of correction factors to take into account both in-plan and elevation irregularities. Displacements are used for in plan, while the drifts are considered for the elevation to evaluate the correction factors for each horizontal direction. These factors are location dependent. In the presence of both in-plan and elevation irregularity, the correction factors are obtained individually and then multiplied to attain the final value.

   a. Application of the correction factor for displacements due to in-plan irregularity: Firstly, the normalized roof displacement is calculated by dividing the roof displacement at a specific location by the displacement at the roof level at CM. Then, the correction factor applied to displacements is computed as the ratio between the normalized roof displacements obtained by elastic modal analysis and pushover analysis. The correction factor is equal to this ratio if the normalized displacement obtained by modal analysis is higher than 1.0. Otherwise, the value of the coefficient is assumed as 1.0.
   b. Application of the correction factor for drifts due to vertical irregularity: Similarly, the correction factor applied to drifts in each horizontal direction is calculated as the ratio of elastic to inelastic normalized story drifts. The reduction factor is not considered if the ratio is lower than 1.0.

3.2.3. Adaptive Pushover Method

Adaptive pushover analysis might be the most advanced nonlinear static procedure. The analysis can be either forced or displacement based. In the present paper, the focus is given to the displacement-based adaptive pushover analysis (DAP). The major improvements that have been implemented herein are that this method accounts for stiffness degradation, redistribution of the inertial forces and enables higher modes to be considered [53]. Thus, it requires an algorithm that has four main stages: (i) carry out eigenvalue analysis using the stiffness matrix at the end of the previous step; (ii) calculate the new displacement from the modal analysis and compute normalized scaling vector; (iii) update and apply the new displacement pattern to the structure; (iv) calculate the new response of the structure and new stiffness matrix due to progressive damage [54]. To demonstrate this, the flow chart of the adaptive pushover analysis is given in Figure 11. The steps can be summarized as follows:

1. Perform an eigenvalue analysis before the next incremental displacement.
2. Based on the modal response, the displacement profile for the current step is calculated by using Equation (7).

$$D_{ij} = \Gamma_j \phi_{ij} S_D(j) \tag{7}$$

   where:

   - i is the story number.
   - j is the mode number.
   - ϕ_j is the modal participation factor for the jth mode.
   - $\Phi_{i,j}$ is the mass normalized mode shape value for the ith story and the jth mode.
   - $S_D(j)$ is the spectral displacement of the jth mode.

3. To keep top displacement proportional to the load factor, displacements obtained by the previous step are normalized (Equation (8)).

$$\overline{D}_i = \frac{D_i}{maxD_i} \tag{8}$$

4. Update the load factor λ, and calculate the displacement vector (Equations (9) and (10)),

$$u_{i,n} = u_{i,n-1} + \left(\Delta\lambda{\cdot}\overline{D}_i{\cdot}u_{i,0}\right) \text{ incremental loading} \tag{9}$$

$$u_{i,n} = \lambda{\cdot}\overline{D}_i{\cdot}u_{i,0} \text{ total loading} \tag{10}$$

where:

- $\Delta\lambda$ is the load increment factor.
- $u_{i,0}$ is the nominal displacement in a story i
- n is the pushover step.

5. Apply the updated displacement to the model and solve the system of equations.
6. Calculate updated stiffness matrix after loading is applied.
7. Return to the first step of the loop to proceed with the next step.

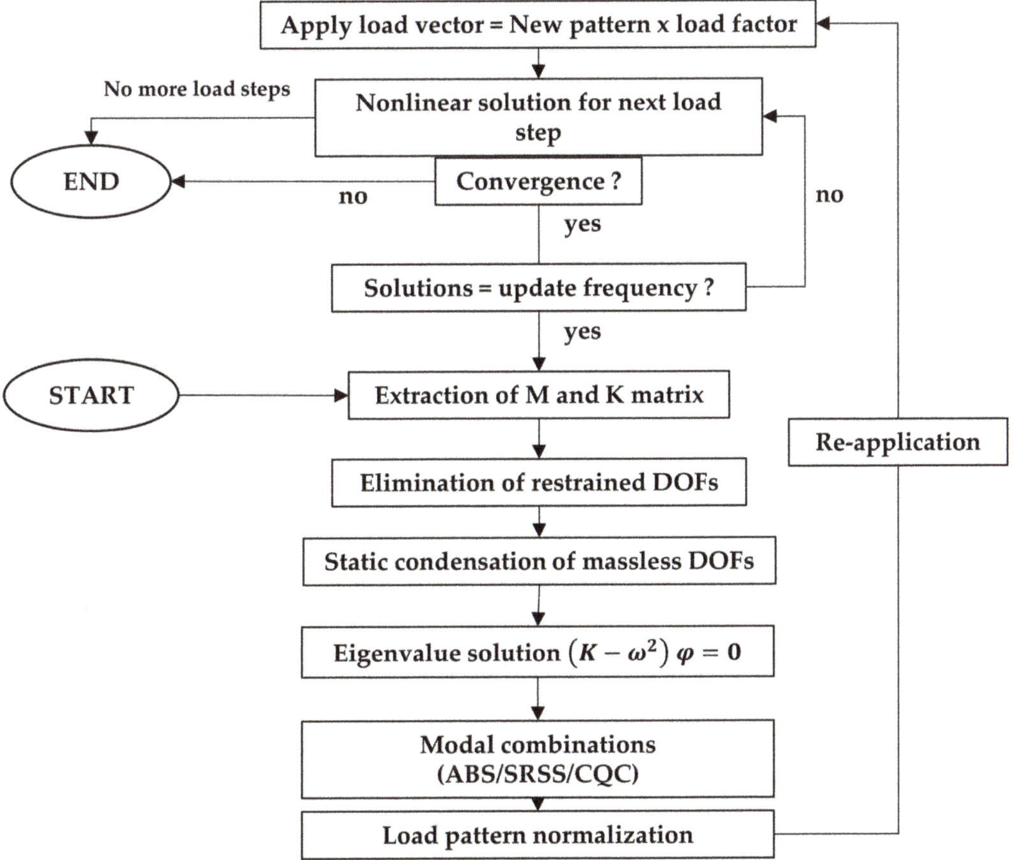

Figure 11. Flow chart of the adaptive pushover method adapted from [54].

4. Applications on Masonry Buildings

It is noticed that the application of nonlinear static procedures to masonry buildings is limited and not common as in reinforced concrete structures. In regard to the modal pushover analysis, it is noticed that it has been particularly applied to high-rise masonry structures, such as chimneys and towers [55–57]. The application of adaptive pushover analysis seems to be limited to framed system structures [58–60]. Nevertheless, there are few cases in which the loading pattern was derived based on the damage accumulation obtained from experimental campaigns carried out on masonry buildings. The dynamic properties (modal response) of the building were also obtained at the end of each step [61]. According to Galasco et al. (2006) [62], adaptive pushover analysis leads to unreliable results if it is performed on masonry buildings without box behavior (mainly with timber floors) due to the lack of redistribution of the forces. It is noted that this issue is not evident in reinforced concrete buildings, which have rigid diaphragms ensuring the load redistribution. Based on results available in the literature, there is also no evidence showing that performing adaptive pushover analysis on masonry buildings with a rigid diaphragm is reasonable or not, and further research on this topic is needed. It is important to notice that the application of these procedures in existing masonry buildings, particularly without box behavior, has limitations due to localized behavior. To overcome this issue, a multi-scale approach was proposed by Lagomarsino et al. (2015) [48].

In addition to the characteristics of masonry buildings, the direction of the seismic action is also a key parameter. Analysis performed in several directions can provide an insight into the least favorable direction that is not evident by simply calculating the eccentricity between CM and CR. This can be considerably relevant due to the high nonlinearity of masonry. In this respect, the concept of capacity dominium was introduced by [63], and an example of its application is found in [14].

Aiming at illustrating the application of NSPs to different types of masonry buildings in more detail, it was decided to present and discuss the results from two different cases studied by [38,64]. It is noticed that, in both cases, extensive research was performed to validate different procedures by comparing static with dynamic responses of masonry buildings. The first case points out the application of traditional procedures, such as the CSM, CM and N2 methods, to masonry buildings with different features, while in the second case study, a comparison between the classical N2 and extended N2 methods is performed. In this section, a discussion is carried out about the limitations in Section 4.2 and improvements in Section 4.3.

4.1. Case Studies Available in the Literature

4.1.1. Case Study 1: Marino et al. (2019)

This study aimed at understanding the nonlinear static procedures and improving their application for the seismic performance assessment of irregular unreinforced masonry buildings. Representative Italian URM buildings with three and four stories, with different irregularity levels in geometry and diaphragm characteristics, were analyzed. A reference model with both in-plan and elevation regularities was compared with others where a set of irregularities was introduced. Buildings with an irregular structural system were developed with (i) in-plan irregularity (in which the classification is made by authors as addressed in Section 2) by simply introducing a change in the distribution, number, or size of the openings at the floor level; (ii) both in-plan and elevation irregularities in which a partial floor area was added above the top level; (iii) a decrease in diaphragm stiffness from rigid to intermediate and flexible (Figure 12). Additionally, the influence of constructive details, such as the tie-rods and ring-beams, on the seismic response was studied. As described in Table 3, 13 numerical models were prepared and analyzed through an equivalent beam-based macro-element approach implemented in TREMURI [65]; see Figure 12.

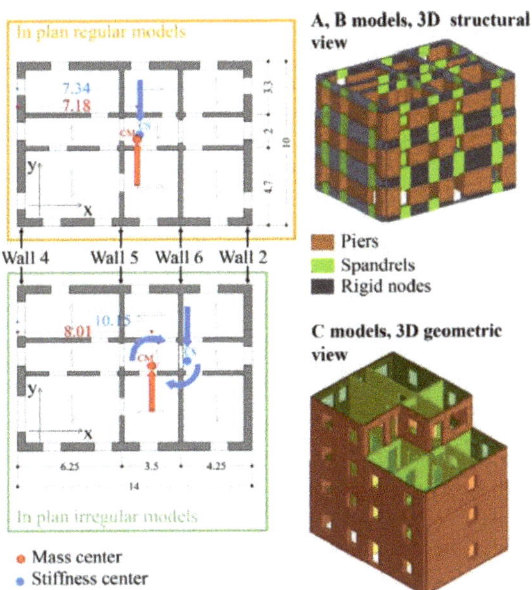

Figure 12. Geometric properties of models (in meters) [38].

Table 3. Acronyms and main characteristics of the buildings [38].

Building	Diaphragm Stiffness	In-Plan Regularity	Elevation Regularity	Constructive Details
$A_{r,rig}$	Rigid	Yes	Yes	Tie-rods
$B_{r,rig}$	Rigid	Yes	Yes	Ring-beams
$A_{irr,rig}$	Rigid	No	Yes	Tie-rods
$B_{irr,rig}$	Rigid	No	Yes	Ring-beams
$C_{irr,rig}$	Rigid	No	No	Tie-rods
$A_{r,int}$	Intermediate	Yes	Yes	Tie-rods
$B_{r,int}$	Intermediate	Yes	Yes	Ring-beams
$A_{irr,int}$	Intermediate	No	Yes	Tie-rods
$B_{irr,int}$	Intermediate	No	Yes	Ring-beams
$C_{irr,int}$	Intermediate	No	No	Tie-rods
$A_{r,flex}$	Flexible	Yes	Yes	Tie-rods
$A_{irr,flex}$	Flexible	No	Yes	Tie-rods
$C_{irr,flex}$	Flexible	No	No	Tie-rods

The work addresses three main issues related to the application of nonlinear static procedures, such as (i) the selection of the load pattern for the obtainment of the pushover curve; (ii) the identification of the damage levels based on deformation; (iii) the calculation of the target displacement and intensity measures. Thus, the work was divided into two main parts: (i) the first part deals with the nonlinear dynamic analyses to achieve an insight into the seismic response of the selected models and to serve as a reference; (ii) the second part is focused on the pushover analyses and on the application of basic nonlinear static procedures to attain the target displacement. Incremental dynamic analysis (IDA) was performed for each model in one direction (Y) by considering ten different seismic inputs. The Y direction was selected for analysis, because in this direction, the response was highly influenced by the torsional effects. Furthermore, it was necessary to avoid the coupled effect resulting from two components applied simultaneously for comparison with pushover analyses results [38]. After the obtainment of the capacity plot, a multiscale approach developed by Lagomarsino et al. (2015) [48] was used to identify the damage limits (DL) of each building. The pushover analyses were performed by firstly considering different load patterns to compare the nonlinear static responses and to select the most accurate loading protocol, for instance, uniform (mass proportional), inversely triangular (mass and height proportional), first mode shape, the SRSS combination of the first modes of each wall and the CQC combination of the first modes of each wall. The SRSS+ combination was only applied to the building with elevation irregularity. Based on the results obtained, it was suggested to use a loading pattern derived by combining load patterns proportional to the relevant first mode of each wall by using the SRSS method. Next, obtained pushover curves were transformed into a capacity curve representing an SDOF system. Accordingly, nonlinear seismic demands for each model, load pattern and damage limits were calculated by using the N2 method, Coefficient Method and Capacity Spectrum Method. Moreover, results obtained from a proposal for improving the N2 method [38,66] were also analyzed. This proposal is based on a new procedure to obtain the bilinear curve equivalent to the capacity curve and will be discussed in more detail in Section 4.3. Intensity measures (IM), which were selected as PGA, were calculated aiming at the development of the Incremental Static Analysis (ISA) curve. It is noticed that the different procedures are compared in terms of PGA rather than displacement demands directly. According to [66], PGA provides more insightful results than spectral displacement within the framework of the IDA and ISA comparison. The displacement demands at certain damage limits are considered to obtain the relevant IM to compare different procedures [66].

For the sake of simplicity, only the DL4 that represents the ultimate displacement demand is discussed here. The comparison of the results among the typology of the buildings is performed in terms of the intensity measure ratio (static over dynamic); see Figure 13. It is found that CSM is the most conservative method, and the reason for this may be associated with the use of the secant stiffness. CM provides a good prediction, but it is very demanding, as the procedure requires an iterative approach. On the contrary, the N2 method seems to be the least suitable for the given case studies. This may be attributed to the disregarding of the strength degradation and change in the damping, which is inevitable in masonry buildings. Moreover, it is important to note that the prediction of the response of the buildings with both in-plan and elevation irregularities is significantly poor, except for the CSM. Likewise, the N2 method, which is mainly recommend in EC8 and NTC, presents unconservative results, even in the case of buildings with in-plan regularity. The comparison of the results obtained from the N2 adaptive method and the new proposal made by the author is discussed in Section 4.3 for the sake of content flow.

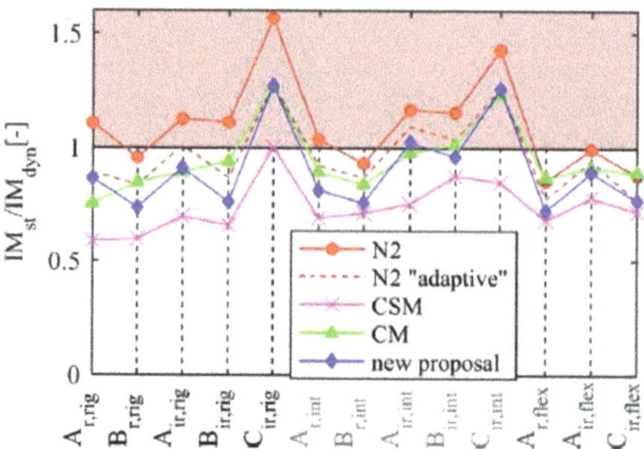

Figure 13. Comparison of the different procedures in terms of IM_{st}/IM_{dyn} ratio found for DL4 [38].

4.1.2. Case Study 2: Azizi (2018)

This case study refers to the assessment of the performance of the extended N2 method in URM buildings (concrete block masonry buildings) with in-plan and elevation irregularities, and rigid diaphragms by performing a set of nonlinear dynamic and pushover analyses. The numerical analyses were performed through the structural component model available in TREMURI software [65]. Three different models with different structural layouts were analyzed, namely, COM and CLM isolated buildings with different levels of irregularity and ACM buildings integrated in an aggregate; see Figure 14. The nonlinear dynamic analyses were executed by considering artificial accelerograms derived based on the EC8 elastic response spectrum, considering seven pairs for the COM and CLM buildings and four pairs for ACM buildings (Figure 14). A set of pushover analyses was also carried out based on a load protocol proportional to the mass and proportional to the shape of main translational mode. The results of the dynamic analysis were taken into account as reference values to compare and choose the most reliable pushover-obtained demands. Accordingly, it was found that the capacity curves obtained from mass proportional pushover analyses were similar to those found from nonlinear dynamic analyses (NLD) and, therefore, were chosen to evaluate the extended N2 method.

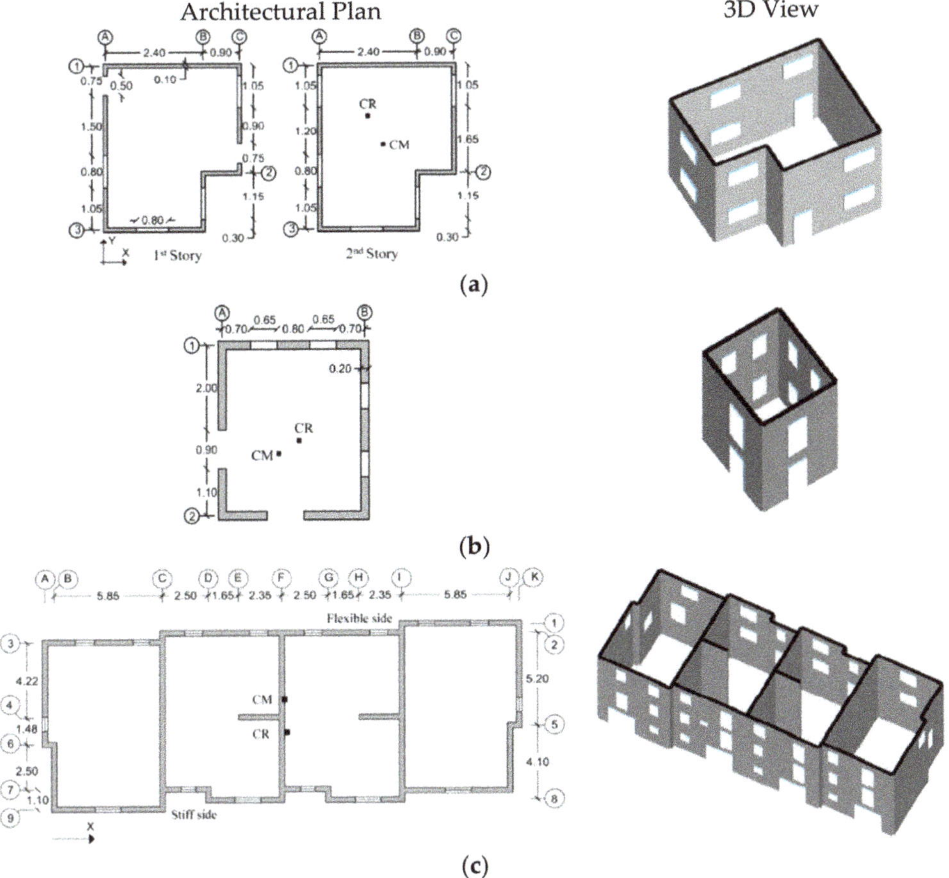

Figure 14. Plan configurations and 3D views of models studied by [64]: (**a**) COM building, (**b**) CLM building and (**c**) ACM building.

In this study, response spectrum analyses (RS) were preferred to attain elastic seismic demand and, therefore, to calculate correction factors that are functions of elastic and inelastic displacement demands. The first four- and five-mode shapes were considered and combined by using the SRSS rule for COM and ACM models, respectively. The CQC rule was applied for CLM buildings taking into account the first five-mode shapes. According to the alteration suggested by Azizi (2018) [64], the combination of RS results for each direction was disregarded. In this sense, elastic displacements in both horizontal directions were found individually at various locations, such as the center of mass (CM), stiff and flexible sides (the side closer to the center of rigidity (CR) is named as a stiff side).

To examine the higher mode effects on the masonry response due to in-plan and elevation irregularities, two different procedures were applied to achieve target displacement by using the extended N2 method as follows.

- First approach: The target displacement was computed by considering the mean value of NLD analyses instead of the basic N2 method. Next, pushover analysis was performed until the target displacement value was equal to that obtained from NLD analysis. The response computed in the previous step was updated employing a correction factor considering both torsional and elevation effects (extended N2 method). This method was applied to all models, namely, COM, CLM and ACM.
- Second approach: The target displacement was calculated by the basic N2 method, and then updated using a correction factor to take into account the effects of in-plan and elevation irregularities on the response (extended N2 method). This approach was utilized for COM buildings only.

The results are compared in terms of absolute roof displacement and inter-story drifts at both flexible and stiff sides of the buildings. The absolute roof displacement is used for indicating in-plan irregularity, while inter-story drift values are considered to calculate in-elevation irregularity. The results show that the extended N2 method is conservative and provides accurate displacement demands which are similar to NLD results at the flexible sides of the models (Figure 15). On the contrary, inter-story drift ratios, which are used as indicators of the higher mode effects in elevation, differ among the procedures (Figure 16). Correction factors are given in the figures indicating the factors for in-plan (upper number) and in-elevation (lower number) irregularities. Overall, the seismic demands calculated using the extended N2 method slightly overestimate displacements in the flexible sides of the buildings.

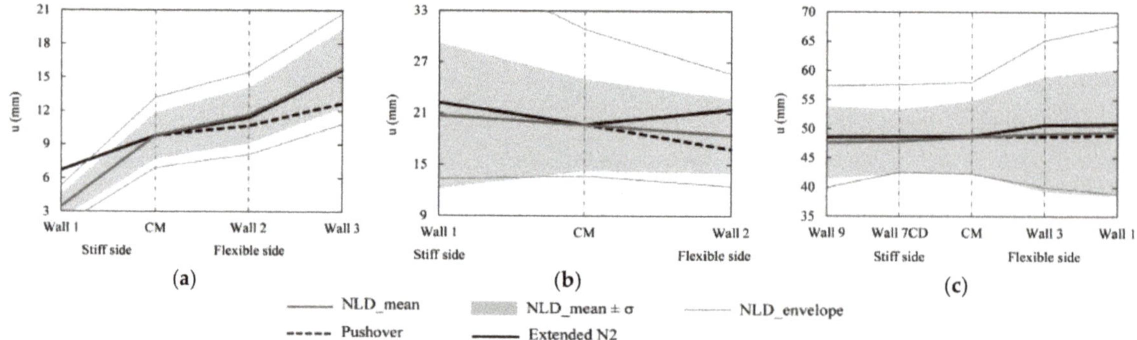

Figure 15. First approach, absolute roof displacements at the highest seismic load in X direction obtained from different methods: (**a**) COM (0.35 × g), (**b**) CLM (0.30 × g) and (**c**) ACM (0.35 × g) [64].

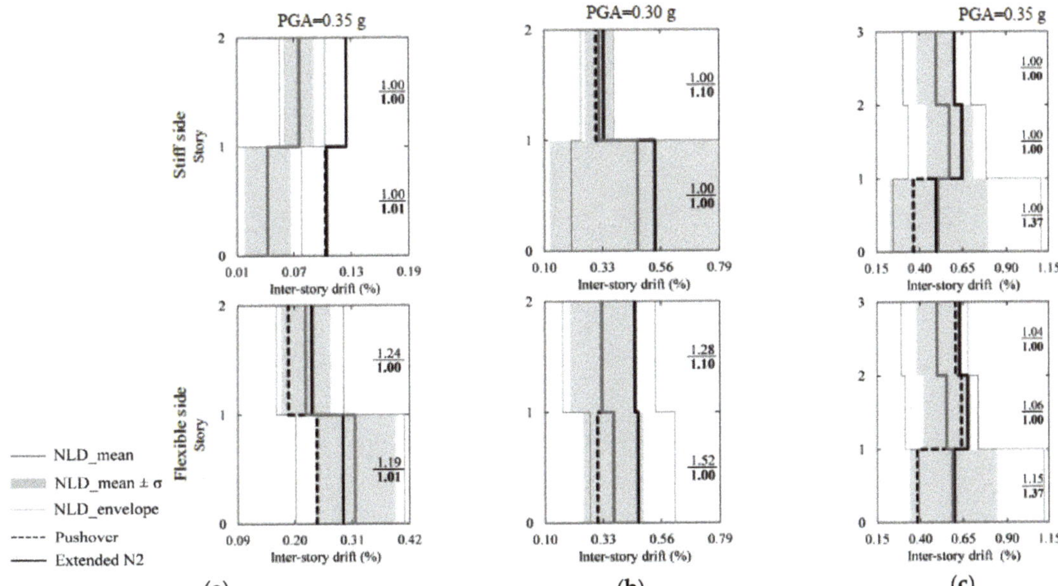

Figure 16. The first approach, inter-story drifts at the highest seismic load in X direction obtained from different methods: (**a**) COM, (**b**) CLM and (**c**) ACM. [64].

The second approach demonstrates that the basic N2 method shows a minor error (%2) for the flexible side, while the extended N2 method overestimates the average target displacement achieved by several NLD analyses by nearly 23%, as depicted in Figure 17. On the other hand, both the pushover analysis and extended N2 method considerably overestimate the displacement in the stiff sides of the buildings (Figure 18). Accordingly, it is suggested that the second procedure, including the combination of the basic and extended N2 methods, is adequate and provides reasonable demands within the present cases. Regardless of the applied approach, it is important to stress that the response revealed to be highly location-dependent due to the torsional behavior is present, as expected. This means that different control points used to compute capacity curves (stiff and flexible sides) result in different displacement values throughout the structure (both in plan and elevation). Consequently, such variation leads to a significant difference in the target displacement obtained.

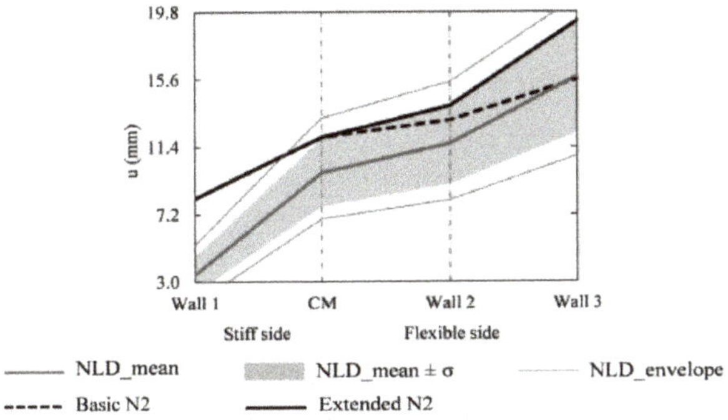

Figure 17. The second approach, absolute roof displacements at the highest seismic load in X direction obtained from different methods for COM building (0.35 × g) [64].

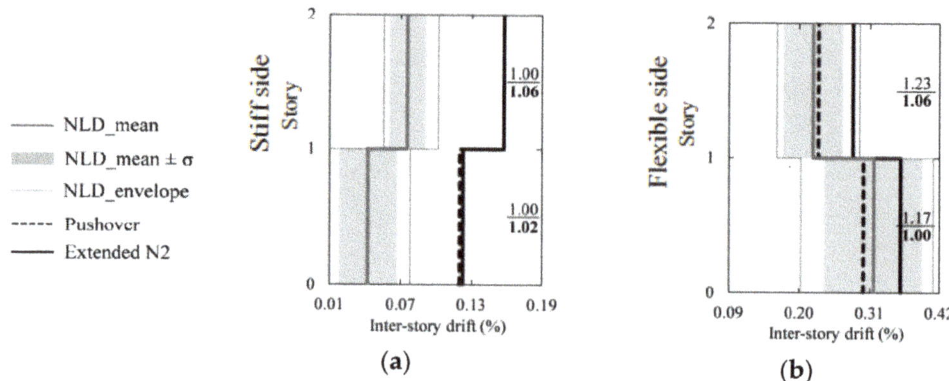

Figure 18. The second approach, inter-story drifts at the highest seismic load in X direction obtained from different methods for COM building (0.35 × g): (**a**) stiff side and (**b**) flexible side. Correction factors are given for in-plan (upper number) and in-elevation (lower number) irregularities [64].

4.2. Limitations of NSPs in Masonry Buildings

The nonlinear static procedures have been successfully used in framed structural systems, such as reinforced concrete and steel. However, there is relatively limited research on the application of the performance-based approach to masonry structures and, in particular, to irregular masonry buildings. In fact, within the case studies presented in the previous section, several limitations can be identified.

- The selection of load patterns is important, and the structural response is highly influenced by the presence of irregularities. According to [38], neither uniform nor triangular load patterns are suitable for buildings with elevation irregularities. The main reason for this is that the damage is concentrated at the top level as a result of dynamic behavior, but in pushover analysis with a uniform or triangular load pattern, a reduction in internal forces is recorded, as the applied force is a function of mass and height. Therefore, in the case of irregularity, it may be expected to have a reduction in mass, which will lead to a reduction in force. In addition, the mode proportional load pattern is only feasible if the response is governed by the so-called box behavior.
- Another important aspect is the identification of damage levels and the corresponding limit states. According to the codes, the definition of limit states is based on drift values associated with the failure mechanism at the building scale, which can be unconservative [36,38,67,68]. To overcome this issue, a multiscale approach, combining global and microelement scale behavior, was suggested by [48], particularly for buildings with an intermediate or flexible diaphragm.
- A considerable difference in displacements is found due to the selection of bilinearization methods, and, therefore, the accuracy of NSP is highly dependent on the bilinearization method [38,64].
- It should be mentioned that although IDA and ISA consider different levels of intensity, the major differences found between them are mostly justified by inherent differences found in the static and dynamic behavior of masonry buildings.

4.3. Improvements Proposed

In this section, adjustments of NSPs to masonry buildings proposed by several researchers are presented and discussed. It is noticed that improvements are generally suggested for classical and extended N2 methods to make them more feasible for masonry buildings. Proposals made by different researchers are summarized and listed in Table 4.

Table 4. NSP formulations to estimate maximum displacement proposed by different sources.

Reference	Improvements Proposed
Graziotti et al. (2014) [69]	$R = \sqrt[\beta]{(\mu - 1) \cdot \frac{T}{T_C}} + 1$ where $\mu = (R - 1)^\beta \cdot \frac{T_C}{T} + 1$ $d_{max} = \frac{d_e}{R} \left[(R - 1)^\beta \left(\frac{T_C}{T} \right) + 1 \right]$
Guerrini et al. (2017) [70]	$d_{max} = \frac{d_e}{R} \left[\dfrac{(R-1)^c}{\left(\frac{T}{T_{hyst}} + a_{hyst} \right) \left(\frac{T}{T_c} \right)^b} + R \right]$ if R > 1 $d_{max} = d_e$ if R < 1 where $\mu_R = \left[\dfrac{(R-1)^c}{\left(\frac{T}{T_{hyst}} + a_{hyst} \right) \left(\frac{T}{T_c} \right)^b} + R \right]$ and $C_R = \frac{\mu_R}{R}$
Marino et al. (2019) [38]	$d^*_{max} = \frac{d^*_{e,max}}{R} R^c = d^*_{e,max} R^{(c-1)}$ where $d^*_y = \frac{d^*_{e,max}}{R}$ $c = \frac{1}{\ln(a)} \cdot \ln \left(1 + (a - 1) b \frac{T_C}{T^*} \right) \geq 1$

Azizi (2018) [64] proposed two major alterations to application of the extended N2 method to masonry buildings: (i) disregarding the superposition principle by the SRSS to the results obtained from pushover analysis and response spectrum analysis; (ii) the application of location-dependent correction factors. The former alteration is justified by the different distribution of the mass and stiffness throughout the masonry building both in plan and elevation regarding framed systems that have lumped mass and stiffness at the floor levels. The second alteration proposes the use of correction factors obtained for each floor, since the in-plan mass and stiffness distribution differ among the floor levels.

Graziotti et al. (2014) [69] pointed out that the application of the classical N2 method to masonry buildings has drawbacks, because the maximum displacement is underestimated for high-ductility buildings, and overestimation is observed for low ductility systems. To overcome this, they propose the introduction of an exponential correction factor β to the regular N2 formulation. The calculation of the strength reduction factor (R) requires an iterative procedure since it is a function of ductility, which is updated by introducing the exponential correction factor (Table 4). The proposed correction factor β is a function of the hysteretic dissipation of the system and the formulation given in Equation (11).

$$\beta = -9.37 \zeta_{hyst} + 3.36 > 1 \text{ for } R \geq 2 \qquad \beta = 1 \text{ for } R < 2 \tag{11}$$

The main objective of using a correction factor is to improve the relationship between strength reduction factor R, ductility μ and the period to corner period ratio (T/T_c). From this improvement, higher accuracy in the estimation of displacement demand for short-period masonry buildings is achieved. The authors noted that a constant value of β equal to 1.8 ensures sufficient agreement for all the structures, in particular if there is no cyclic behavior information available.

Guerrini et al. (2017) [70] pointed out that the application of the classical N2 method may not be accurate in masonry buildings. The main reason for this is that code NSPs only take into account the structural period and ductility. Nevertheless, it is stressed that demand is highly dependent on the hysteretic behavior of an equivalent SDOF system. In fact, an SDOF system with a short period demands higher displacement due to low dissipation capacity, and, therefore, the use of code NSPs may underestimate the inelastic displacement demand for short period structures. With this aim, Guerrini et al. (2017) [70] derived an improved formulation in which hysteretic energy dissipation is introduced to the relation between elastic and inelastic displacement demands, as given in Table 4. The parameters a_{hyst}, b, c and T_{hyst} are adopted based on the hysteretic dissipation range, assuming that these depend on the dominant resisting mechanisms, namely, flexure or shear (Table 5). However, if there is no available information regarding to the hysteretic dissipation, i.e., cyclic pushover analysis, it is suggested to consider geometrical and mechanical properties, such as axial load and aspect ratio of piers, for the selection of the parameters required.

Table 5. Calibrated parameters for the proposed formulation given in [70].

Case	a_{hyst} (-)	b (-)	c (-)	T_{hyst} (s)
Mainly FD * $13\% \leq \xi_{hyst} < 15\%$	0.7	2.3	2.1	0.055
Intermediate $15\% \leq \xi_{hyst} \leq 18\%$	0.2	2.3	2.1	0.030
Mainly SD * $18\% < \xi_{hyst} \leq 20\%$	0.0	2.3	2.1	0.022

* FD, flexure-dominated; SD, shear-dominated.

Recently, further research was developed by Marino (2018) [66] to improve the classical N2 method used in Eurocode 8 and NTC 2018. The author proposes (a) an adaptive N2 method by considering an adaptive bilinear diagram (N2 adaptive), (b) an improved N2 method by the combination of the adaptive N2 method with a new formulation to calculate the target displacement (new proposal). An adaptive bilinear idealization was proposed to take into account strength and stiffness reduction to achieve a more precise prediction of the target displacement. The main reason for this is that the classical N2 method is limited to one equivalent bilinear system considering a constant effective period T^*. Thus, the inelastic displacement demand is evaluated based on the elastic range and only updated by R. Figure 19 depicts an example of the procedure to obtain bilinear curves by different approaches. For instance, based on the pre-peak phase, as given in Figure 19a, V_y^* is equal to the shear force obtained at each step of the pushover curve until the maximum force is achieved. Accordingly, the stiffness at each loading step is obtained by considering the area under the pushover curve and bilinear approximation. Once the system exceeds the peak base shear capacity, instead of the stiffness remaining constant (Figure 19b), the author proposes the use of constant yield strength equal to the maximum base shear strength in the post-peak, and to update the stiffness by using the equivalence of the energy (Figure 19c). In this way, stiffness degradation is taken into account. Indeed, the accuracy of the prediction by adopting adaptive bilinear curves improves considerably, and results are closer to the dynamic ones, as illustrated in Figure 13.

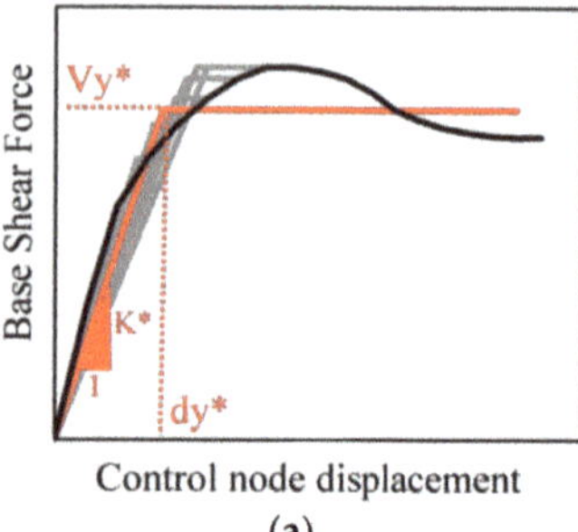

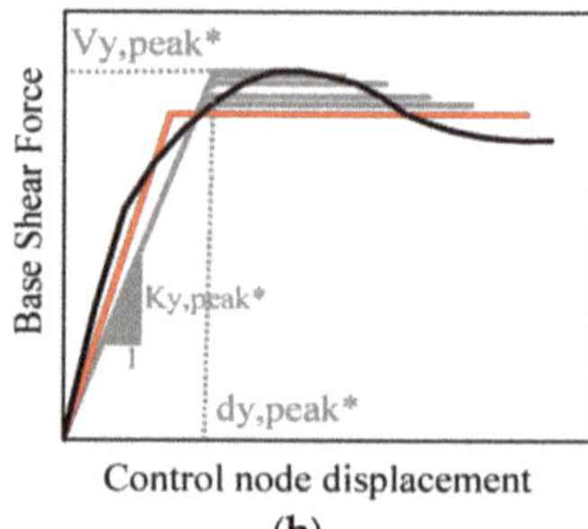

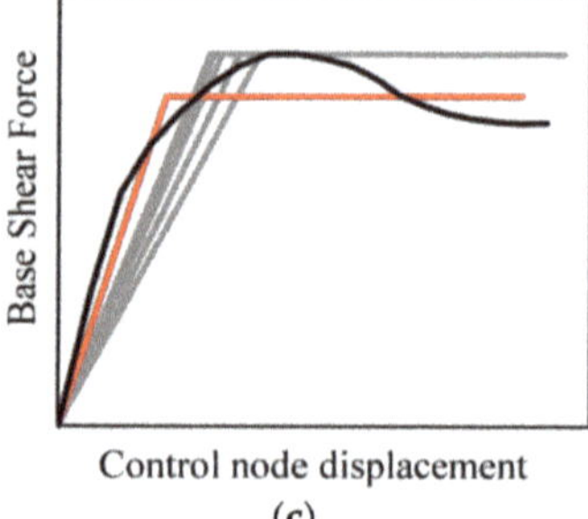

Figure 19. Examples of equivalent bilinear adapted from [66]: in red, the equivalent bilinear as proposed in NTC 2018 [71]; in grey, proposed adaptive bilinear; (**a**) before the peak; (**b**) after the peak; (**c**) post-peak phase; final proposal for the equivalent bilinear.

The new formulation for the calculation of the target displacement intends to take into account issues related to hysteretic behavior [66]. Within this framework, the proposed formulation is an improved version of the N2 method, where d_y^* is the displacement corresponding to the yielding point of the equivalent bilinear curve, while the coefficient

a is the ductility demand (Table 4). Moreover, a coefficient b is introduced to take into account the dissipative capacity and strength degradation of the system. The new equation is applicable for systems with $T^* > T_C$ and a coefficient c equal to or greater than 1. It is noticed that the new proposal provides a better prediction than the classical N2 method (Figure 13), but it is interesting to note that the new proposal underestimates slightly more than the adaptive N2 method.

5. Discussion

Based on the literature review carried out in the present paper, it is noticed that several aspects of PBD for masonry buildings have not yet been clarified. It is found that structural irregularities have various definitions given by different codes, and, therefore, there is not a systematic and uniform procedure for their characterization. In particular, for masonry buildings, a comprehensive description is needed in the design codes (Eurocode 8, NTC 2018, TEC 2019 and ASCE 7-16). Exceptionally, only a few research studies have been carried out to provide quantitative metrics to identify the irregularities for masonry buildings in elevation. Indeed, particular attention should be given to irregularities in elevation, since there is almost no such classification made for masonry buildings in the standards. Defining the correct irregularities plays a crucial role, because structural irregularities play an important role in the structural behavior and accuracy of the performance-based assessment procedure. Hence, NSPs have been improved so that the effects of irregularities on the response can be included. The proposed improvements are mainly limited to existing code formulations, namely, the N2, CSM or DCM methods, which were developed for regular frame systems. Some authors have intended to make some changes to the basic N2 method to improve its applicability to masonry buildings [66]. Indeed, there is still a research gap concerning the applicability of NSPs, in particular to irregular masonry buildings, and only a few works are available in the literature. This shows, to a great extent, that further studies are needed.

6. Conclusions

The present paper provides an overview of nonlinear static procedures for the seismic performance-based assessment and design of masonry buildings and, in particular, of irregular masonry buildings. It is observed that a great amount of work is devoted to the evaluation of the seismic vulnerability of regular buildings, regardless of construction type. It is not surprising that nonlinear static procedures have been developed for symmetric and regular buildings due to the fact that the design codes discourage irregularities in the load-bearing systems. The characterization of regularity is indeed straightforward for frame systems as provided by different seismic design codes, but such classifications may not always be applicable in the case of masonry buildings. The development of quantitative definitions for masonry buildings is needed, since the structural system also serves as an architectural component, and it can be characterized by geometrical complexity (dimensions and distribution of openings in masonry façades). Under this perspective, some researchers proposed quantitative indicators to characterize irregularities based on the geometry and alignment of the openings located in the load-bearing walls. These aspects are important to define the eccentricity at each floor level, which is not commonly taken into account. Considering the influence of the irregularity of building on seismic behavior, it is important to assess the reliability of the existing nonlinear static procedures for the seismic performance-based assessment/design of masonry buildings. Indeed, the reliability of the extended NSPs applied to irregular masonry buildings remains uncertain and requires further research. It is strongly believed that the seismic performance evaluation of an irregular masonry building supported with a systematic irregularity classification and reliable NSPs will ensure higher accuracy in the design and assessment of masonry buildings.

Author Contributions: Conceptualization, A.A., G.V. and P.B.L.; investigation, A.A.; writing—original draft preparation, A.A.; writing—review and editing, G.V. and P.B.L.; visualization, A.A.; supervision, G.V. and P.B.L.; project administration, G.V. and P.B.L.; funding acquisition, A.A., G.V. and P.B.L. All authors have read and agreed to the published version of the manuscript.

Funding: This research was funded by the Portuguese Foundation for Science and Technology FCT, grant number SFRH/BD/143949/2019 and PTDC/ECI-EGC/29010/2017.

Institutional Review Board Statement: Not applicable.

Informed Consent Statement: Not applicable.

Data Availability Statement: The data presented in this study are openly available in Universidade do Minho Repository, http://hdl.handle.net/1822/55845 (accessed on 13 February 2021). The data presented in this study are available in article [37] Lagomarsino, S.; Camilletti, D.; Cattari, S.; Marino, S. Seismic Assessment of Existing Irregular Masonry Buildings by Non-linear Static and Dynamic Analyses. Eurocode-Compliant Seism. *Anal. Design of R/C Build.* **2018**, *46*, 123–151, doi:10.1007/978-3-319-75741-4_5.

Acknowledgments: The first author acknowledges the financial support from the Portuguese Foundation for Science and Technology (FCT) through the Ph.D. Grant SFRH/BD/143949/2019. This work is financed by national funds through FCT, in the scope of the research project "Experimental and Numerical Pushover Analysis of Masonry Buildings" (PUMA) (PTDC/ECI-EGC/29010/2017).

Conflicts of Interest: The authors declare no conflict of interest. The funders had no role in the design of the study; in the collection, analyses, or interpretation of data; in the writing of the manuscript, or in the decision to publish the results.

Abbreviations

ADRS	Acceleration–displacement response spectrum
CM	Center of mass
CQC	Complete-Quadratic-Combination
CR	Center of rigidity
CSM	Capacity spectrum method
DAP	Displacement-based adaptive pushover analysis
DCM	Displacement coefficient method
DL	Damage limit
IDA	Incremental dynamic analysis
IM	Intensity measure
ISA	Incremental static analysis
MDOF	Multi-degree-of-freedom
MMP	Multi-modal pushover analysis
MMPA	Modified modal pushover analysis
MPA	Modal pushover analysis
NLD	Nonlinear dynamic analysis
NSP	Nonlinear static procedures
PBD	Performance based design
PGA	Peak ground acceleration
RS	Response spectrum
SDOF	Single-degree-of-freedom
SRSS	Square-root-of-sum-of-square

Appendix A

Table A1. Horizontal irregularity indexes are given by different design codes. Figures from [72].

Irregularity Type	EC 8 [11]	TEC 2019 [33]	ASCE/SEI 7-16 [72]	NTC 2018 [71]
Torsional $\frac{\Delta_{max}}{\Delta_{avg}}$ [72]	N.A	>1.2	>1.2 >1.4 (extreme)	N.A
Setback $\frac{X_p}{X}$, $\frac{Y_p}{Y}$; $\frac{A_{set}}{A_t}$ (EC8, NTC 2018)	>0.05	>0.2	>0.15	>0.05
Diaphragm discontinuity $\frac{A_{open}}{X*Y}$	N.A	$>\frac{1}{3}$	>0.5	N.A
Out-of-plane offset	N.A	N.A	QL	Not allowed
Nonparallel system	N.A	N.A	QL	N.A
Plan shape regularity $\frac{L_{max}}{L_{min}}$ where L_{max} is larger, L_{min} is smaller dimensions of the plan	< 4.0	N.A	N.A	< 4.0

N.A: not available; no definition is mentioned. QL: qualitative.

Table A2. Vertical irregularity indexes given by different design codes. Figures from [72].

Irregularity Type	EC 8 [11]	TEC 2019 [33]	ASCE/SEI 7-16 [72]	NTC 2018 [71]
Soft story (lateral stiffness) 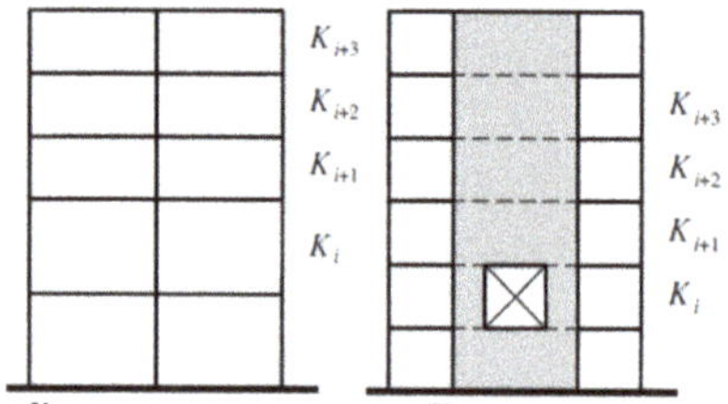 $\frac{K_i}{K_{i+1}} = a$ or $\frac{K_i}{(K_{i+1}+K_{i+2}+K_{i+3})} = b$ * TEC 2019 considers inter-story drift Δd as a parameter instead of *Str*. $\frac{\Delta d_i}{\Delta d_{i+1}}$ or $\frac{\Delta d_i}{\Delta d_{i-1}}$ Φ NTC 2018 consider reduction or increase from one level to its above. $\frac{(M_i - M_{i+1})}{M_i} = a$	QL	>2.0 *	a < 0.7 or b < $\frac{0.8}{3}$ Extreme: a < 0.6 or b < $\frac{0.7}{3}$	Reduction: a < 30 Φ% Increase: a < 10 Φ%
Weak story (lateral strength) 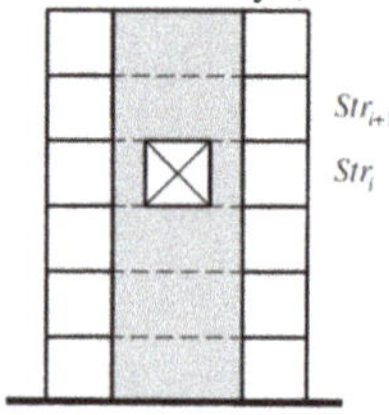 $\frac{Str_i}{Str_{i+1}}$ (ASCE) $\frac{(\Sigma A_e)_i}{(\Sigma A_e)_{i+1}}$ (TEC) * TEC 2019 considers effective shear area A_e as a parameter instead of *Str*. Φ Eurocode 8 considers the difference in shear area between two adjacent stories, specifically defined for masonry. $\frac{(A_i - A_{i+1})}{A_i}$	<20 Φ%	<0.80 *	<0.8 <0.65 (extreme)	N.A
Weight (Mass) 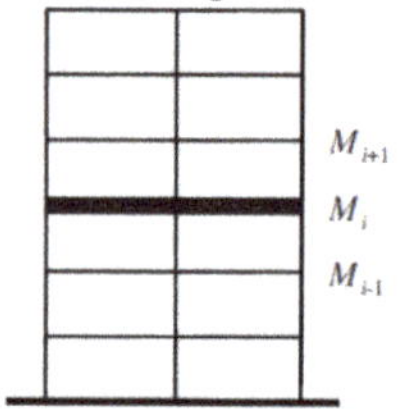 $\frac{M_i}{M_{i+1}}$ or $\frac{M_i}{M_{i-1}}$ * Eurocode 8 and NTC 2018 consider the difference in mass between two adjacent stories. $\frac{(M_i - M_{i+1})}{M_i}$	<20 *%	N.A	>1.5	<25%

Table A2. Cont.

Irregularity Type	EC 8 [11]	TEC 2019 [33]	ASCE/SEI 7-16 [72]	NTC 2018 [71]
Geometric ASCE/SEI 7-16 (Figure from [72]) 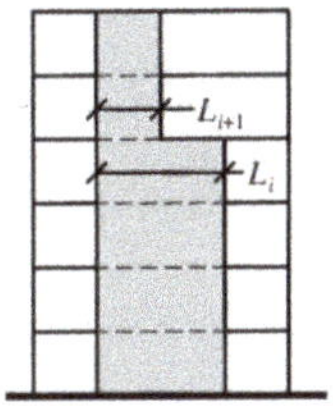 $\dfrac{L_i}{L_{i+1}}$ Eurocode 8 (Figures given below from [11]) 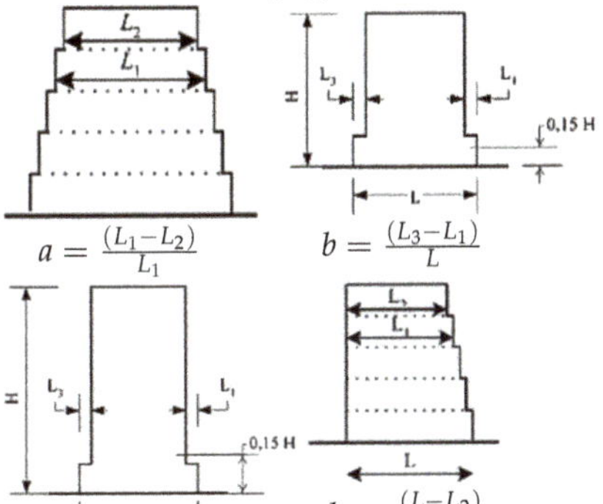 $a = \dfrac{(L_1-L_2)}{L_1}$ $b = \dfrac{(L_3-L_1)}{L}$ $c = \dfrac{(L_3-L_1)}{L}$ $d_1 = \dfrac{(L-L_2)}{L}$ $d_2 = \dfrac{(L_1-L_2)}{L}$ NTC 2018 [71] Setbacks are considered in terms of the plan area. The difference between the levels should be;	$a \leq 0.20$ $b \leq 0.20$ $c \leq 0.50$ $d_1 \leq 0.30$ $d_2 \leq 0.10$	N.A	>1.3	<30% at the first level <10% at other levels
In-plane discontinuity of lateral force resisting elements (figures from [72]) 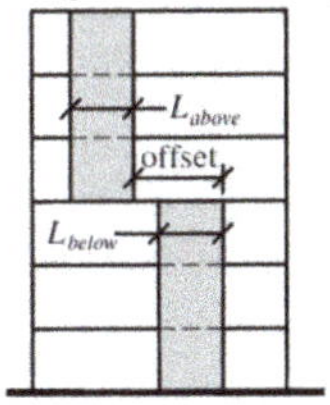 $\dfrac{offset}{L_{above}}$ or $\dfrac{offset}{L_{below}}$ 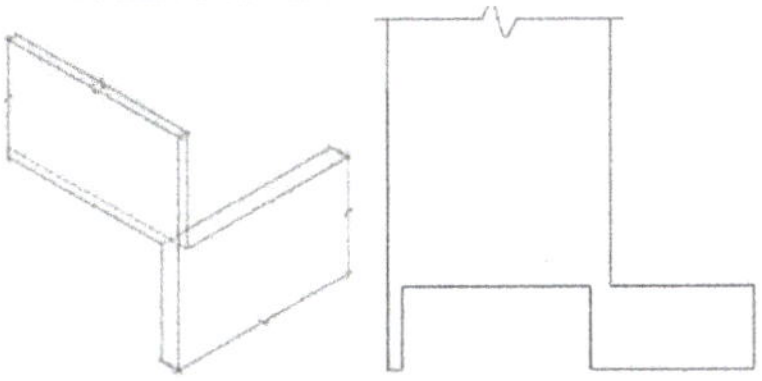 Perpendicular walls, walls with offset	QL	QL	>1.0	QL

N.A: not available; no definition is mentioned. QL: qualitative.

Appendix B. Nomenclature

Symbols in Figure 3 and Table 1

D	Inter-story height
G	Centroid
H_a	Height of the higher opening
H_b	Height of the lower opening
H_{max}	Maximum height of the opening
H_{min}	Minimum height of the opening
i	Irregularity index
L_{max}	Maximum opening length
L_{min}	Minimum opening length
L_w	Overall length of the wall
t_f	Thickness of the slab
X_G	Distance of the centroid G
Δ_0	Distance between the upper edges of the two openings
ΔH_a	Distance between upper opening edge
ΔH_b	Distance between lower opening edge
ΔL	Total irregularity, distances between right and left opening edges
ΔH	Total irregularity, difference between maximum and minimum height
N_{min}	Minimum number of openings per story
N_{max}	Maximum number of openings per story

Symbols in Figure 4 and Table 1

$\overline{h}_j$	Regularized opening height at j-th story
$\overline{X}_{G,i}$	Regularized horizontal alignment at i-th vertical alignment
$\overline{X}_{G,i+1}$	Subsequent regularized horizontal alignment at i-th vertical alignment
$\overline{X}_{G,i-1}$	Preceding regularized horizontal alignment at i-th vertical alignment
$X_{G,ij}$	Centroid ordinate of an opening at i-th level j-th opening
$\overline{Y}_{G,j}$	Regularized vertical alignment at j-th story
$\overline{b}_i$	Regularized opening width at i-th vertical alignment
b_{ij}	Opening width
H	Total height of the wall
h_{ij}	Opening height
L	Total length of the wall
$Y_{G,ij}$	Centroid ordinate of an opening
ΔHj	Inter-story height

References

1. Battaglia, L.; Ferreira, T.M.; Lourenço, P.B. Seismic fragility assessment of masonry building aggregates: A case study in the old city Centre of Seixal, Portugal. *Earthq. Eng. Struct. Dyn.* **2021**, *50*, 1358–1377. [CrossRef]
2. Valente, M.; Milani, G.; Grande, E.; Formisano, A. Historical masonry building aggregates: Advanced numerical insight for an effective seismic assessment on two row housing compounds. *Eng. Struct.* **2019**, *190*, 360–379. [CrossRef]
3. Formisano, A.; Florio, G.; Landolfo, R.; Mazzolani, F.M. Numerical calibration of an easy method for seismic behaviour assessment on large scale of masonry building aggregates. *Adv. Eng. Softw.* **2015**, *80*, 116–138. [CrossRef]
4. Lourenço, P.; Marques, R. Design of masonry structures (General rules): Highlights of the new European masonry code. In *Brick and Block Masonry-From Historical to Sustainable Masonry, Proceedings of the 17th International Brick/Block Masonry Conference, Kraków, Poland, 5–8 July 2020*; CRC Press: Boca Raton, FL, USA, 2020; p. 3.
5. Priestley, M.J.N.; Calvi, G.M.; Kowalsky, M.J. *Displacement-Based Seismic Design of Structures*; IUSS Press: Pavia, Italy, 2007; ISBN 978-88-6198-0006.
6. Fajfar, P. A Nonlinear Analysis Method for Performance-Based Seismic Design. *Earthq. Spectra* **2000**, *16*, 573–592. [CrossRef]
7. Fajfar, P.; Marušić, D.; Perus, I. Torsional effects in the pushover-based seismic analysis of buildings. *J. Earthq. Eng.* **2005**, *9*, 831–854. [CrossRef]
8. Kreslin, M.; Fajfar, P. The extended N2 method considering higher mode effects in both plan and elevation. *Bull. Earthq. Eng.* **2012**, *10*, 695–715. [CrossRef]
9. Silva, V.; Crowley, H.; Pinho, R.; Varum, H. Extending displacement-based earthquake loss assessment (DBELA) for the computation of fragility curves. *Eng. Struct.* **2013**, *56*, 343–356. [CrossRef]
10. Eurocode 6. *Eurocode 6–Design of Masonry Structures–Part 1-1: General Rules for Reinforced and Unreinforced Masonry Structures*; European Committee for Standardization: Brussels, Belgium, 2018.
11. Eurocode 8. *EN 1998-1: Design of Structures for Earthquake Resistance–Part 1: General Rules, Seismic Actions and Rules for Buildings*; European Committee for Standardization: Brussels, Belgium, 2004.

12. De Stefano, M.; Pintucchi, B. A review of research on seismic behaviour of irregular building structures since 2002. *Bull. Earthq. Eng.* **2007**, *6*, 285–308. [CrossRef]
13. D'Altri, A.; Sarhosis, V.; Milani, G.; Rots, J.; Cattari, S.; Lagomarsino, S.; Sacco, E.; Tralli, A.; Castellazzi, G.; De Miranda, S. *A Review of Numerical Models for Masonry Structures*; Numerical Modeling of Masonry and Historical Structures; Woodhead Publishing Series in Civil and Structural Engineering; Woodhead Publishing: Cambridge, UK, 2019; pp. 3–53. ISBN 9780081024393.
14. Aşıkoğlu, A.; Vasconcelos, G.; Lourenço, P.B.; Pantò, B. Pushover analysis of unreinforced irregular masonry buildings: Lessons from different modeling approaches. *Eng. Struct.* **2020**, *218*, 110830. [CrossRef]
15. Silva, L.C.; Lourenço, P.B.; Milani, G. Numerical homogenization-based seismic assessment of an English-bond masonry prototype: Structural level application. *Earthq. Eng. Struct. Dyn.* **2020**, *49*, 841–862. [CrossRef]
16. Lourenço, P.B.; Silva, L.C. computational applications in masonry structures: From the meso-scale to the super-large/super-complex. *Int. J. Multiscale Comput. Eng.* **2020**, *18*, 1–30. [CrossRef]
17. Abbati, S.D.; D'Altri, A.M.; Ottonelli, D.; Castellazzi, G.; Cattari, S.; de Miranda, S.; Lagomarsino, S. Seismic assessment of interacting structural units in complex historic masonry constructions by nonlinear static analyses. *Comput. Struct.* **2019**, *213*, 51–71. [CrossRef]
18. Shehu, R. Implementation of Pushover Analysis for Seismic Assessment of Masonry Towers: Issues and Practical Recommendations. *Buildings* **2021**, *11*, 71. [CrossRef]
19. Clementi, F. Failure Analysis of Apennine Masonry Churches Severely Damaged during the 2016 Central Italy Seismic Sequence. *Buildings* **2021**, *11*, 58. [CrossRef]
20. Saloustros, S.; Pelà, L.; Roca, P. Nonlinear Numerical Modeling of Complex Masonry Heritage Structures Considering History-Related Phenomena in Staged Construction Analysis and Material Uncertainty in Seismic Assessment. *J. Perform. Constr. Facil.* **2020**, *34*, 04020096. [CrossRef]
21. Croce, P.; Landi, F.; Formichi, P. Probabilistic Seismic Assessment of Existing Masonry Buildings. *Buildings* **2019**, *9*, 237. [CrossRef]
22. Rodrigues, H.; Šipoš, T.K. Masonry Buildings: Research and Practice. *Buildings* **2019**, *9*, 162. [CrossRef]
23. Simões, A. Evaluation of the Seismic Vulnerability of the Unreinforced Masonry Buildings Constructed in the Transition between the 19 th and 20 th Centuries in Lisbon, Portugal. Ph.D. Thesis, Universidade de Lisboa Instituto Superior Tecnico, Lisbon, Portugal, 2018.
24. Cundari, G.; Milani, G.; Failla, G. Seismic vulnerability evaluation of historical masonry churches: Proposal for a general and comprehensive numerical approach to cross-check results. *Eng. Fail. Anal.* **2017**, *82*, 208–228. [CrossRef]
25. Tiberti, S.; Acito, M.; Milani, G. Comprehensive FE numerical insight into Finale Emilia Castle behavior under 2012 Emilia Romagna seismic sequence: Damage causes and seismic vulnerability mitigation hypothesis. *Eng. Struct.* **2016**, *117*, 397–421. [CrossRef]
26. Parisse, F.; Cattari, S.; Marques, R.; Lourenço, P.; Magenes, G.; Beyer, K.; Calderoni, B.; Camata, G.; Cordasco, E.; Erberik, M.; et al. Benchmarking the seismic assessment of unreinforced masonry buildings from a blind prediction test. *Structures* **2021**, *31*, 982–1005. [CrossRef]
27. Asteris, P.; Chronopoulos, M.; Chrysostomou, C.; Varum, H.; Plevris, V.; Kyriakides, N.; Silva, V. Seismic vulnerability assessment of historical masonry structural systems. *Eng. Struct.* **2014**, *62–63*, 118–134. [CrossRef]
28. Parisi, F.; Augenti, N. Seismic capacity of irregular unreinforced masonry walls with openings. *Earthq. Eng. Struct. Dyn.* **2013**, *42*, 101–121. [CrossRef]
29. Augenti, N.; Parisi, F. Learning from Construction Failures due to the 2009 L'Aquila, Italy, Earthquake. *J. Perform. Constr. Facil.* **2010**, *24*, 536–555. [CrossRef]
30. Giordano, A.; Guadagnuolo, M.; Faella, G. Pushover analysis of plan irregular masonry buildings. In Proceedings of the 14th World Conference on Earthquake Engineering, Beijing, China, 12–17 October 2008.
31. Drysdale, R.G.; Hamid, A.A. *Masonry Structures Behavior and Design*, 3rd ed.; The Masonry Society: Englewood Cliffs, NJ, USA, 2008.
32. De Stefano, M.; Mariani, V. Pushover analysis for plan irregular building structures. In *Perspectives on European Earthquake Engineering and Seismology Geotechnical, Geological and Earthquake Engineering*; Springer: Berlin/Heidelberg, Germany, 2014; pp. 8–11.
33. TEC. *Turkish Earthquake Code: Specifications for Building Design under Earthquake Effects*; Official Newspaper No 30364; TEC: Ankara, Turkey, 2019.
34. Abrams, D.P. Response of Unreinforced Masonry Buildings. *J. Earthq. Eng.* **1997**, *1*, 257–273. [CrossRef]
35. Bairrão, R.; Silva, M.F. Shaking table tests of two different reinforcement techniques using polymeric grids on an asymmetric limestone full-scaled structure. *Eng. Struct.* **2009**, *31*, 1321–1330. [CrossRef]
36. Kallioras, S.; Guerrini, G.; Tomassetti, U.; Marchesi, B.; Penna, A.; Graziotti, F.; Magenes, G. Experimental seismic performance of a full-scale unreinforced clay-masonry building with flexible timber diaphragms. *Eng. Struct.* **2018**, *161*, 231–249. [CrossRef]
37. Lagomarsino, S.; Camilletti, D.; Cattari, S.; Marino, S. Seismic Assessment of Existing Irregular Masonry Buildings by Nonlinear Static and Dynamic Analyses. *Eurocode Compliant Seism. Anal. Des. R/C Build.* **2018**, *46*, 123–151. [CrossRef]
38. Marino, S.; Cattari, S.; Lagomarsino, S. Are the nonlinear static procedures feasible for the seismic assessment of irregular existing masonry buildings? *Eng. Struct.* **2019**, *200*, 109700. [CrossRef]

39. Berti, M.; Salvatori, L.; Orlando, M.; Spinelli, P. Unreinforced masonry walls with irregular opening layouts: Reliability of equivalent-frame modelling for seismic vulnerability assessment. *Bull. Earthq. Eng.* **2017**, *15*, 1213–1239. [CrossRef]
40. Grünthal, G. *European Macroseismic Scale*; European Seismological Commission: Luxembourg, 1998.
41. SEAOC. *Performance Based Seismic Engineering of Buildings*; Vision 2000; Structural Engineers Association of California: Sacramento, CA, USA, 1995.
42. Bilgin, H.; Hysenlliu, M. Comparison of near and far-fault ground motion effects on low and mid-rise masonry buildings. *J. Build. Eng.* **2020**, *30*, 101248. [CrossRef]
43. FEMA 440. *Improvement of Nonlinear Static Seismic Analysis Procedures*; Department of Homeland Security Federal Emergency Management Agency: Washington, DC, USA, 2005.
44. Fajfar, P.; Fischinger, M. N2—A Method for Nonlinear seismic analysis of regular buildings. In Proceedings of the Ninth World Conference Earthquake Engineering, Tokyo, Japan, 2–6 August 1988.
45. Freeman, S.; Nicoletti, J.; Tyrell, J. Evaluation of existing buildings for seismic risk—A case study of Puget Sound Naval Shipyard. In Proceedings of the U.S. National Conference on Earthquake, Bremerton, WA, USA, 18–20 June 1975; pp. 113–122.
46. FEMA 356. *Prestandard and Commentary for the Seismic Rehabilitation of Buildings*; Department of Homeland Security Federal Emergency Management Agency: Washington, DC, USA, 2000.
47. Calvi, G.M. A Displacement-Based Approach for Vulnerability Evaluation of Classes of Buildings. *J. Earthq. Eng.* **1999**, *3*, 411–438. [CrossRef]
48. Lagomarsino, S.; Cattari, S. PERPETUATE guidelines for seismic performance-based assessment of cultural heritage masonry structures. *Bull. Earthq. Eng.* **2015**, *13*, 13–47. [CrossRef]
49. Paret, T.; Sasaki, K.; Elibeck, D.; Freeman, S. Approximate inelastic procedures to identify failure mechanism from higher modes effects. In Proceedings of the 11th World Conference Earthquake Engineering, Acapulco, Mexico, 23–28 June 1996.
50. Chopra, A.K.; Goel, R.K. A modal pushover analysis procedure for estimating seismic demands for buildings. *Earthq. Eng. Struct. Dyn.* **2002**, *31*, 561–582. [CrossRef]
51. Chopra, A.K.; Goel, R.K. A modal pushover analysis procedure to estimate seismic demands for unsymmetric-plan buildings. *Earthq. Eng. Struct. Dyn.* **2004**, *33*, 903–927. [CrossRef]
52. Fajfar, P.; Kilar, V.; Marusic, D.; Perus, I.; Magliulo, G. The extension of the N2 method to asymmetric buildings. In Proceedings of the 4th European Workshop on the Seismic Behaviour of Irregular and Complex Structures, Thessaloniki, Greece, 26–27 August 2005.
53. Antoniou, S. Advanced Inelastic Static Analysis for Seismic Assessment of Structures. PhD. Thesis, Imperial College London (University of London), London, UK, 2002.
54. Papanikolaou, V.K.; Elnashai, A.S. Evaluation of conventional and adaptive pushover analysis I: Methodology. *J. Earthq. Eng.* **2005**, *9*, 923–941. [CrossRef]
55. Minghini, F.; Bertolesi, E.; Del Grosso, A.; Milani, G.; Tralli, A. Modal pushover and response history analyses of a masonry chimney before and after shortening. *Eng. Struct.* **2016**, *110*, 307–324. [CrossRef]
56. Minghini, F.; Milani, G.; Tralli, A. Seismic risk assessment of a 50m high masonry chimney using advanced analysis techniques. *Eng. Struct.* **2014**, *69*, 255–270. [CrossRef]
57. Peña, F.; Lourenço, P.B.; Mendes, N.; Oliveira, D.V. Numerical models for the seismic assessment of an old masonry tower. *Eng. Struct.* **2010**, *32*, 1466–1478. [CrossRef]
58. Ferracuti, B.; Pinho, R.; Savoia, M.; Francia, R. Verification of displacement-based adaptive pushover through multi-ground motion incremental dynamic analyses. *Eng. Struct.* **2009**, *31*, 1789–1799. [CrossRef]
59. Silva, V.; Crowley, H.; Varum, H.; Pinho, R.; Sousa, L. Investigation of the characteristics of Portuguese regular moment-frame RC buildings and development of a vulnerability model. *Bull. Earthq. Eng.* **2015**, *13*, 1455–1490. [CrossRef]
60. Antoniou, S.; Rovithakis, A.; Pinho, R. Development and Verification of a Fully Adaptive Pushover Procedure. In Proceedings of the 12th European Conference Earthquake Engineering, London, UK, 9–13 September 2002.
61. Yi, T.; Moon, F.L.; Leon, R.T.; Kahn, L.F. Lateral Load Tests on a Two-Story Unreinforced Masonry Building. *J. Struct. Eng.* **2006**, *132*, 643–652. [CrossRef]
62. Galasco, A.; Lagomarsino, S.; Penna, A. On the use of pushover analysis for existing masonry buildings. In Proceedings of the 1st European Conference on Earthquake and Seismology, Geneva, Switzerland, 3–8 September 2006.
63. Caliò, I.; Cannizzaro, F.; D'Amore, E.; Marletta, M.; Pantò, B. A new discrete-element approach for the assessment of the seismic resistance of composite reinforced concrete-masonry buildings. *Fourth Huntsville Gamma Ray Burst Symp.* **2008**, *1020*, 832–839. [CrossRef]
64. Azizi, H. Analytical and Empirical Seismic Fragility Analysis of Irregular URM Buildings with Box Behavior. Ph.D. Thesis, University of Minho, Guimarães, Portugal, 2018.
65. Lagomarsino, S.; Penna, A.; Galasco, A.; Cattari, S. *TREMURI Program: Seismic Analyses of 3D Masonry Buildings, Release 2.0*; University of Genoa: Genoa, Italy, 2012.
66. Marino, S. Nonlinear Static Procedures for the Seismic Assessment of Irregular URM Buildings. Ph.D. Thesis, University of Genoa, Genoa, Italy, 2018.
67. Kallioras, S.; Correia, A.A.; Graziotti, F.; Penna, A.; Magenes, G. Collapse shake-table testing of a clay-URM building with chimneys. *Bull. Earthq. Eng.* **2020**, *18*, 1009–1048. [CrossRef]

68. Graziotti, F.; Tomassetti, U.; Kallioras, S.; Penna, A.; Magenes, G. Shaking table test on a full scale URM cavity wall building. *Bull. Earthq. Eng.* **2017**, *15*, 5329–5364. [CrossRef]
69. Graziotti, F.; Penna, A.; Bossi, E.; Magenes, G. Evaluation of displacement demand for unreinforced masonry buildings by equivalent SDOF systems. In Proceedings of the 9th International Conference on Structural Dynamics, EURODYN 2014, Porto, Portugal, 30 June–2 July 2014; pp. 365–372.
70. Guerrini, G.; Graziotti, F.; Penna, A.; Magenes, G. Improved evaluation of inelastic displacement demands for short-period masonry structures. *Earthq. Eng. Struct. Dyn.* **2017**, *46*, 1411–1430. [CrossRef]
71. NTC. *Norme Tecniche per le Costruzioni*; DM 17/1/2018; Gazzetta Ufficiale della Repubblica Italiana n. 42 del 20 febbraio 2018; NTC: Roma, Italy, 2018.
72. ASCE/SEI 7-16. *Minimum Design Loads and Associated Criteria for Buildings and Other Structures*; American Society of Civil Engineers: Reston, VA, USA, 2017.

Article

Displacement Demand for Nonlinear Static Analyses of Masonry Structures: Critical Review and Improved Formulations

Gabriele Guerrini [1,2,*], **Stylianos Kallioras** [1,2], **Stefano Bracchi** [1,2], **Francesco Graziotti** [1,2] and **Andrea Penna** [1,2]

[1] Department of Civil Engineering and Architecture (DICAr), University of Pavia, Via Ferrata 3, 27100 Pavia, PV, Italy; stylianos.kallioras@unipv.it (S.K.); stefano.bracchi@unipv.it (S.B.); francesco.graziotti@unipv.it (F.G.); andrea.penna@unipv.it (A.P.)

[2] European Centre for Training and Research in Earthquake Engineering (EUCENTRE), Via Ferrata 1, 27100 Pavia, PV, Italy

* Correspondence: gabriele.guerrini@unipv.it

Abstract: This paper discusses different formulations for calculating earthquake-induced displacement demands to be associated with nonlinear static analysis procedures for the assessment of masonry structures. Focus is placed on systems with fundamental periods between 0.1 and 0.5 s, for which the inelastic displacement amplification is usually more pronounced. The accuracy of the predictive equations is assessed based on the results from nonlinear time-history analyses, carried out on single-degree-of-freedom oscillators with hysteretic force–displacement relationships representative of masonry structures. First, the study demonstrates some limitations of two established approaches based on the equivalent linearization concept: the capacity spectrum method of the Dutch guidelines NPR 9998-18, and its version outlined in FEMA 440, both of which overpredict maximum displacements. Two codified formulations relying on inelastic displacement spectra are also evaluated, namely the N2 method of Eurocode 8 and the displacement coefficient method of ASCE 41-17: the former proves to be significantly unconservative, while the latter is affected by excessive dispersion. A non-iterative procedure, using an equivalent linear system with calibrated optimal stiffness and equivalent viscous damping, is then proposed to overcome some of the problems identified earlier. A recently developed modified N2 formulation is shown to improve accuracy while limiting the dispersion of the predictions.

Keywords: capacity spectrum method; equivalent linear system; inelastic displacement spectra; masonry structure assessment; nonlinear static analysis; seismic displacement demand; single-degree-of-freedom oscillator

Citation: Guerrini, G.; Kallioras, S.; Bracchi, S.; Graziotti, F.; Penna, A. Displacement Demand for Nonlinear Static Analyses of Masonry Structures: Critical Review and Improved Formulations. *Buildings* **2021**, *11*, 118. https://doi.org/10.3390/buildings11030118

Academic Editor: Rita Bento and Ana Simões

Received: 15 February 2021
Accepted: 9 March 2021
Published: 16 March 2021

Publisher's Note: MDPI stays neutral with regard to jurisdictional claims in published maps and institutional affiliations.

1. Introduction

Nonlinear static procedures (NSPs) have gained popularity in the professional practice for the seismic performance assessment of existing masonry structures. In fact, they can provide good predictions of local and global earthquake-induced deformations directly related to structural and non-structural damage [1]. At the same time, NSPs are not affected by some of the hurdles of nonlinear time-history analyses (NLTHA), namely the definition of cyclic constitutive models, the adoption of viscous damping models, and the selection of representative ground motions. Various NSPs require determining first the capacity curve of a single-degree-of-freedom (SDOF) oscillator, equivalent to the multi-degree-of-freedom (MDOF) structure, through pushover analyses [2–8]. It is then necessary to determine the inelastic displacement demands on the SDOF system due to certain seismic hazards and compare them with displacement thresholds identified on the capacity curve, corresponding to meaningful limit states. Over recent decades, several methods have been developed for the evaluation of the displacement demands, which can be classified into two main families and form the object of this study.

The first family includes methods based on the concept of an "equivalent linear system". The seismic demand on a nonlinear structure is estimated using overdamped elastic spectra and a substitute linear-elastic SDOF system, with reduced effective stiffness and equivalent viscous damping to account for the effects of period elongation and hysteretic energy dissipation due to yielding [9–12]. With these techniques, one generally achieves a solution through an iterative process. The "capacity spectrum method", initially proposed by Freeman et al. [13] and documented thoroughly in the American ATC-40 and FEMA 274 guidelines [14,15], is the most prominent example of these analysis procedures. The method has been recently revamped in the Italian building code NTC-18 [16,17], in the Dutch NPR 9998-18 [18] derived from the New Zealand code for the seismic assessment of existing buildings [19], and in some displacement-based procedures [20,21]. Modified versions of this approach have been proposed in the FEMA 440 guidelines [22] and other seismic assessment procedures.

Methods that employ inelastic response spectra, referring to the initial elastic stiffness and viscous damping of a first-mode-equivalent SDOF oscillator, belong to the second family. Pioneer studies in the development of such approaches were conducted in the 1960s [23,24], illustrating the "equal displacement rule" for medium and long-period systems and the significant amplification of displacement demands for inelastic systems in the short-period range. During the following decades, several researchers confirmed these observations and addressed the influence of oscillators relative strength, hysteretic rules, supplemental viscous damping, P-Δ effects, soil conditions, and ground motion characteristics on constant-ductility or constant-relative-strength inelastic response spectra [1,25–30]. Some of these efforts led to the development of the so-called "N2 method" [31–34], included in Eurocode 8 [35] and the Italian building code NTC-18 [16,17], and of the "displacement coefficient method" of FEMA 273 and FEMA 274 guidelines [15,36], adopted by the ASCE 41-17 code [37].

Recently, criticism has been raised against the first family of methods because of accuracy issues, convergence issues, and lack of mechanical correlation between viscous damping and hysteretic energy dissipation [4,38–42]. Nevertheless, building codes worldwide still include formulations based on the equivalent linear system concept. Moreover, inelastic displacement demands depend on the oscillator hysteretic behavior, being generally larger for less dissipative systems with shorter elastic periods [26,38,42]. However, current building codes propose NSP formulations that do not address this hysteresis dependence but give only explicit consideration of structural period and ductility. Research has been conducted in recent years to optimize methods belonging to the second family [43–46]. Applications of NSP to probabilistic seismic assessment and urban-scale risk evaluation have also been proposed [47–49]. Statistics-based, rather than mechanics-based approaches, such as those relying on surrogate models, could also be employed to calibrate these relationships, as has been done for other engineering applications [50].

For these reasons, this paper first discusses the accuracy of two established methods per family to calculate the inelastic seismic displacement demand on short-period masonry buildings, highlighting their shortcomings. Then, it presents an improved formulation for each family, the "optimal stiffness method" and the "modified N2 method" [44], respectively, which account for the typical hysteretic dissipation of masonry structures. The assessment and calibration of the predictive equations are based on the results from extensive NLTHA on nonlinear SDOF oscillators, with fundamental periods ranging from 0.05 to 0.5 s and hysteretic behavior representative of masonry structures, performed with TREMURI [51,52]. Two independent databases of real earthquake records were used, as well as two sets of oscillators. A total number of 3,434,900 analyses supported the evaluation and calibration processes.

2. Evaluation and Calibration with Nonlinear Time-History Analyses

2.1. Nonlinear SDOF Oscillators

A large number of NLTHA was performed on inelastic SDOF oscillators, covering a comprehensive range of structural parameters representative of masonry buildings as detailed in previous work by the authors [44,53,54]. The oscillators were analyzed in TREMURI [51,52]: this software allows modeling structures through macroelements with constitutive relationships compatible with masonry in-plane flexural and shear behavior. The monotonic acceleration-displacement (AD) response of each oscillator, obtained from the software, was idealized into a bilinear elastoplastic relationship (Figure 1a). The idealized elastic stiffness, k, was first established as the slope of the secant line through 70% of the maximum base shear. Then, the idealized yield pseudo-acceleration, a_y, and displacement, d_y, were calculated by equating the areas below the curves between the origin and the ultimate displacement, d_u, identified at a base-shear drop equal to 20% of the maximum strength [16,17]. The idealized elastic period was determined as $T = 2\pi\sqrt{m/k}$, where m was the mass associated with each SDOF system.

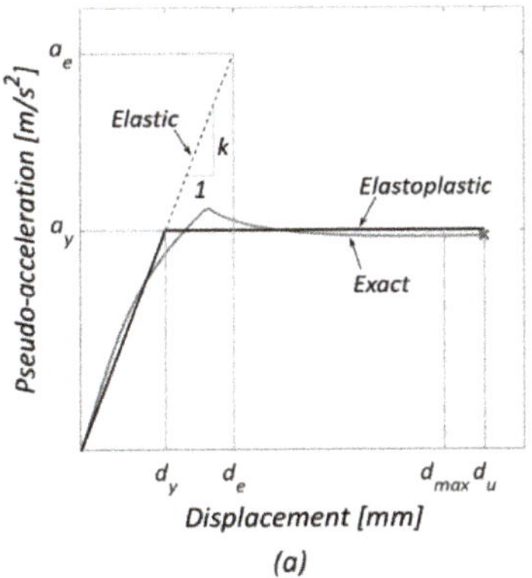

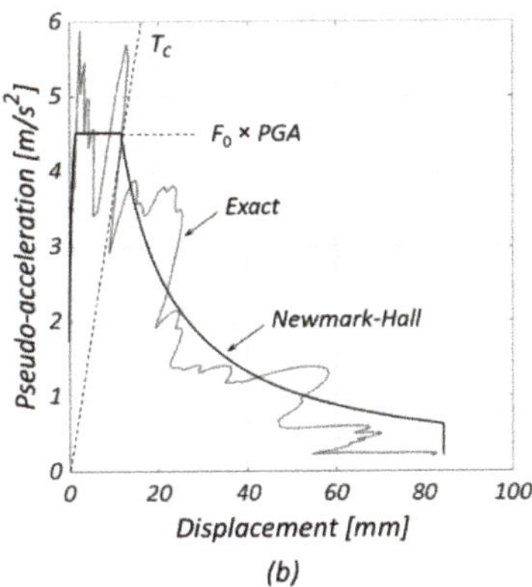

Figure 1. Oscillators and input ground motions for nonlinear time-history analyses (NLTHA): (**a**) elastoplastic idealization of monotonic acceleration-displacement (AD) response (adapted from [44]); (**b**) elastic response spectra in AD format of the Niijima Island earthquake (1 July 2000).

The displacement ductility demand, μ, the strength ratio, R (also termed force-reduction factor, response-modification factor, or behavior factor), and the inelastic displacement ratio C were defined on the elastoplastic backbone curve as follows Equations (1)–(3):

$$\mu = \frac{d_{max}}{d_y} \tag{1}$$

$$R = \frac{a_e}{a_y} = \frac{d_e}{d_y} \tag{2}$$

$$C = \frac{d_{max}}{d_e} \tag{3}$$

where d_{max} is the maximum inelastic displacement demand, $d_e = a_e \cdot (2\pi/T)^2$ is the elastic displacement demand, and $a_e(T)$ is the elastic pseudo-spectral acceleration for idealized elastic period T and 5% viscous damping ratio. The following relationship between μ, R, and C can be derived from Equation (1) through Equation (3):

$$\mu = R \cdot C \tag{4}$$

meaning that for a given R one can determine C by calculating μ or, conversely, for a given μ, one can determine C by calculating R. In what follows, the symbols μ_R and C_R will denote the ductility demand and inelastic displacement ratio for a given R, as opposed to R_μ and C_μ, which would indicate the strength ratio and inelastic displacement ratio for a given μ.

Two sets of target SDOF oscillators were defined, starting from seven hysteretic reference models with Jacobsen's equivalent viscous damping ratio, ξ_{hyst}, between 13.8% and 19.9%, and targeting ten different idealized elastic periods T between 0.05 and 0.5 s [44,54]. The first set of oscillators (Set 1) was then generated assuming ten values of the idealized yield strength a_y between 0.5 and 5.0 m/s^2, resulting in 700 oscillators employed to evaluate current formulations and calibrate new equations. The second set of SDOF systems (Set 2) was instead obtained considering five values of strength ratio R between 1.0 and 5.0, resulting in a group of 350 oscillators used to determine $C_R - R - T$ and $\mu_R - R - T$ relationships.

2.2. Ground Motion Records and Response Spectra

Two databases of earthquake records were selected to conduct NLTHA. The first group (Database A) included 467 pairs of accelerograms from the third release of the SIMBAD database [55]. The second group (Database B) comprised 1753 pairs of records [56], combining tectonic ground motions from the NGA1 [57] and the RESORCE [58] databases with induced-seismicity motions from the Groningen gas field [59]. Both databases cover wide ranges of site conditions and seismological parameters.

The actual elastic response spectra of the signals were approximated by Newmark-Hall's spectral shapes [25,44] (Figure 1b). The approximated spectra were anchored to the actual peak ground acceleration (PGA) of the records. A least-square regression was performed on each pseudo-acceleration spectrum within the period range between 0 and 4.0 s to select parameter F_0, which quantifies the plateau acceleration as a multiple of PGA, and corner period T_C, which identifies the transition from the constant-acceleration to the constant-velocity branch of the idealized spectrum. Elastic displacement spectra were derived from the Newmark-Hall pseudo-acceleration spectra, multiplying each ordinate by $(2\pi/T)^2$. Figure 1b shows an example of actual and approximated elastic response spectrum in AD format.

Each SDOF oscillator from Set 1 was subjected to the records of both databases, resulting in 653,800 (Suite 1-A) and 2,454,200 (Suite 1-B) earthquake simulations using Database A and Database B, respectively. Both suites of simulations were used to evaluate current approaches, while Suite 1-A served for the calibration of new equations and Suite 1-B for their validation. Instead, the oscillators of Set 2 were analyzed only with the ground motions of Database A, resulting in additional 326,900 simulations (Suite 2-A), which were employed to produce constant-relative-strength inelastic response spectra. A total number of 3,434,900 analyses formed the basis of this study.

2.3. Evaluation and Calibration Procedures

Statistical analysis was performed with MATLAB (MathWorks, version R2019a) considering pairs of equation-predicted and NLTHA ductility demands obtained from Suite 1-A or Suite 1-B analyses, represented by gray dots in Figure 2 [44]. The points were assigned to diagonal bins, with boundaries orthogonal to the bisector of the first quadrant. The median distance from the bisector $d_{m,i}$, the 16th percentile distance $d_{16,i}$, and the 84th percentile distance $d_{84,i}$ were calculated for the data points within the ith bin; points for these percentiles were determined for each bin, associating those distances with the bin central value, as plotted in Figure 2. Similarly, the points corresponding to the 5th and 95th percentile were also determined. Median, 5th, 16th, 84th, and 95th percentile lines were then drawn by connecting these points.

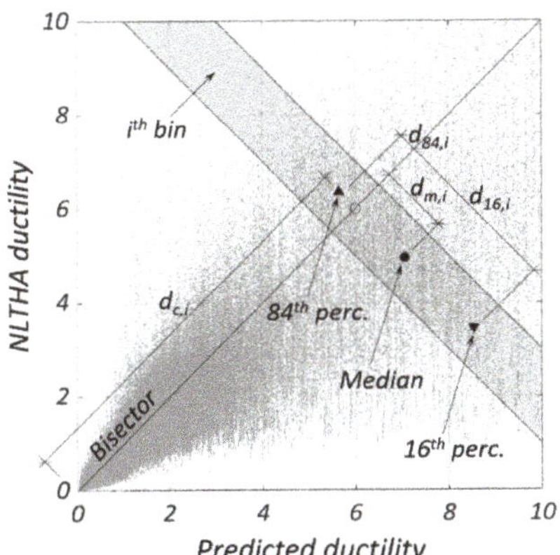

Figure 2. Comparison between predicted and NLTHA ductility demands: definition of diagonal bins and percentile distances from the bisector.

Parameters of the proposed equations were calibrated with the dynamic responses from Suite 1-A, applying an orthogonal regression between the predictions and the results of NLTHA to minimize the error on the median and the scatter (i.e., the 16th and 84th percentiles) by diagonal bins [44]. The accuracy of the calibrated equations was then evaluated following the same approach based on Suite 1-B analysis results.

The evaluation and calibration procedures by diagonal bins were limited to displacement ductility demands up to 10, considered as a limit value for most structures to which these methods would be applied. Ideally, an accurate and precise method would result in the median line coinciding with the bisector, with upper and lower percentiles as close as possible to it. Due to the high rate of divergence towards the infinity of the ductility demand from both NLTHA and predictions, it was not possible to obtain meaningful results for oscillators with periods of 0.05 s, which were consequently excluded from the statistical analysis [44].

The NLTHA results from Suite 2-A were instead used to determine $C_R - R - T$ and $\mu_R - R - T$ relationships, calculating for each idealized period T and strength ratio R the median values of inelastic displacement ratio C_R and ductility μ_R. In this case, also ductility demands greater than 10 were included in the determination of the median values.

3. Established Formulations Based on Equivalent Linearization

3.1. Capacity Spectrum Method (NPR 9998-18)

The capacity spectrum method (CSM) was initially adopted by the ATC-40 guidelines [14] for the seismic evaluation of existing concrete buildings. The same version of the method has been proposed by the 2018 edition of the Italian building code NTC-18 [16,17] as one of the two available methods to estimate displacement demands. Recently, the CSM has also been adopted as the preferred method for the nonlinear static analysis of masonry buildings by the Dutch code NPR 9998-18 [18], derived from the New Zealand code [19].

For oscillators that remain elastic (i.e., with $R \leq 1$), simply $d_{max} = d_e$. Based on the equivalent linearization approach, the CSM approximates the response of an SDOF oscillator undergoing inelastic deformations (i.e., with $R > 1$) through a substitute linear-elastic system, with reduced stiffness and increased viscous damping to account for nonlinear effects. The procedure implies a relationship between the inelastic excursion, expressed in terms of ductility μ, and an equivalent viscous damping ratio ξ_{eff}, which is used to adjust the initial elastic demand spectrum. In NPR 9998-18 [18], ξ_{eff} is first related to μ; then, a spectral reduction factor η is calculated, as follows Equations (5)–(7):

$$\xi_{eff} = \xi_{hyst} + \xi_{soil} + 0.05 \leq 0.40 \tag{5}$$

$$\xi_{hyst} = 0.42 \left(1 - \frac{0.9}{\sqrt{\mu}} - 0.1\sqrt{\mu}\right) \leq 0.15 \tag{6}$$

$$\eta = \sqrt{\frac{0.07}{0.02 + \xi_{eff}}} \geq 0.55 \tag{7}$$

In this study, the effect of soil–structure interaction (ξ_{soil}) on the equivalent viscous damping is ignored. In fact, NPR 9998-18 [18] allows ignoring ξ_{soil} for buildings up to two stories, which include most masonry structures. In any case, imposing $\eta \geq 0.55$ limits the effectiveness of additional damping sources for $\mu \geq 4.3$. The method assumes that the period of the equivalent linear system T_{eff} corresponds to the secant stiffness at the maximum displacement; for elastoplastic systems without hardening, this is Equation (8):

$$T_{eff} = T\sqrt{\mu} \tag{8}$$

Since μ is the unknown of the problem, the solution requires iterations that end when the spectral displacement demand at $T_{eff}(\mu)$, obtained from the elastic spectrum reduced by $\eta(\mu)$, is equal to $\mu \cdot d_y$. Graphically, the seismic demand on the nonlinear oscillator results from the intersection of its AD capacity curve with the elastic response spectrum reduced by $\eta(\mu)$ to account for hysteretic energy dissipation.

Figure 3 compares the displacement ductility demands resulting from the CSM procedure by NPR 9998-18 [18] with the ones obtained from NLTHA of Suite 1-A and Suite 1-B. The NPR method results in a significant overestimation of the demand, as the median line falls below the bisector, especially for large μ, while the percentiles are scattered away. Similar trends are obtained using both ground motion databases.

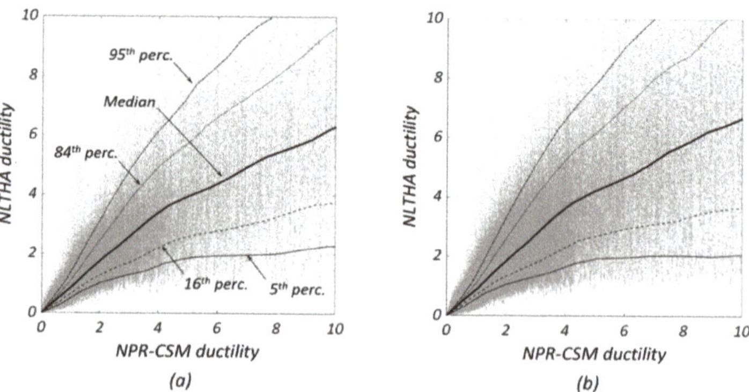

Figure 3. Comparison between ductility demands from the capacity spectrum method (CSM) by NPR 9998-18 and from NLTHA: (**a**) Suite 1-A, and (**b**) Suite 1-B.

Figure 4 compares the median $C_R - R - T$ and $\mu_R - R - T$ spectra calculated with the CSM by NPR 9998-18 [18] (solid lines) with those derived from NLTHA (dashed lines) for analysis Suite 2-A. The curves are limited to periods T between 0.1 and 0.5 s and strength ratios R between 1.0 and 5.0. One can observe that, generally, the inelastic displacement ratios C_R from NLTHA tend to infinity as T approaches zero, and approximate 1.0 as T goes to infinity. Consistently, the ductility demand μ_R from NLTHA approaches the R factor for long periods according to Equation (4).

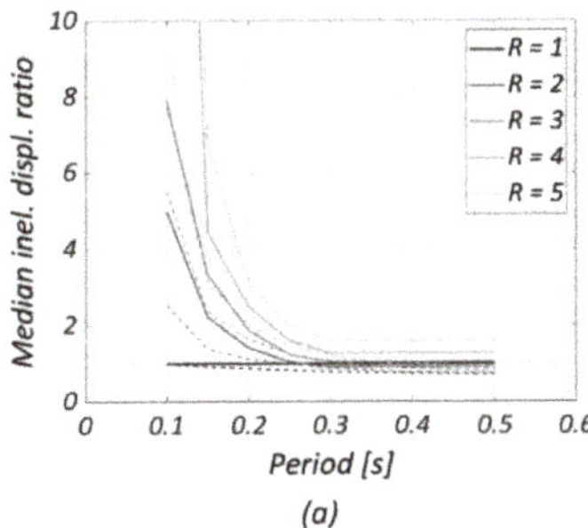
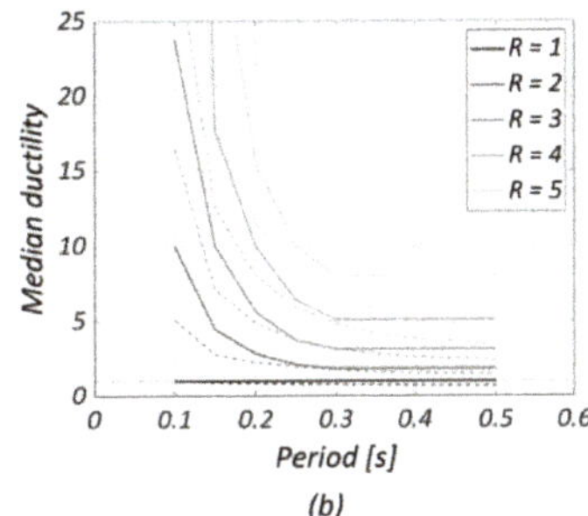

Figure 4. Comparison between the predictions by the CSM of NPR 9998-18 (solid lines) and the results from NLTHA of Suite 2-A (dashed lines): (**a**) median $C_R - R - T$ curves, and (**b**) median $\mu_R - R - T$ curves.

The CSM by NPR 9998-18 [18] overestimates systematically median C_R and μ_R for all R factors across all periods T, when compared to the NLTHA results. Specifically, for periods approaching zero (i.e., very stiff structures), the estimated displacement demands tend to infinity with a faster rate compared to the NLTHA results. Instead, for periods $T > 0.3$ s (i.e., more flexible structures), predicted C_R and μ_R become constant but higher than 1.0 and R, respectively.

3.2. Modified Capacity Spectrum Method (FEMA-440)

An improved version of the CSM has been proposed in the FEMA-440 guidelines [22], where the equivalent linear system is defined by an optimal effective period T_{eff} and effective viscous damping ζ_{eff}. Similar to the NPR 9998-18 [18] procedure, T_{eff} and ζ_{eff} depend on the unknown ductility μ: consequently, an iterative process is required to calculate the displacement demand. For SDOF oscillators that remain elastic (i.e., with $R \leq 1$), simply $d_{max} = d_e$. For oscillators undergoing inelastic deformations (i.e., with $R > 1$) with degrading stiffness and no hardening, compatible with masonry behavior, the spectral reduction parameters are computed as Equations (9) and (10):

$$\zeta_{eff} = \begin{cases} 5.1(\mu - 1)^2 - 1.1(\mu - 1)^3 + 5, & 1.0 < \mu < 4.0 \\ 12 + 1.4(\mu - 1) + 5, & 4.0 \leq \mu \leq 6.5 \\ 20\left\{\dfrac{0.62(\mu-1)-1}{[0.62(\mu-1)]^2}\right\}\left(\dfrac{T_{eff}}{T}\right)^2 + 5, & \mu > 6.5 \end{cases} \tag{9}$$

$$\eta = 0.25\left(5.6 - \ln \zeta_{eff}\right) \tag{10}$$

There is no lower bound limit applied to the reduction factor of Equation (10), as opposed to Equation (7). Unlike the CSM by NPR 9998-18 [18], this method does not require the intersection between the capacity curve and demand spectra. The resulting effective period for stiffness-degrading, not hardening systems, is given by Equation (11):

$$T_{eff} = \begin{cases} \left[0.17(\mu - 1)^2 - 0.032(\mu - 1)^3 + 1\right]T, & 1.0 < \mu < 4.0 \\ [0.10 + 0.19(\mu - 1) + 1]T, & 4.0 \leq \mu \leq 6.5 \\ \left\{0.85\left[\sqrt{(\mu - 1)} - 1\right] + 1\right\}T, & \mu > 6.5 \end{cases} \tag{11}$$

Figure 5 compares the ductility demands resulting from the CSM procedure of FEMA 440 [22] with the ones obtained from NLTHA of Suite 1-A and Suite 1-B. Despite some accuracy improvements compared to the NPR 9998-18 [18] formulation, especially for larger ductility, this approach is still affected by overestimation and high scatter issues. Similar trends are observed for both earthquake record databases.

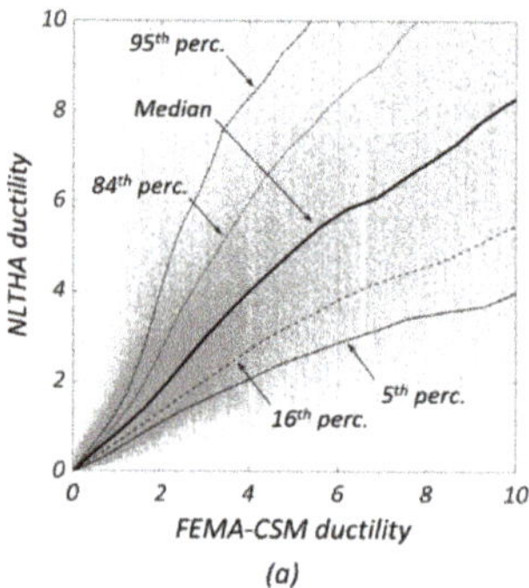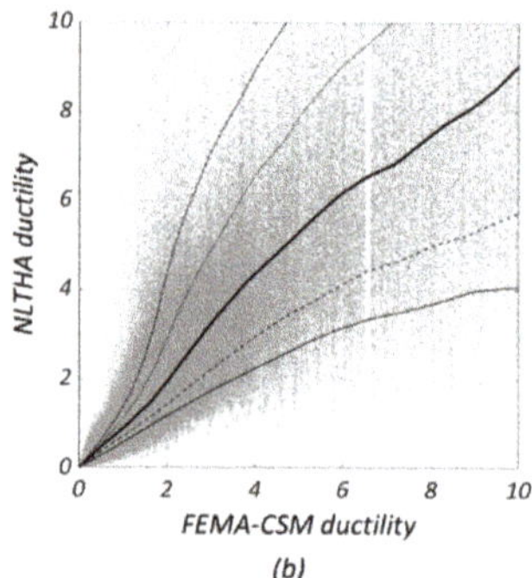

Figure 5. Comparison between ductility demands from the CSM by FEMA 440 and from NLTHA: (**a**) Suite 1-A, and (**b**) Suite 1-B.

Figure 6 illustrates the comparison of the median $C_R - R - T$ and $\mu_R - R - T$ curves. Different from NPR 9998-18 [18], FEMA 440 [22] offers predictions that align with the NLTHA results for short periods. Deviations are noticed only for oscillators with $T \leq 0.2$ s and $R \leq 2.5$, for which the predicted median C_R and μ_R remain constant instead of going to infinity. Similar to the CSM of NPR 9998-18 [18], this method overpredicts demands for systems with periods $T > 0.3$ s, even though to a lesser extent.

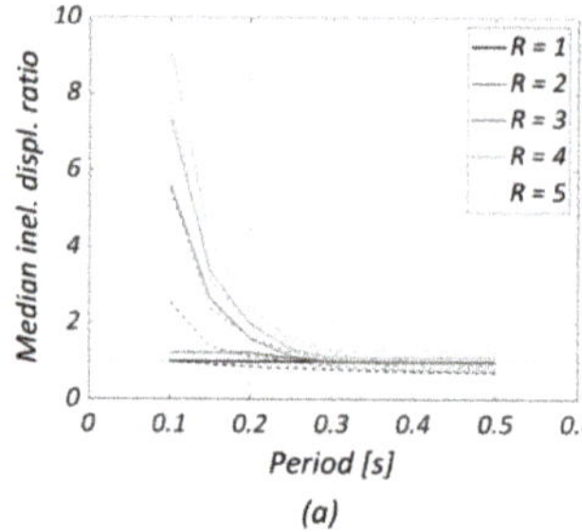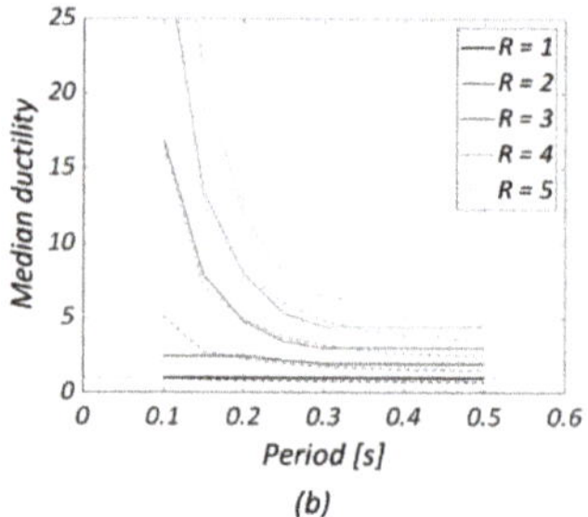

Figure 6. Comparison between the predictions by the CSM of FEMA 440 (solid lines) and the results from NLTHA of Suite 2-A (dashed lines): (**a**) median $C_R - R - T$ curves, and (**b**) median $\mu_R - R - T$ curves.

3.3. Issues with Methods Based on Equivalent Linearization

The overestimation observed for both NPR 9998-18 [18] and FEMA 440 [22] capacity spectrum methods is particularly evident (Figure 7a,b) considering only oscillators with a secant period T_{NLTHA}, defined as the effective period at the displacement demand from NLTHA (stars on Figure 7e,f), shorter than the corner period T_C of the demand spectrum. Three main causes may be responsible for the origin of the observed behavior.

The first issue affects all capacity spectrum method formulations, which define an effective period corresponding to the intersection between capacity and demand diagrams, such as the original ATC–40 [14] and the approaches followed by the Italian NTC-18 [16,17], the New Zealand guidelines [19], and the Dutch NPR 9998-18 [18]. The problem is due to the dependence of the spectral reduction factor η on the ductility demand μ, which tends to saturate as μ increases, and in some formulations is limited to a minimum value (Figure 7c,d).

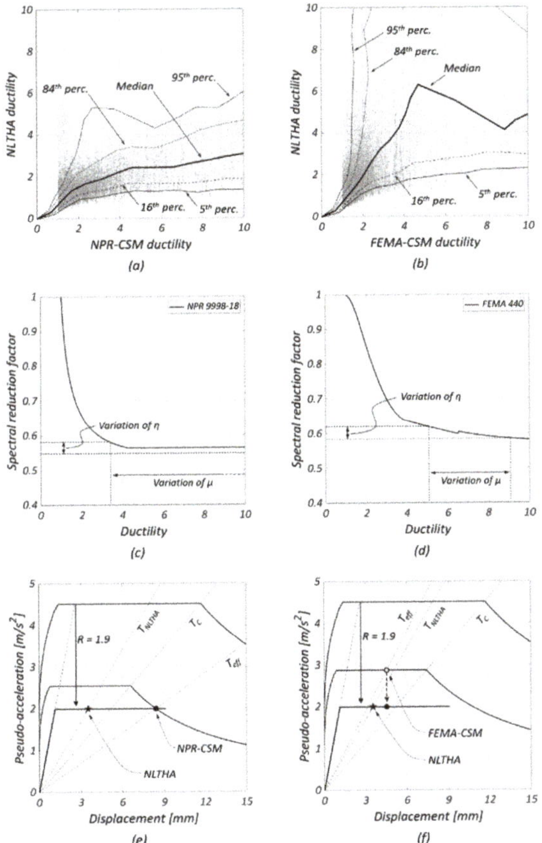

Figure 7. Comparison between ductility demand predictions from CSM and NLTHA on systems with $T_{NLTHA} < T_C$ from Suite 1-A: (**a**) NPR 9998-18, and (**b**) FEMA 440. Relationships between spectral reduction factor and ductility according to (**c**) NPR 9998-18 and (**d**) FEMA 440. Application of the CSM to a system with $T_{NLTHA} < T_C$: (**e**) NPR 9998-18, and (**f**) FEMA 440.

In particular, the NPR 9998-18 [18] method limits $\xi_{hyst} \leq 0.15$, then imposing $\eta \geq 0.56$ (excluding soil–structure interaction damping). If a system is characterized by $R > 1/\eta = 1.77$, the only possible intersection is with the constant-velocity or constant-displacement branch of the demand spectrum (solid dot on Figure 7e), resulting necessarily in $T_{eff} > T_C$ as opposed to the results of NLTHA.

Consequently, this formulation cannot be used to check limit states associated with displacement capacities corresponding to secant periods shorter than T_C: in fact, it would automatically result in a violation of such displacement capacity thresholds. This problem does not affect the improved CSM by FEMA 440 [22] because it does not seek convergence through the direct intersection between capacity and demand diagrams (hollow dot in Figure 7f).

The second source of inaccuracy is identified in the relationship between η and μ for any value of T_{NLTHA}. As η tends to level off for increasing μ, small variations in η strongly affect the predicted μ, with a bias towards overestimation of μ when η is slightly underestimated (Figure 7c,d) [60]. This problem affects both NPR 9998-18 [18] and FEMA 440 [22] capacity spectrum methods, as they seek convergence on μ. However, it is amplified in the Dutch formulation due to the lower bound imposed on η.

The third issue depends on the statistical distribution of μ values associated with a certain η, which is non-symmetrical with positive skewness for medium- and long-period systems, as inferred from the data shown by Pennucci et al. [61]. This means that, for given η, values of μ exceeding the median ductility will more likely result in larger errors than values falling below it. This happens with both iterative formulations.

Differently from NPR 9998-18 [18], the CSM formulation by FEMA 440 [22] significantly underestimates displacement demands in the low-ductility range (Figure 7b). This problem can be explained by looking at the $C_R - R - T$ and $\mu_R - R - T$ curves of Figure 6: for any system with period $T \leq 0.2$ s and strength ratio $R \leq 2.5$ the method predicts constant ductility demands, below the values obtained from NLTHA.

4. Established Formulations Based on Inelastic Response Spectra

4.1. N2 Method (Eurocode 8 and NTC-18)

The current NSP formulations by Eurocode 8 [35] and by the Italian building code NTC-18 [16,17] descend from the N2 method [31–34], which relates maximum inelastic and elastic displacement demands on an SDOF oscillator, when $R > 1$, with Equation (12):

$$d_{max} = \frac{1}{R}\left[(R-1)\left(\frac{T_C}{T}\right) + 1\right]d_e \geq d_e \qquad (12)$$

while $d_{max} = d_e$ when $R \leq 1$. The lower-bound limit of Equation (12) is necessary because for $T \geq T_C$ the inelastic displacement demand should not be taken as less than the elastic one (equal displacement rule).

Figure 8 compares the ductility demands predicted by the equation with the ones obtained from the NLTHA of Suite 1-A and Suite 1-B. In both cases, the comparison reveals that the current code formulation underestimates inelastic displacement demands significantly when they exceed a ductility of 4, as indicated by the median line found above the bisector. Nevertheless, the N2 approach results in limited dispersion compared to other existing methods.

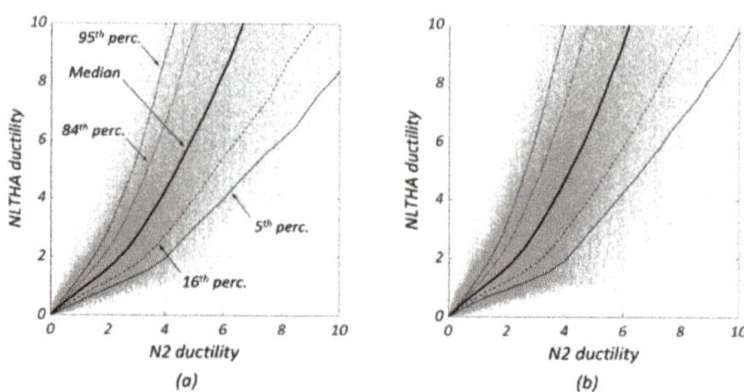

Figure 8. Comparison between ductility demands from the N2 method and from NLTHA: (**a**) Suite 1-A, and (**b**) Suite 1-B.

The tendency of the method to underestimate displacement demands is also visible in the $C_R - R - T$ and $\mu_R - R - T$ relationships in Figure 9. Differences between the predictions by the N2 equations and the results from NLTHA are more pronounced for systems of short period and low relative strength (high R factors).

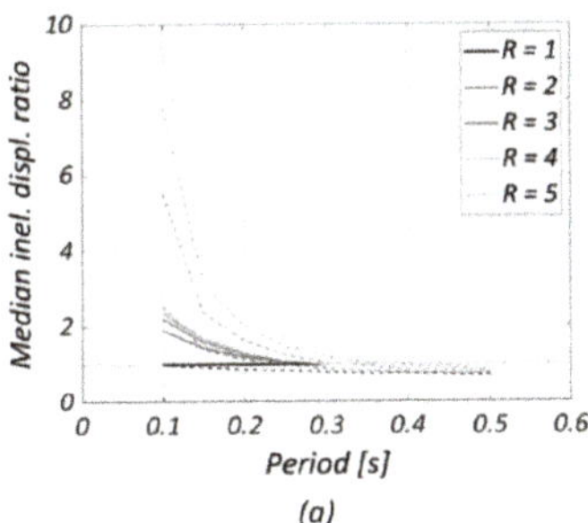
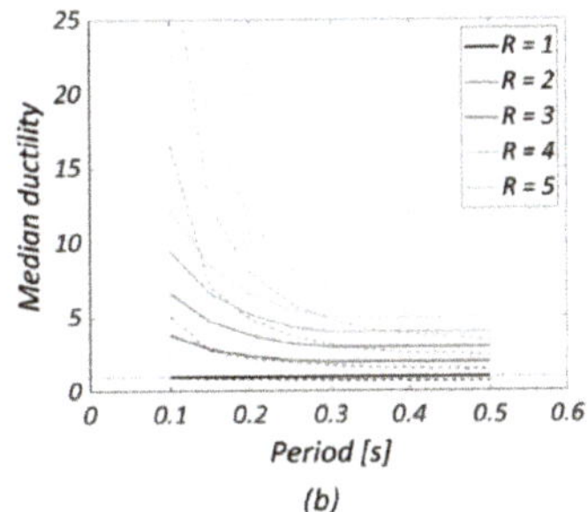

Figure 9. Comparison between the predictions by the N2 method (solid lines) and the results from NLTHA of Suite 2-A (dashed lines): (**a**) median $C_R - R - T$ curves, and (**b**) median $\mu_R - R - T$ curves.

4.2. Displacement Coefficient Method (ASCE 41-17)

The displacement coefficient method (DCM) is presented in the ASCE 41-17 code [37]. When $R > 1$, the method calculates the maximum displacement demand on nonlinear SDOF system as in Equation (13):

$$d_{max} = C_1 C_2 d_e \tag{13}$$

while $d_{max} = d_e$ when $R \leq 1$. Coefficients C_1 and C_2 are empirical modification factors given by the following Equations (14) and (15):

$$C_1 = \begin{cases} 1 + \frac{R-1}{0.04a}, & T \leq 0.2 \text{ s} \\ 1 + \frac{R-1}{aT^2}, & 0.2 \text{ s} < T \leq 1.0 \text{ s} \\ 1.0, & T > 1.0 \text{ s} \end{cases} \tag{14}$$

$$C_2 = \begin{cases} 1 + \frac{1}{800}\left(\frac{R-1}{T}\right)^2, & T \leq 0.7 \text{ s} \\ 1.0, & T > 0.7 \text{ s} \end{cases} \tag{15}$$

where a is a site-dependent parameter associated with the known site class [62] of each recording station: $a = 130$ for site classes A and B, $a = 90$ for site class C, and $a = 60$ for site classes D, E, and F. Figure 10 shows that, despite good accuracy in the median prediction, these equations result in large scatter, as indicated by the distance of the lower and upper percentile lines from the median.

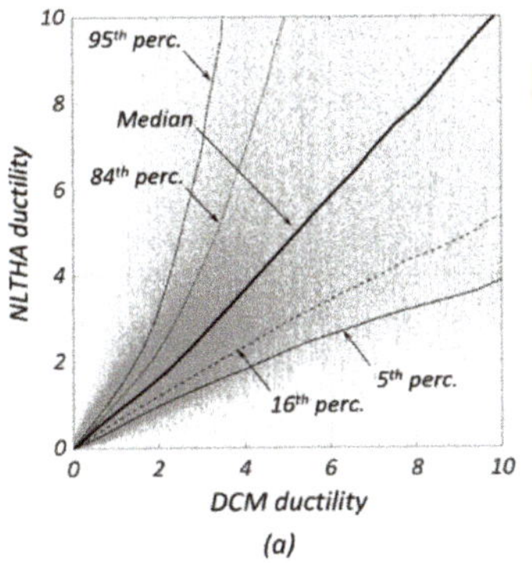

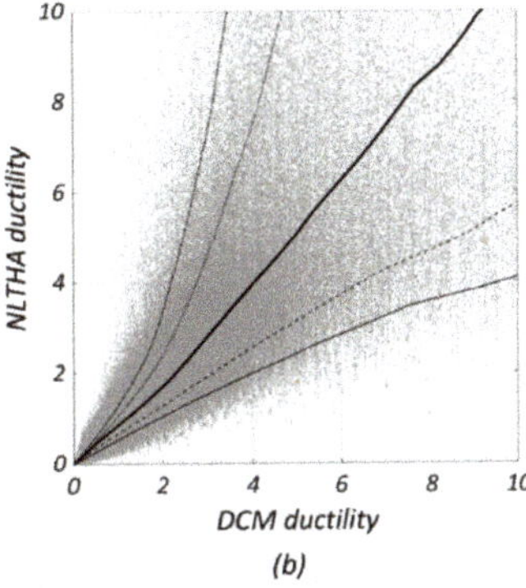

Figure 10. Comparison between ductility demands from the displacement coefficient method (DCM) by ASCE 41-17 and from NLTHA: (**a**) Suite 1-A, and (**b**) Suite 1-B.

Figure 11 illustrates the comparison between the $C_R - R - T$ and $\mu_R - R - T$ spectra by the DCM and those obtained from NLTHA results. One can notice that the method provides considerably lower median values of C_R and μ_R for systems with periods $T < 0.15$ s regardless of the strength ratio R. Instead, the method overpredicts demands for longer periods and for all strength ratios R. This confirms the overall large scatter affecting the DCM predictions, as pointed out by FEMA 440 [22] and Ruiz-García and Miranda [1]. The transition of Equation (14) at $T = 0.2$ s explains the change of slope of the curves in Figure 11: this happens because for periods $T \leq 0.2$ s coefficient C_1 remains constant with the period.

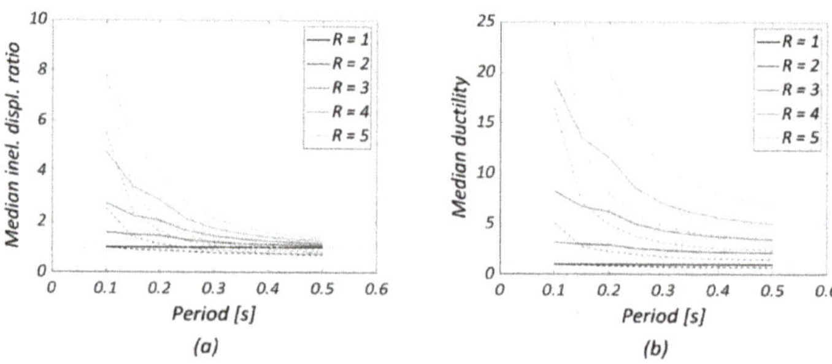

Figure 11. Comparison between the predictions by the DCM of ASCE 41-17 (solid lines) and the results from NLTHA of Suite 2-A (dashed lines): (**a**) median $C_R - R - T$ curves, and (**b**) median $\mu_R - R - T$ curves.

4.3. Issues with Methods Based on Inelastic Response Spectra

Figure 8 through Figure 11 show that both the N2 method of Eurocode 8 and NTC-18 [16,17,35] and the DCM of ASCE 41-17 [37] can result in significant errors when applied to short-period masonry-type oscillators. Both formulations were originally derived for steel and concrete frame structures, characterized by longer fundamental periods and higher hysteretic dissipation capacity. In fact, Guerrini et al. [44] demonstrated that the inefficiency of the N2 equation is more evident for oscillators with $T < 0.5$ s, and that errors are more pronounced for systems with low hysteretic dissipation. They suggested that the accuracy of the method would benefit from explicit consideration of the hysteretic behavior, especially at short periods, where inelastic displacement demand amplification is more sensitive to this parameter.

Another source of inaccuracy for the N2 formulation [16,17,35] lies in the fact that it belongs to the so-called "indirect methods" [63], where the $\mu_R - R - T$ relationship (i.e., for given R) results from the inversion of the calibrated $R_\mu - \mu - T$ equation (i.e., for a given μ) by Vidic et al. [25]. This process can introduce systematic errors that tend to underestimate the maximum inelastic displacement demands, with a greater error for increasing ductility [1].

The N2 method and the DCM correctly predict ductility demands that tend towards infinity for very short periods and approach the R value for long periods. Nevertheless, both methods display problems with the rate of convergence to these two limits. In particular, the DCM approaches the two boundaries with a lower rate than the one obtained from NLTHA, while the N2 method significantly underestimates the displacement amplification at short periods.

5. Proposed Formulations

5.1. Optimal Stiffness Method

This section presents an improved equivalent linearization procedure, named optimal stiffness method (OSM), as it defines an optimal stiffness T_{opt} and the corresponding equivalent viscous damping ratio ζ_{opt} in terms of the idealized elastic period T and the strength ratio R of the oscillator. Adapting the equation form discussed by Lin and Miranda [64], the following relationships were calibrated based on the results from NLTHA on masonry-type oscillators. For oscillators that remain elastic (i.e., with $R \leq 1$), simply $d_{max} = d_e$. The optimal linear parameters for SDOF oscillators with $R > 1$ are given by Equations (16) and (17):

$$\zeta_{opt} = 0.05 + n_{hyst}(R - 1)^2 \tag{16}$$

$$T_{opt} = T + m_{hyst}(R - 1)^2 \tag{17}$$

The coefficients m_{hyst} and n_{hyst} in these equations were calibrated with dynamic responses of SDOF oscillators from Suite 1-A by the orthogonal regression algorithm mentioned earlier. The calibration of the parameters was performed separately for systems of low ($13\% \leq \zeta_{hyst} < 15\%$), intermediate ($15\% \leq \zeta_{hyst} \leq 18\%$), and high ($18\% < \zeta_{hyst} \leq 20\%$) hysteretic dissipation. The resulting values are summarized in Table 1.

Table 1. Calibrated parameters for the proposed OSM equations.

Hysteresis Case	m_{hyst} (s)	n_{hyst} (-)
$13\% \leq \zeta_{hyst} < 15\%$	0.067	0.040
$15\% \leq \zeta_{hyst} \leq 18\%$	0.065	0.059
$18\% < \zeta_{hyst} \leq 20\%$	0.061	0.077

The relationship between spectral reduction factor η and equivalent viscous damping ratio ζ_{opt} was taken by Eurocode 8, without lower-bound limitations, according to Equation (18):

$$\eta = \sqrt{\frac{0.10}{0.05 + \zeta_{opt}}} \tag{18}$$

Owing to the dependence of the equivalent system properties on the known strength ratio R, rather than on the unknown ductility μ, the method offers the advantage of a direct non-iterative solution. In this way, the method overcomes the problems associated with the relationship between η and μ of NPR 9998-18 [18] and FEMA 440 [22]. The proposed equation generally provides an accurate estimate of the ductility demand, as demonstrated by the median line approaching the bisector in Figure 12. Moreover, predictions are characterized by low dispersion for both Suite 1-A and 1-B analyses.

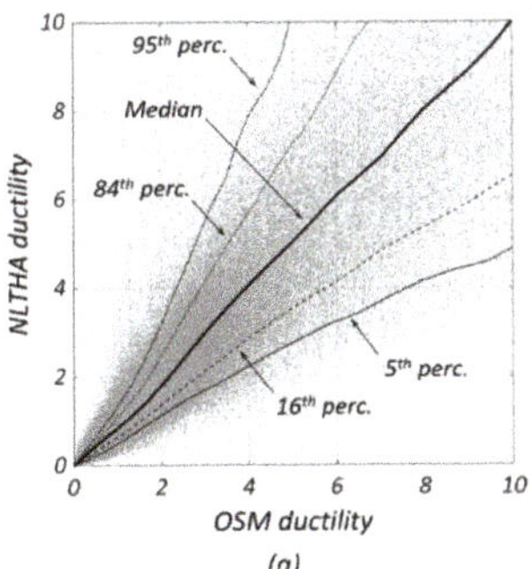

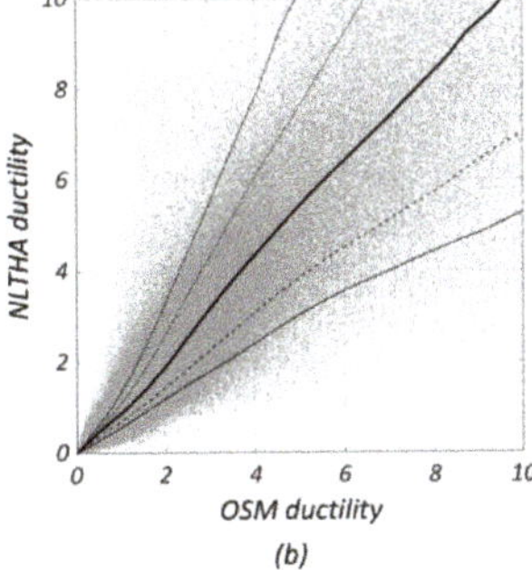

Figure 12. Comparison between ductility demands from the proposed optimal stiffness method (OSM) and from NLTHA: (**a**) Suite 1-A, and (**b**) Suite 1-B.

Good agreement was also observed between the predicted and the NLTHA-derived median $C_R - R - T$ and $\mu_R - R - T$ spectra, as shown in Figure 13. It is noteworthy that, for periods $T > 0.3$ s, the predicted curves approach those obtained by NLTHA, as opposed to the equivalent linearization procedures by NPR 9998-18 [18] and FEMA 440 [22], which deviate significantly.

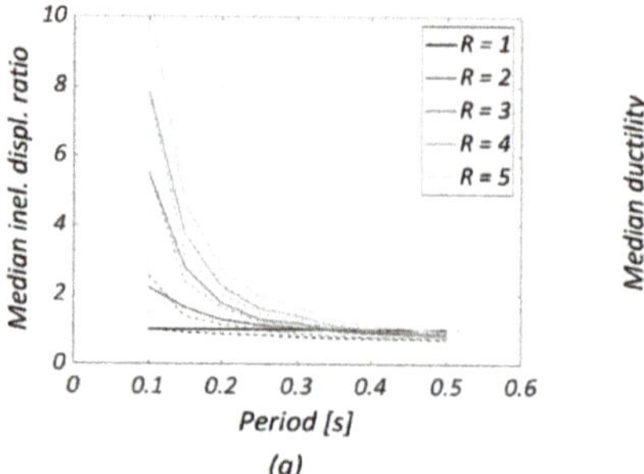
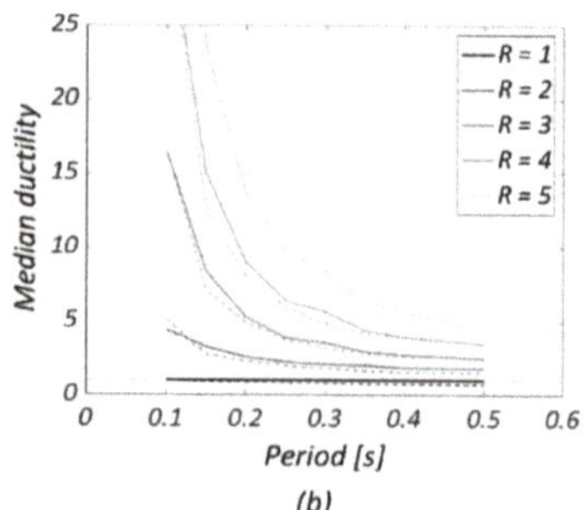

Figure 13. Comparison between the predictions by the proposed OSM (solid lines) and the results from NLTHA of Suite 2-A (dashed lines): (**a**) median $C_R - R - T$ curves, and (**b**) median $\mu_R - R - T$ curves.

5.2. Modified-N2 Method

Motivated by the limitations of the N2 method of Eurocode 8 and NTC-18 [16,17,35], when applied to masonry-like systems, a modified-N2 (MN2) method has been formulated and calibrated against NLTHA results to relate inelastic and elastic seismic displacement demands for this kind of oscillators [44]. For SDOF systems with $R > 1$, the formulation results in Equation (19):

$$d_{max} = \frac{1}{R}\left[\frac{(R-1)^{2.1}}{\left(\frac{T}{T_{hyst}} + a_{hyst}\right)\left(\frac{T}{T_C}\right)^{2.3}} + R\right] d_e \tag{19}$$

while $d_{max} = d_e$ when $R \leq 1$. Unlike the original N2 formulation, this equation tends asymptotically to the elastic displacement demand as T approaches infinity, without the need for a lower-bound limit. The coefficients T_{hyst} and a_{hyst} were calibrated with the results from NLTHA on SDOF oscillators (Suite 1-A) using the same orthogonal regression algorithm discussed above [44]. Similar to the OSM equations, Equation (19) was calibrated separately for three ranges of hysteretic dissipation capacity; the resulting values for the parameters are listed in Table 2.

Table 2. Calibrated parameters for the proposed modified-N2 (MN2) equation [44].

Hysteresis Case	a_{hyst} (-)	T_{hyst} (s)
$13\% \leq \xi_{hyst} < 15\%$	0.7	0.055
$15\% \leq \xi_{hyst} \leq 18\%$	0.2	0.030
$18\% < \xi_{hyst} \leq 20\%$	0.0	0.022

The ductility demands predicted by the MN2 method are shown in Figure 14. The equation exhibits significantly improved accuracy and reduced dispersion compared to the original N2 method, considering both Site 1-A and 1-B analysis results. The improved performance of this method for periods ranging between 0.1 and 0.5 s is also reflected in the predicted $C_R - R - T$ and $\mu_R - R - T$ spectra, which accurately approximate the NLTHA-derived spectra (Figure 15).

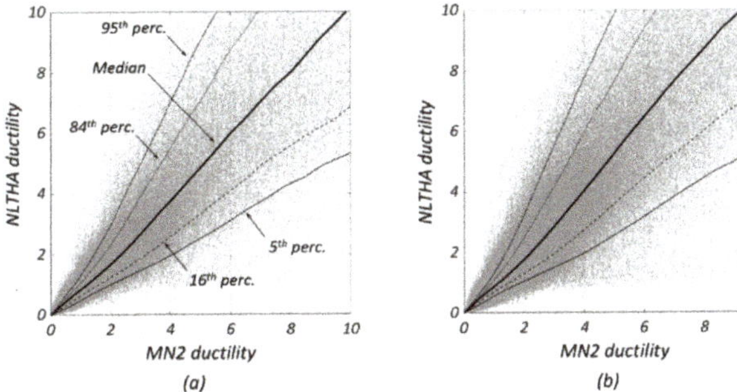

Figure 14. Comparison between ductility demands from the MN2 method and from NLTHA: (**a**) Suite 1-A, and (**b**) Suite 1-B.

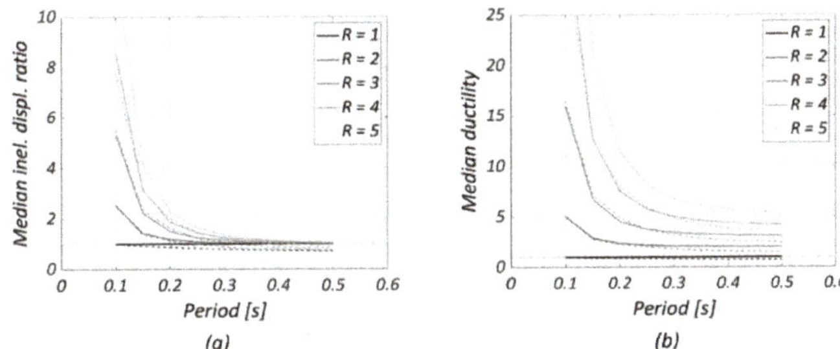

Figure 15. Comparison between the predictions by the MN2 method (solid lines) and the results from NLTHA of Suite 2-A (dashed lines): (**a**) median $C_R - R - T$ curves, and (**b**) median $\mu_R - R - T$ curves.

5.3. Accuracy and Scatter of OSM and MN2 Methods

Figures 12 and 14 show that median and percentile distances from the bisector can be approximated well by straight lines on the plane of ductility demands predicted by the proposed equations versus the NLTHA. This means that the ratio of the percentile to the central ductility value of each bin remains nearly constant over the considered ductility range (up to 10). Percentile factors γ_p were evaluated for different percentiles p over the entire set of oscillators and for individual periods T, and were taken to be equal to the slope of the corresponding percentile line. The inelastic displacement demand $d_{max,p}$, corresponding to percentile p, can be expressed as Equation (20):

$$d_{max,p} = \gamma_p d_{max} \tag{20}$$

Tables 3 and 4 summarize the values of γ_p for the 50th (median), 70th, 84th and 95th percentiles for both proposed methods, showing that they provide accurate estimates of the median inelastic displacement demand ($\gamma_{50} \approx 1.0$). It can also be observed that the OSM is somehow less precise than the MN2 method in its predictions because it is characterized by a larger scatter: this is reflected by larger values of γ_p for higher percentiles.

Table 3. Percentile factors of displacement predicted by the proposed OSM for various percentiles.

Percentile	Idealized Elastic Period (s)					Overall
	0.1	0.2	0.3	0.4	0.5	
50th	1.1	1.0	1.0	1.0	1.1	1.0
70th	1.5	1.2	1.2	1.2	1.4	1.2
84th	2.3	1.5	1.4	1.5	1.7	1.5
95th	6.4	2.1	1.7	1.9	2.4	2.3

Table 4. Percentile factors of displacement demands predicted by the MN2 method for various percentiles.

Percentile	Idealized Elastic Period (s)					Overall
	0.1	0.2	0.3	0.4	0.5	
50th	0.9	1.1	1.0	0.9	0.8	1.0
70th	1.2	1.3	1.2	1.1	1.0	1.2
84th	1.4	1.5	1.4	1.2	1.1	1.4
95th	1.8	1.9	1.7	1.5	1.4	1.7

Lacking detailed information on the amount of hysteretic dissipation for a specific masonry structure, the parameters of the intermediate case ($15\% \leq \zeta_{hyst} \leq 18\%$) could be generally adopted with both methods. This simplification would have a minimal effect on the accuracy of the results. In most cases, only the second decimal figure would change in factors γ_p for higher percentiles. No appreciable effect of the focal distance was observed on the accuracy and dispersion of the predictions, when the ground motion records were treated in two groups considering distance thresholds of 5, 10, 15 or 20 km.

Finally, it has been observed that for oscillators with a secant period T_{NLTHA} from NLTHA shorter than the corner period T_C of the demand spectrum, both methods provide improved estimations of the median inelastic displacement demand compared to established formulations. The MN2 method appears more accurate than the OSM in these situations, despite a slight tendency towards overestimations. On the other hand, the OSM results in a more pronounced underestimation of the median displacement demand with somewhat larger dispersion for these oscillators.

6. Discussion and Conclusions

This paper discussed the implementation of six methods for estimating inelastic seismic displacement demands to be used in nonlinear static analysis procedures for the assessment of existing masonry structures. The predictive accuracy of the methods was assessed based on the results from nonlinear time-history analyses (NLTHA) on single-degree-of-freedom (SDOF) oscillators with hysteretic relationships typical of masonry buildings and periods between 0.05 and 0.5 s. Two independent ground motion databases, as well as two sets of oscillators, were used for this scope, resulting in a total number of 3,434,900 analyses.

The study first demonstrated some limitations of two established iterative methods based on the equivalent linearization concept: the capacity spectrum method recommended by the Dutch code NPR 9998-18, and its modified version outlined in the American guidelines FEMA 440. Both methods provide general overprediction of inelastic displacement demands because of issues in the relationships between ductility demand, effective viscous damping, and spectral reduction factors. Overestimations are more pronounced for methods seeking convergence through direct intersection between capacity and demand diagrams, like in NPR 9998-18, especially if the NLTHA results in a maximum displacement corresponding to an effective period shorter than the corner period of the demand spectrum. This problem can be found in all similar methods, such as the ones of the Italian NTC-18 and of the New Zealand guidelines.

Two established formulations based on inelastic displacement spectra, oscillator idealized elastic period, and strength ratio were also evaluated. The N2 method of Eurocode 8 and Italian building code NTC-18 proved to underestimate the displacement demand in spite of a limited scatter. In contrast, the displacement coefficient method of ASCE 41-17 was shown to better predict the median ductility but with excessively high dispersion. Both methods have problems with the rate of convergence to infinite and elastic displacement demands as the elastic period approaches zero and infinity, respectively. These issues are mainly due to lack of calibration with the dynamic responses of masonry structures.

An alternative approach based on the equivalent linearization concept, termed optimal stiffness method (OSM), was then proposed. This procedure defines an optimal stiffness and equivalent viscous damping in terms of idealized elastic period and strength ratio, which are both known in the assessment of an existing building. This formulation does not need iterations on the ductility demand and overcomes the problems with the relationships between ductility demand, effective viscous damping, and spectral reduction factors. A modified version of the N2 method (MN2), directly calibrated with the dynamic response of short-period oscillators, was finally presented.

Both proposed formulations predict the median ductility demand accurately while limiting the dispersion of the results. However, the MN2 method is somehow more precise than the OSM because it is characterized by a smaller scatter. The MN2 method appears more accurate than the OSM when the NLTHA results in a maximum displacement corresponding to an effective period shorter than the corner period of the demand spectrum, despite a slight tendency towards overestimations. On the other hand, in these cases, the OSM results in a more pronounced underestimation of the median displacement demand with somewhat larger dispersion for these oscillators. Neither of the proposed methods appeared sensitive to the focal distance of the earthquake records. Due to the high rate of divergence towards the infinity of the ductility demand from both NLTHA and predictions, it was not possible to obtain meaningful results for oscillators with periods shorter than 0.1 s.

The general equations of the OSM and MN2 methods can be calibrated with the NLTHA responses of SDOF oscillators with other hysteretic rules. Different sets of parameters can then be derived, allowing the application of these formulations to other structural systems.

Author Contributions: G.G.: conceptualization, investigation, methodology, software, formal analysis, data curation, writing—original draft preparation, visualization; S.K.: investigation, software, formal analysis, data curation, validation, writing—original draft preparation, visualization; S.B.: investigation, validation, writing—review and editing; F.G.: conceptualization, methodology, writing—review and editing, supervision, project administration, funding acquisition; A.P.: resources, supervision, project administration, funding acquisition. All authors have read and agreed to the published version of the manuscript.

Funding: This research was funded by ReLUIS-DPC project 2019-2021 WP10 "Contributi normativi relativi a costruzioni esistenti in muratura", funded by the Italian Department of Civil Protection and the UNIPV-TU Delft project "Development and verification of software for the seismic assessment of masonry buildings according to Annex G of NPR9998:2018", funded by Nationaal Coördinator Groningen (NCG), Centrum Veilig Wonen (CVW), and Econstruct BV.

Data Availability Statement: Data can be made available upon request to the corresponding author.

Acknowledgments: The authors would like to express their gratitude to G. Magenes for his valuable comments and to V. Bonura, E. Bossi, and C. Rossi who contributed to the elaboration of the SDOF analysis results. The authors are also grateful to H. Crowley and R. Pinho for providing part of the seismic input for nonlinear time-history analyses.

Conflicts of Interest: The authors declare no conflict of interest. The funders had no role in the design of the study; in the collection, analyses, or interpretation of data; in the writing of the manuscript, or in the decision to publish the results.

References

1. Ruiz-García, J.; Miranda, E. Inelastic displacement ratios for evaluation of existing structures. *Earthq. Eng. Struct. Dyn.* **2003**, *32*, 1237–1258. [CrossRef]
2. Saiidi, M.; Sozen, M.A. Simple Nonlinear Seismic Analysis of R/C Structures. *J. Struct. Div.* **1981**, *107*, 937–953. [CrossRef]
3. Miranda, E. Approximate Seismic Lateral Deformation Demands in Multistory Buildings. *J. Struct. Eng.* **1999**, *125*, 417–425. [CrossRef]
4. Chopra, A.K.; Goel, R.K. Evaluation of NSP to Estimate Seismic Deformation: SDF Systems. *J. Struct. Eng.* **2000**, *126*, 482–490. [CrossRef]

5. Chopra, A.K.; Goel, R.K. A modal pushover analysis procedure for estimating seismic demands for buildings. *Earthq. Eng. Struct. Dyn.* **2002**, *31*, 561–582. [CrossRef]

6. Antoniou, S.; Pinho, R. Advantages and limitations of adaptive and non-adaptive force-based pushover procedures. *J. Earthq. Eng.* **2004**, *8*, 497–522. [CrossRef]

7. Antoniou, S.; Pinho, R. Development and verification of a displacement-based adaptive pushover procedure. *J. Earthq. Eng.* **2004**, *8*, 643–661. [CrossRef]

8. Brozovič, M.; Dolšek, M. Envelope-based pushover analysis procedure for the approximate seismic response analysis of buildings. *Earthq. Eng. Struct. Dyn.* **2013**, *43*, 77–96. [CrossRef]

9. Jacobsen, L.S. Steady Forced Vibrations as Influenced by Damping. *Trans. ASME-APM* **1930**, *52*, 169–181.

10. Hudson, D.E. Equivalent Viscous Friction for Hysteretic Systems with Earthquake-Like Excitations. In Proceedings of the 3rd World Conference on Earthquake Engineering, New Zealand National Committee on Earthquake Engineering, Wellington, New Zealand, 22 January–1 February 1965; Volume 2, pp. 185–202.

11. Jennings, P.C. Equivalent Viscous Damping for Yielding Structures. *J. Eng. Mech. Div.* **1968**, *94*, 103–116. [CrossRef]

12. Shibata, A.; Sözen, M.A. Substitute-Structure Method for Seismic Design in R/C. *J. Struct. Div. ASCE* **1976**, *102*, 1–18. [CrossRef]

13. Freeman, S.A.; Nicoletti, J.P.; Tyrrell, J.V. Evaluation of Existing Buildings for Seismic Risk—A Case Study of Puget Sound Naval Shipyard, Bremerton, Washington. In Proceedings of the 1st US National Conference on Earthquake Engineering, Oakridge, CA, USA, 18–20 June 1975; pp. 113–122.

14. Applied Technology Council (ATC). *Seismic Evaluation and Retrofit of Concrete Buildings, ATC–40*; Applied Technology Council: Redwood City, CA, USA, 1996.

15. Federal Emergency Management Agency (FEMA). *NEHRP Commentary on the Guidelines for the Seismic Rehabilitation of Buildings, FEMA Publication 274*; Federal Emergency Management Agency: Washington, DC, USA, 1997.

16. Ministry of Infrastructures and Transport (MIT). *Norme Tecniche per le Costruzioni (NTC-18), DM 17/01/2018*; Ministry of Infrastructures and Transport: Rome, Italy, 2018. (In Italian)

17. Ministry of Infrastructures and Transport (MIT). *Istruzioni per l'Applicazione dell'Aggiornamento delle "Norme Tecniche per le Costruzioni (NTC-18)", Circ. 7 of 21/01/2019*; Ministry of Infrastructures and Transport: Rome, Italy, 2019. (In Italian)

18. Netherlands Standardization Institute (NEN). *Assessment of Structural Safety of Buildings in Case of Erection, Reconstruction, and Disapproval—Induced Earthquakes—Basis of Design, Actions and Resistances, NPR 9998*; Netherlands Standardization Institute: Delft, The Netherlands, 2018. (In Dutch)

19. New Zealand Society for Earthquake Engineering (NZSEE). *The Seismic Assessment of Existing Buildings, Part C: Detailed Seismic Assessment*; MBIE, EQC, SESOC, NZSEE, NZGS: Wellington, New Zealand, 2016.

20. Dwairi, H.M.; Kowalsky, M.J.; Nau, J.M. Equivalent Damping in Support of Direct Displacement-Based Design. *J. Earthq. Eng.* **2007**, *11*, 512–530. [CrossRef]

21. Priestley, M.J.N.; Calvi, G.M.; Kowalsky, M.J. *Displacement-Based Seismic Design of Structures*; IUSS Press: Pavia, Italy, 2007.

22. Federal Emergency Management Agency (FEMA). *Improvement of Nonlinear Static Seismic Analysis Procedures, FEMA Publication 440*; Federal Emergency Management Agency: Washington, DC, USA, 2005.

23. Veletsos, A.S.; Newmark, N.M. Effect of Inelastic Behavior on the Response of Simple Systems to Earthquake Motions. In Proceedings of the 2nd World Conference on Earthquake Engineering, Tokyo, Japan, 11–18 July 1960; Science Council of Japan: Tokyo, Japan, 1960; Volume 2, pp. 895–912.

24. Veletsos, A.S.; Newmark, N.M.; Chelapati, C.V. Deformation Spectra for Elastic and Elastoplastic Systems Subjected to Ground Shock and Earthquake Motions. In Proceedings of the 3rd World Conference on Earthquake Engineering, Wellington, New Zealand, 22 January–1 February 1965; Volume 2, pp. 663–682.

25. Newmark, N.M.; Hall, W.J. *Earthquake Spectra and Design*; Earthquake Engineering Research Institute: Berkeley, CA, USA, 1982.

26. Riddell, R.; Hidalgo, P.; Cruz, E. Response Modification Factors for Earthquake Resistant Design of Short Period Buildings. *Earthq. Spectra* **1989**, *5*, 571–590. [CrossRef]

27. Rahnama, M.; Krawinkler, H. *Effects of Soft Soil and Hysteresis Model on Seismic Demands*; Report No., 108; The John A. Blume Earthquake Engineering Center, Stanford University: Stanford, CA, USA, 1993.

28. Miranda, E. Evaluation of Site-Dependent Inelastic Seismic Design Spectra. *J. Struct. Eng.* **1993**, *119*, 1319–1338. [CrossRef]

29. Miranda, E. Inelastic Displacement Ratios for Structures on Firm Sites. *J. Struct. Eng.* **2000**, *126*, 1150–1159. [CrossRef]

30. Ramirez, O.M.; Constantinou, M.C.; Whittaker, A.S.; Kircher, C.A.; Chrysostomou, C.Z. Elastic and Inelastic Seismic Response of Buildings with Damping Systems. *Earthq. Spectra* **2002**, *18*, 531–547. [CrossRef]

31. Fajfar, P.; Fischinger, M. N2–A Method for Nonlinear Seismic Analysis of Regular Buildings. In Proceedings of the 9th World Conference on Earthquake Engineering, Tokyo, Japan, 2–6 August 1988; Science Council of Japan: Tokyo, Japan, 1988; Volume 5, pp. 111–116.

32. Vidic, T.; Fajfar, P.; Fischinger, M. Consistent inelastic design spectra: Strength and displacement. *Earthq. Eng. Struct. Dyn.* **1994**, *23*, 507–521. [CrossRef]

33. Fajfar, P. Capacity Spectrum Method Based on Inelastic Demand Spectra. *Earthq. Eng. Struct. Dyn.* **1999**, *28*, 979–993. [CrossRef]

34. Fajfar, P. A Nonlinear Analysis Method for Performance-Based Seismic Design. *Earthq. Spectra* **2000**, *16*, 573–592. [CrossRef]

35. European Committee for Standardization (CEN). *Eurocode 8: Design of Structures for Earthquake Resistance—Part 1: General Rules, Seismic Actions, and Rules for Buildings, EN 1998-1*; European Committee for Standardization: Brussels, Belgium, 2004.

36. Federal Emergency Management Agency (FEMA). *NEHRP Guidelines for the Seismic Rehabilitation of Buildings, FEMA Publication 273*; Federal Emergency Management Agency: Washington, DC, USA, 1997.

37. American Society of Civil Engineers (ASCE). *Seismic Evaluation and Retrofit of Existing Buildings*; ASCE/SEI American Society of Civil Engineers: Reston, VG, USA, 2017.

38. Bertero, V.V.; Anderson, J.C.; Krawinkler, H.; Miranda, E. *Design Guidelines for Ductility and Drift Limits*; Report No. UCB/EERC-91/15; Earthquake Engineering Research Center, University of California at Berkeley: Berkeley, CA, USA, 1991.

39. Krawinkler, H. New Trends in Seismic Design Methodology. In Proceedings of the 10th European Conference on Earthquake Engineering, Vienna, Austria, 28 August–2 September 1994; Balkema: Rotterdam, The Netherlands, 1995; Volume 2, pp. 821–830.

40. Chopra, A.K.; Goel, R.K. Direct Displacement-Based Design: Use of Inelastic vs. Elastic Design Spectra. *Earthq. Spectra* **2001**, *17*, 47–64. [CrossRef]

41. Martinelli, E.; Faella, C. Nonlinear static analyses based on either inelastic or elastic spectra with equivalent viscous damping: A parametric comparison. *Eng. Struct.* **2015**, *88*, 241–250. [CrossRef]

42. Amadio, C.; Rinaldin, G.; Fragiacomo, M. Investigation on the accuracy of the N2 method and the equivalent linearization procedure for different hysteretic models. *Soil Dyn. Earthq. Eng.* **2016**, *83*, 69–80. [CrossRef]

43. Michel, C.; Lestuzzi, P.; Lacave, C. Simplified non-linear seismic displacement demand prediction for low period structures. *Bull. Earthq. Eng.* **2014**, *12*, 1563–1581. [CrossRef]

44. Guerrini, G.; Graziotti, F.; Penna, A.; Magenes, G. Improved evaluation of inelastic displacement demands for short-period masonry structures. *Earthq. Eng. Struct. Dyn.* **2017**, *46*, 1411–1430. [CrossRef]

45. Diana, L.; Manno, A.; Lestuzzi, P. Seismic displacement demand prediction in non-linear domain: Optimization of the N2 method. *Earthq. Eng. Eng. Vib.* **2019**, *18*, 141–158. [CrossRef]

46. Lestuzzi, P.; Diana, L. Accuracy Assessment of Nonlinear Seismic Displacement Demand Predicted by Simplified Methods for the Plateau Range of Design Response Spectra. *Adv. Civ. Eng.* **2019**, *2019*, 1–16. [CrossRef]

47. Dolšek, M.; Fajfar, P. Simplified probabilistic seismic performance assessment of plan-asymmetric buildings. *Earthq. Eng. Struct. Dyn.* **2007**, *36*, 2021–2041. [CrossRef]

48. Diana, L.; Lestuzzi, P.; Podestà, S.; Luchini, C. Improved Urban Seismic Vulnerability Assessment Using Typological Curves and Accurate Displacement Demand Prediction. *J. Earthq. Eng.* **2019**, 1–23. [CrossRef]

49. Snoj, J.; Dolšek, M. Pushover-based seismic risk assessment and loss estimation of masonry buildings. *Earthq. Eng. Struct. Dyn.* **2020**, *49*, 567–588. [CrossRef]

50. Shishegaran, A.; Khalili, M.R.; Karami, B.; Rabczuk, T.; Shishegaran, A. Computational predictions for estimating the maximum deflection of reinforced concrete panels subjected to the blast load. *Int. J. Impact Eng.* **2020**, *139*, 103527. [CrossRef]

51. Lagomarsino, S.; Penna, A.; Galasco, A.; Cattari, S. TREMURI program: An equivalent frame model for the nonlinear seismic analysis of masonry buildings. *Eng. Struct.* **2013**, *56*, 1787–1799. [CrossRef]

52. Penna, A.; Lagomarsino, S.; Galasco, A. A nonlinear macroelement model for the seismic analysis of masonry buildings. *Earthq. Eng. Struct. Dyn.* **2014**, *43*, 159–179. [CrossRef]

53. Graziotti, F.; Penna, A.; Bossi, E.; Magenes, G. Evaluation of Displacement Demand for Unreinforced Masonry Buildings by Equivalent SDOF Systems. In Proceedings of the 9th International Conference on Structural Dynamics, Porto, Portugal, 30 June–2 July 2014; pp. 365–372.

54. Graziotti, F.; Penna, A.; Magenes, G. A nonlinear SDOF model for the simplified evaluation of the displacement demand of low-rise URM buildings. *Bull. Earthq. Eng.* **2016**, *14*, 1589–1612. [CrossRef]

55. Smerzini, C.; Galasso, C.; Iervolino, I.; Paolucci, R. Ground Motion Record Selection Based on Broadband Spectral Compatibility. *Earthq. Spectra* **2014**, *30*, 1427–1448. [CrossRef]

56. Crowley, H.; Pinho, R. Report on the v5 Fragility and Consequence Models for the Groningen Field. In *Report on Groningen Field Seismic Hazard and Risk Assessment Project*; Van Elk, J., Doornhof, D., Eds.; Nederlandse Aardolie Maatschappij (NAM): Assen, The Netherlands, 2017; Available online: www.nam.nl/feiten-en-cijfers (accessed on 13 March 2021).

57. Chiou, B.S.J.; Darragh, R.; Gregor, N.; Silva, W.J. NGA Project Strong-Motion Database. *Earthq. Spectra* **2008**, *24*, 23–44. [CrossRef]

58. Akkar, S.; Sandikkaya, M.A.; Şenyurt, M.; Sisi, A.A.; Ay, B.O.; Traversa, P.; Douglas, J.H.; Cotton, F.; Luzi, L.; Hernandez, B.M.; et al. Reference database for seismic ground-motion in Europe (RESORCE). *Bull. Earthq. Eng.* **2014**, *12*, 311–339. [CrossRef]

59. Bommer, J.J.; Dost, B.; Edwards, B.; Stafford, P.J.; Van Elk, J.; Doornhof, D.; Ntinalexis, M. Developing an Application-Specific Ground-Motion Model for Induced Seismicity. *Bull. Seism. Soc. Am.* **2015**, *106*, 158–173. [CrossRef]

60. Graziotti, F. Contributions towards a Displacement-Based Seismic Assessment of Masonry Structures. Ph.D. Thesis, University School for Advanced Studies IUSS, Pavia, Italy, 2013.

61. Pennucci, D.; Sullivan, T.J.; Calvi, G.M. Displacement Reduction Factors for the Design of Medium and Long Period Structures. *J. Earthq. Eng.* **2011**, *15*, 1–29. [CrossRef]

62. American Society of Civil Engineers (ASCE). *Minimum Design Loads and Associated Criteria for Buildings and Other Structures, ASCE/SEI 7-16*; American Society of Civil Engineers: Reston, VG, USA, 2016; pp. 17–41.

63. Miranda, E. Estimation of Inelastic Deformation Demands of SDOF Systems. *J. Struct. Eng.* **2001**, *127*, 1005–1012. [CrossRef]

64. Lin, Y.-Y.; Miranda, E. Noniterative Equivalent Linear Method for Evaluation of Existing Structures. *J. Struct. Eng.* **2008**, *134*, 1685–1695. [CrossRef]

Article

Numerical Simulation of Unreinforced Masonry Buildings with Timber Diaphragms

Igor Tomić [1], Francesco Vanin [2], Ivana Božulić [1] and Katrin Beyer [1,*]

[1] École Polytechnique Fédérale de Lausanne (EPFL), School of Architecture, Civil and Environmental Engineering (ENAC), Earthquake Engineering and Structural Dynamics Laboratory (EESD), 1015 Lausanne, Switzerland; igor.tomic@epfl.ch (I.T.); ivana.bozulic@epfl.ch (I.B.)

[2] Résonance Ingénieurs-Conseils SA, 1227 Carouge, Switzerland; francesco.vanin@resonance.ch

[*] Correspondence: katrin.beyer@epfl.ch

Abstract: Though flexible diaphragms play a role in the seismic behaviour of unreinforced masonry buildings, the effect of the connections between floors and walls is rarely discussed or explicitly modelled when simulating the response of such buildings. These flexible diaphragms are most commonly timber floors made of planks and beams, which are supported on recesses in the masonry walls and can slide when the friction resistance is reached. Using equivalent frame models, we capture the effects of both the diaphragm stiffness and the finite strength of wall-to-diaphragm connections on the seismic behaviour of unreinforced masonry buildings. To do this, we use a newly developed macro-element able to simulate both in-plane and out-of-plane behaviour of the masonry walls and non-linear springs to simulate wall-to-wall and wall-to-diaphragm connections. As an unretrofitted case study, we model a building on a shake table, which developed large in-plane and out-of-plane displacements. We then simulate three retrofit interventions: Retrofitted diaphragms, connections, and diaphragms and connections. We show that strengthening the diaphragm alone is ineffective when the friction capacity of the wall-to-diaphragm connection is exceeded. This also means that modelling an unstrengthened wall-to-diaphragm connection as having infinite stiffness and strength leads to unrealistic box-type behaviour. This is particularly important if the equivalent frame model should capture both global in-plane and local out-of-plane failure modes.

Keywords: unreinforced masonry; seismic assessment; equivalent frame models; incremental dynamic analysis; timber floors; flexible diaphragms; retrofitting

Citation: Tomić, I.; Vanin, F.; Božulić, I.; Beyer, K. Numerical Simulation of Unreinforced Masonry Buildings with Timber Diaphragms. *Buildings* **2021**, *11*, 205. https://doi.org/10.3390/buildings11050205

Academic Editors: Rita Bento and Ana Simões

Received: 5 April 2021
Accepted: 22 April 2021
Published: 14 May 2021

Publisher's Note: MDPI stays neutral with regard to jurisdictional claims in published maps and institutional affiliations.

1. Introduction

Historical unreinforced masonry buildings have proven to be particularly susceptible to earthquakes (e.g., [1–5]). To establish effective seismic risk management strategies and design appropriate retrofitting schemes, simulation tools are required that can reproduce the behaviour of historical unreinforced masonry buildings in their unstrengthened and strengthened configurations.

Different modelling techniques have been adopted to simulate the seismic behaviour of unreinforced masonry buildings, which differ regarding both the level of detail at which the building is modelled and the computational costs of the simulations (e.g. [6,7]). While more detailed techniques, such as the ones used in [8–12], simulate masonry behaviour at a micro-scale, the computational cost limits at present still either the size of the model analysed or the number of simulations. For the simulations in this paper, we chose to use the equivalent frame model approach, which we consider a good compromise between level of detail and computational cost if a large number of analyses are performed [13]. It is also a modelling approach that is widely used in engineering practice, making our findings highly applicable [14]. In equivalent frame models, building facades are idealised as frames consisting of vertical pier elements, horizontal spandrel elements, and nodes [13,15]. This

107

frame idealisation is applicable to buildings with a relatively regular opening layout, such as the layout of many residential masonry buildings [16–18].

In equivalent frame models, the response of individual piers and spandrels is captured through macro-elements, which phenomenologically reproduce the force-displacement response of the piers and spandrels. A number of such macro-element models have been proposed for unreinforced masonry elements [15,19–33]; a recent review is included in [14]. The simplicity of this modelling approach allows multiple static and dynamic analyses to be performed in a short time, and a large number of performed analyses can address aleatory and epistemic uncertainties [13,34–37]. However, because all but the most recent macro-elements for unreinforced masonry elements [38] capture only the in-plane and not the out-of-plane response, equivalent frame model analyses were restricted to the global response.

Timber floors and their wall-to-diaphragm connections affect the global response of unreinforced masonry buildings and the formation of local out-of-plane failure modes [1,2,39]. Solarino et al. [40] reviewed wall-to-diaphragm connections common in unreinforced masonry buildings as well as classical and innovative strengthening solutions. The effect of timber floors and their wall-to-diaphragm connections has been investigated experimentally by several research groups through large-scale shake table tests on masonry buildings [41–55]. Other studies numerically investigated the effect of diaphragm stiffness on the global nonlinear seismic response of unreinforced masonry buildings by modelling the diaphragms as elastic membranes and using equal DOF (Degree of Freedom) constraints for the wall-to-diaphragm connection [56–62]. Recent works by Mirra [63] and Trutalli et al. [64] proposed modelling timber floors by an assemblage of elastic truss elements and nonlinear springs, which are assigned as a uniaxial material using *Pinching4* of OpenSEES [65] that represents a 'pinched' load-deformation response and degrades under cyclic loading. The necessary parameters for obtaining an accurate pinching cycle were then calibrated against experimental results.

The effect of the quality of the wall-to-diaphragm connection on the seismic response of vernacular masonry buildings was addressed by Ortega et al. through an investigation of the influence of several floor parameters, including the diaphragm stiffness, the beam stiffness and the wall-to-diaphragm and wall-to-beam connections [66]. The masonry was modelled using solid 3D elements and an isotropic total strain rotating crack model. Because this method is computationally expensive, pushover analyses rather than time-history analyses were carried out. The wall-to-beam connections were modelled by imposing equal DOF conditions in combination with different embedment lengths for the beams, and the wall-to-diaphragm connections were modelled either with equal DOF conditions or without any connection. The friction connection was therefore idealised as either infinitely strong or non-existent. The results showed that if a proper connection was lacking, a stiffened diaphragm did not have the expected benefits. First equivalent frame models that used the macro-element by Vanin et al. [38] showed that this new formulation can capture out-of-plane mechanisms of single walls and parts of buildings that involve one-way bending of single elements [67]. Common post-earthquake, out-of-plane damage patterns, such as those shown in Figure 1 illustrate the necessity of correctly modelling this phenomenon.

Figure 1. Examples of out-of-plane damage patterns from L'Aquila 2009 earthquake: (**a**) Out-of-plane mechanism in long walls. (**b**) Global overturning of external walls. (**c**) Overturning due to the lack of anchorage between walls and horizontal diaphragms. (**d**) Corner out-of-plane mechanism [66]. (Sources: Dr. Javier Ortega, Prof. Hugo Rodrigues)

The objectives of this paper are to show that the latest equivalent frame modelling approach can be used for studying the effects of diaphragm stiffness and wall-to-diaphragm connections and to highlight the importance of explicitly modelling the wall-to-diaphragm connection. More specifically, we make the following two contributions:

- Equivalent frame model for Building 1 of the Pavia test series on stone masonry buildings [47,48]: This test series comprised uni-directional shake-table tests on three stone masonry buildings. Building 1 had a weak diaphragm and wall-to-diaphragm connections that relied only on friction between beams and walls. It developed significant nonlinear in-plane deformations but eventually succumbed to out-of-plane failure. Building 1 has not yet been modelled by an equivalent frame approach, so we close this gap by developing an equivalent frame model for Building 1 and validating it against the experimental results. Buildings 2 and 3 had strengthened diaphragms and wall-to-diaphragm connections. They did not develop any out-of-plane mechanisms and their in-plane response was modelled successfully by Penna et al. [68] using Tremuri [27] and the macro-element by Penna et al. [28].
- Interplay between diaphragm stiffness and unstrengthened wall-to-diaphragm connections: We model the unstrengthened wall-to-diaphragm connection of Building 1 and analyse various configurations using a nonlinear spring with a force capacity that is limited by Coulomb friction. We confirm the finding by Ortega et al. [66] that when a proper connection is lacking, a stiffened diaphragm lacks its beneficial effects. By modelling the connection through a friction connection rather than as fully connected (equal DOF) or disconnected, we show that there is a threshold PGA (Peak Ground Acceleration) value for which the wall-to-diaphragm connections start to slide. For higher PGA values, stiffened diaphragms lose their beneficial effect.

In this paper, we first outline our modelling strategy for unreinforced masonry buildings (Section 2) and establish the equivalent frame model for Building 1 (Section 3). We compare the results of the analyses to the experimental results. We then simulate three simple strengthening interventions that highlight the interplay of diaphragm stiffness and wall-to-diaphragm connection strength (Section 4). Based on these simulations, we formulate recommendations for modelling timber slabs in unreinforced masonry buildings in the final section (Section 5). Finally, we conclude as much as these case studies permit on the effect of the retrofit techniques on the seismic response of unreinforced masonry buildings and formulate future research needs regarding the modelling of timber slabs in equivalent frame models.

2. Equivalent Frame Models for Unreinforced Masonry Buildings with Timber Slabs

Here, we describe the equivalent frame model that we adopted for simulating the seismic behaviour of stone masonry building with timber floors. As our goal is to investigate the role of the timber diaphragm and wall-to-diaphragm connection on the seismic response of unreinforced masonry buildings, we discuss modelling assumptions with regard to these two points in particular detail.

2.1. A Macro-Element for Modelling the in-Plane and out-of-Plane Response of Unreinforced Masonry Piers and Spandrels

In this study, we use the newly developed macro-element by Vanin et al. [38], which is implemented in OpenSEES [65]. It is the first macro-element for equivalent frame models that can capture the in-plane and out-of-plane behaviour of piers and spandrels by modelling each as a three-node element in three-dimensional space (Figure 2). The element is formulated as a system of two panels, deformable only in shear, rotating around three end sections and where flexural deformations are lumped. The exact equilibrium is ensured at all sections in the deformed configuration using an approximated $P - \Delta$ formulation. The in-plane response of this macro-element is based on the formulation by Penna et al. [28].

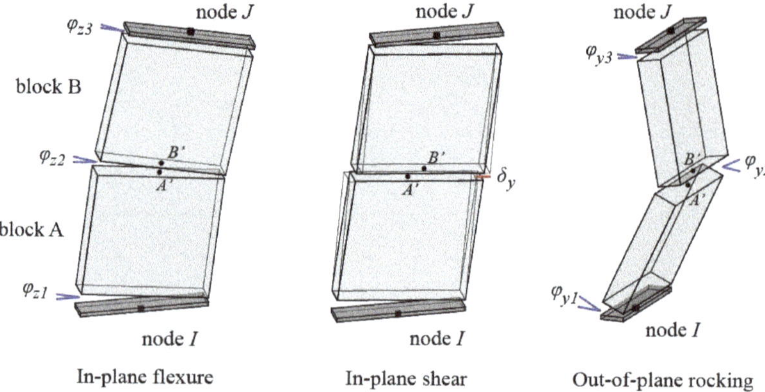

Figure 2. Macro-element by Vanin et al. [38]: deformation modes.

With this approach, the in-plane flexural and shear failures and the out-of-plane overturning of the panel can be modelled [38]. The shear model depends explicitly on the axial load applied to the section. The shear strength of the panel is defined by a Mohr-Coulomb failure criterion imposed by a damage-plasticity model describing residual displacements, stiffness degradation, and post-peak strength degradation. The flexural response, both in-plane and out-of-plane, depends directly on the applied section model. In the following, an analytical section model is used, assuming a material without tensile strength and with limited compressive strength with no post-peak degradation. When large lateral displacements are attained, in-plane failure of the panel is imposed.

2.2. Modelling Assumptions for Masonry Walls and Wall-to-Wall Connections

The strength of the wall-to-wall connection depends on the material properties and the level of interlocking in the corners, which also depends on the skills of the builders and modifications of the structure in its lifetime. We model masonry wall-to-wall connections in the equivalent frame model using a 1D material model that is linear elastic in compression with no crushing and with finite tensile strength with exponential softening [38], shown in Figure 3. The strength of the connections is calculated according to [69].

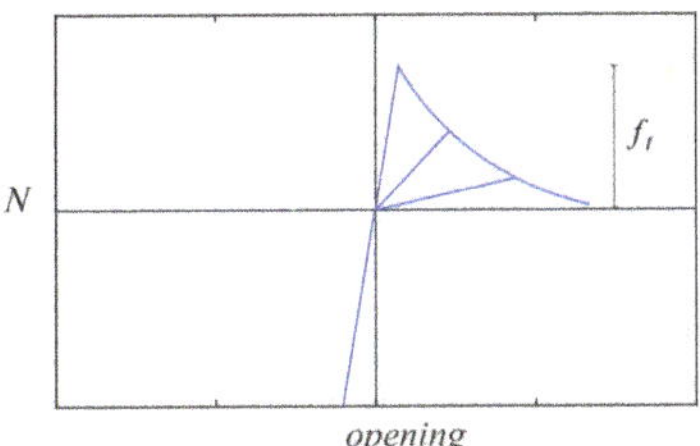

Figure 3. Equivalent frame model of wall-to-wall interface [67].

2.3. Modelling Assumptions for Timber Floors

In equivalent frame models, floors are usually modelled by elastic membrane elements with perfect connections between the floors and walls [14,27,28,49,67,68]. Such simplified assumptions are justified if the main goal is to describe the force redistribution and the floor-provided coupling of the response between different façades. However, if out-of-plane failure modes that span more than one floor are to be captured, the nonlinear in-plane response of the floor needs to be modelled. Since, in general, little distributed damage is observed on timber floors in post-earthquake surveys, concentrating the non-linearity in the connection between floors and walls is a reasonable approach [67]. The floor diaphragm is therefore modelled as linear elastic orthotropic membrane with a larger axial stiffness in the direction of the beams and a lower axial stiffness in the direction orthogonal to the beams [70,71]. To do this, estimates of timber floor properties are needed to describe the in-plane axial stiffness in both directions as well as the shear stiffness.

The diaphragm axial stiffness in the strong and weak direction of the timber floor is based on the timber stiffness both parallel and perpendicular to the grain. The diaphragm shear stiffness is computed according to the approach by Brignola et al. [70,71]. This shear stiffness, which is given in Equation (1), accounts for the (i) rigid rotation of the planks due to the slip of nails; (ii) flexural deformation of timber planks; and (iii) shear deformation of timber planks.

$$G_{eq} = \chi / A (l/(k_{ser}s_n^2) + \chi/(GA) + l^2/(12EI))^{-1}, \tag{1}$$

where χ is the shear correction factor (normally 5/6 for the rectangular cross-section), A is the area of a single plank section, l is the distance between the nail pairs on the opposite sides of a plank, k_{ser} is the nail stiffness per shear plane per fastener provided by codes [72], s_n is the nails spacing, G is the shear modulus of timber planks, E is the flexural modulus parallel to the grains of timber planks, and I is the moment of inertia of the plank section. The nail stiffness is calculated according to EC 1995-1-1-2004 [72] using the equation:

$$k_{ser} = \rho_m^{1.5} d^{0.8} / 30, \tag{2}$$

where ρ_m is the nail density in kg/m^3 and d is the nail diameter in mm. The value of k_{ser} is then in N/mm [72].

To model the increase in floor stiffness when an additional layer of timber planks is added as retrofit measure, the thickness and the equivalent shear stiffness are increased. The thickness of the retrofitted diaphragm corresponds to the thickness of the original planks plus the thickness of the new planks. The nails that are needed to fix the additional planks on the original planks increase the shear stiffness. Therefore, for the retrofitted configuration, we rewrite the equation as:

$$G_{eq} = \chi / A (l/(4k_{ser}s_n^2) + \chi/(GA) + l^2/(12EI))^{-1}. \tag{3}$$

2.4. Modelling Assumptions for Wall-to-Diaphragm Connections

As outlined in the introduction, we model the limited force capacity of this connection explicitly for the unstrengthened wall-to-diaphragm connection, meaning when the force transfer from the floor to the wall relies on a Coulomb friction mechanism [73]. The values for the friction coefficient were derived from a series of friction tests between both timber and timber and timber and mortar [73]. As shown in Figure 4, we model this connection by a nonlinear spring, coupling axial and shear force. The model allows for loading in the positive direction through sliding (beam pulled off the support) and in the negative direction through pounding (beam pounding against the wall).

Wall-to-diaphragm connections are typically reinforced by anchoring the floor beam to the wall [40], which can make a rather stiff wall-to-diaphragm connection [74,75]. For the purpose of this study, these retrofitted connections are therefore assumed as infinitely stiff and strong and are modelled with the EqualDOF command in OpenSEES, which constructs a multi-point constraint between nodes [67]. In the future, additional simulations with a limited anchor capacity could be envisaged. First strength models for the anchor capacity in stone masonry and numerical simulations of the anchors were put forward by several groups [74,76,77].

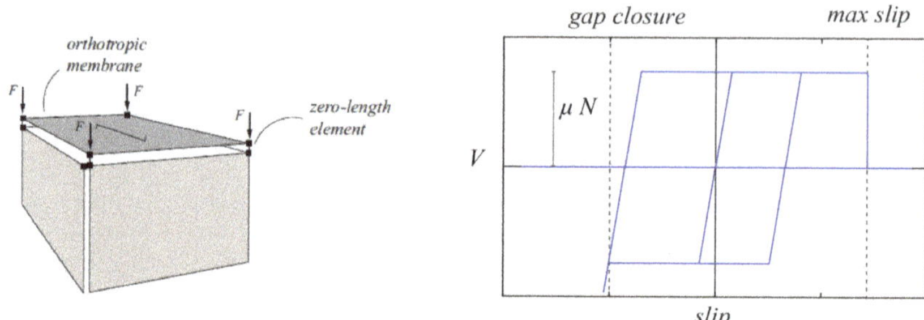

Figure 4. Equivalent frame model of wall-to-diaphragm connection [38].

2.5. Damping Model for Dynamic Analyses

Here, we conduct nonlinear time-history simulations using the equivalent frame model. Vanin et al. [78] showed that the overdamping that occurs with the classical Rayleigh damping model becomes especially relevant for out-of-plane behaviour. To avoid both the overdamping attributed to initial-stiffness proportional damping and the numerical problems stemming from tangential-stiffness proportional damping, we used the newly developed secant-stiffness proportional damping model in this study [78]. The secant-stiffness proportional damping model defines a correction term for the initial stiffness-matrix proportional damping in the classical Rayleigh damping matrix, thus approximating the secant-stiffness proportional damping model. This correction term is updated at each converged analysis step. The use of secant-stiffness proportional damping simulates out-of-plane rocking with acceptable accuracy when compared to the experimental data and classical rocking formulations, such as those of Housner et al. [79], and experimental evidence [78].

3. Case-Study Building

The case-study building is Building 1 of an experimental campaign by Magenes et al. comprising a full-scale unretrofitted stone masonry building (Pavia Building 1) and two retrofitted configurations of the same building (Pavia Building 2 and 3) [47–49]. The building was tested on the shake table at the EUCENTRE, Pavia, Italy. As outlined in the introduction, the two retrofitted configurations have already been successfully modelled by Penna et al. [68]. In this paper, we model Building 1, which developed significant in-plane deformations and then an out-of-plane failure mode. We directly base this model on the equivalent frame models of Buildings 2 and 3 by Penna et al., modified so the new equivalent frame model can capture the out-of-plane response developed by Building 1. In the following, we describe the unretrofitted building as well as the obtained experimental and numerical input data and the seismic record.

3.1. Experimental Campaign

The Pavia Building 1 [47], shown in Figure 5, is representative of an existing stone masonry building without any aseismic detailing. The building was 5.8 m long and 4.4 m wide. It had two storeys and a roof; the total height from the base to the top of the gable was 6.0 m. The walls were 32–cm-thick double-leaf stone masonry, without throughstones except in the corners and in the vicinity of openings. Two leaves of undressed stones were simply built adjacent to each other with smaller stones and mortar filling the irregular gaps. The four facades had different opening layouts such that some rotation in the building was expected. It is therefore a suitable case study of the effects of the floor diaphragms and wall-to-diaphragm connections on the seismic response.

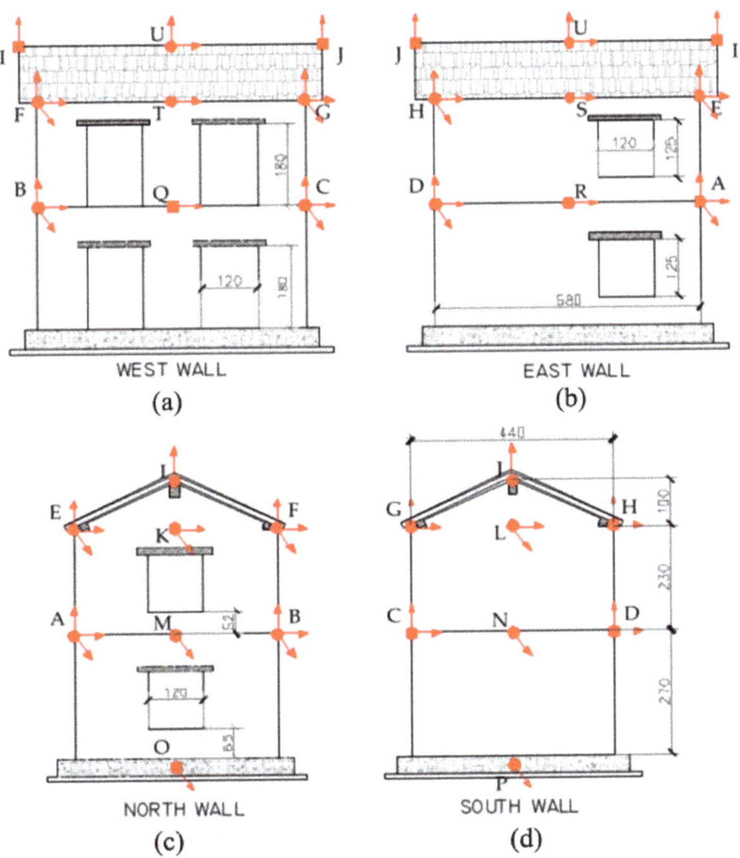

Figure 5. Drawings of Pavia Building 1 with the positions of accelerometers: (**a**) West wall. (**b**) East wall. (**c**) North wall. (**d**) South wall [47].

The floor was composed of timber beams that were 12 cm wide and 16 cm thick with planks that were 30 mm thick simply nailed on top of the beam [47]. The roof was composed of a 20 cm × 32 cm ridge beam, two 32 cm × 12 cm spreader beams and 8 cm × 12 cm purlins. The 30–mm-thick roof planks were again simply nailed on top of the purlins. The details of the masonry walls, floors and connections are shown in Figure 6. Additional masses were evenly distributed onto the floors, for a total amount of 3.2 tons.

Figure 6. Building 1 details: (**a**) masonry wall. (**b,c**) timber floor. (**d,e**) timber roof [47–49,68].

3.2. Numerical Model

All three buildings had the same overall geometry, differing only with regard to the floor and roof details. The previous model of Buildings 2 and 3 [68] included a sensitivity study regarding the discretisation of the equivalent frame model. We built here on their work and use the "MOD" discretisation, which they concluded to be the most appropriate for capturing the force capacity and damage mechanism observed during testing. In the "MOD" discretisation, the height of the piers was equal to the height of the adjacent openings.

Our material parameters for masonry in the numerical model are based on those used in the equivalent frame models for Buildings 2 and 3 [68], which were analysed using the software Tremuri [27] with the macro-element by Penna et al. [28]. As outlined in Section 2, we used the macro-element by Vanin et al. [38], which builds the in-plane response on the macro-element by Penna et al. [28]. For this reason, the macro-element parameters were based on the values used in the original modelling of Buildings 2 and 3 [68]. The chosen set of material parameters for the macro-element simulating masonry piers and spandrels is shown in Table 1. For each model, the modal properties were calculated first and then the Rayleigh damping model parameters were computed such that the damping ratios at the first and sixth mode corresponded to the damping ratio of this model. We built upon the existing models by using the ability of the macro-element [38] to explicitly model the out-of-plane behaviour. As shown in Tomić et al. [37], when the out-of-plane behaviour is accounted for, the influence of non-linear connections can be highlighted. Therefore, unlike the Tremuri model where the floors were assumed to be perfectly connected to the walls and the stiffness was calibrated accordingly, here we explicitly model the non-linear connections to base the stiffness of the floor diaphragm on material properties and mechanical formulation. The OpenSEES Building 1 model is shown in Figure 7.

Table 1. Pavia Building 1: Material parameters assumed for masonry elements [68].

E (MPa)	G (MPa)	ρ (kg/m)^3	f_c (MPa)	τ (MPa)	μ
1900	300	2200	4.50	0.175	0.20

Floors and roofs were modelled as orthotropic membranes, with the parameters calibrated according to Brignola et al. [70,71]. The floor-wall connection was modelled using a frictional interface calibrated according to the experimental tests performed by Almeida et al. [73]. The floor parameters and floor-wall connection parameters are summarised in Table 2. E_1 and E_2 represent the membrane axial moduli in the strong and weak direction, respectively, and G represents the shear modulus of the membrane. The first row contains stiffness values that estimate the timber properties of Building 1. For evaluating the influence of the retrofitting strategies on the building response, a diaphragm is first modelled considering a retrofit of an additional layer of planks and nails. Then, the retrofitted floor-wall connection is modelled followed by the seismic performance of the building for both retrofitting techniques applied simultaneously.

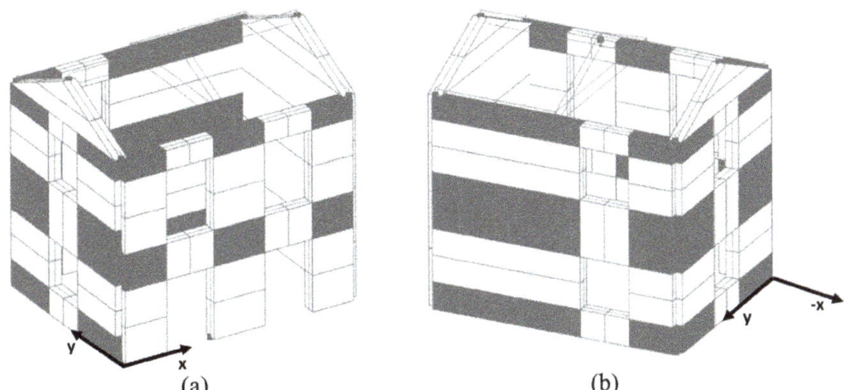

Figure 7. OpenSEES model of Pavia Building 1: (**a**) View from the northwest corner. (**b**) View from the northeast corner.

Table 2. Pavia Building 1: Material parameters of diaphragms and the wall-to-diaphragm connections.

	E_1 (GPa)	E_2 (GPa)	G (MPa)	t (m)	μ_{w-to-d}
Unretrofitted	10	0.5	10.3	0.03	1.0
Diaphragms retrofitted	10	0.5	19.6	0.06	1.0
Wall-to-diaphragm connections retrofitted	10	0.5	10.3	0.03	fixed
Diaphragms and connections retrofitted	10	0.5	19.6	0.06	fixed

3.3. Seismic Excitation

In the experimental campaign, Building 1 was subjected to the east-west component of the seismic record of the Montenegro 1979 earthquake at the Albatros station [47]. The record was scaled to nominal PGAs (Peak Ground Accelerations) between 0.05–0.40 g, and Building 1 was subjected to five runs of increasing intensities (Table 3). The actual applied PGAs were between 0.07–0.63 g. The numerical simulations used the actual applied ground motion as recorded during the test. To capture the damage evolution, we performed one long analysis that comprised all five runs. In between runs, we included zero ground acceleration records to again reach near zero building vibrations at the start of the next run.

Table 3. Building 1: Summary of shake-table runs [47].

Test Run	Nominal PGA (g)	Actual PGA (g)
1	0.05	0.07
2	0.10	0.14
3	0.20	0.31
4	0.30	0.50
5	0.40	0.63

For the second part of the analyses where we studied the interplay of diaphragm- and connection-strengthening interventions, incremental dynamic analyses (IDAs) were performed. The applied ground motions comprised both horizontal components of the Montenegro 1979 earthquake at the Albatros station, with the east-west component applied in the x-direction and the north-south component in the y-direction. For these analyses, we used the record as downloaded from Engineering Strong Motion Database [80]. The acceleration records and the response spectra are shown in Figures 8 and 9, respectively. More details on the IDAs are given in Section 4.2.

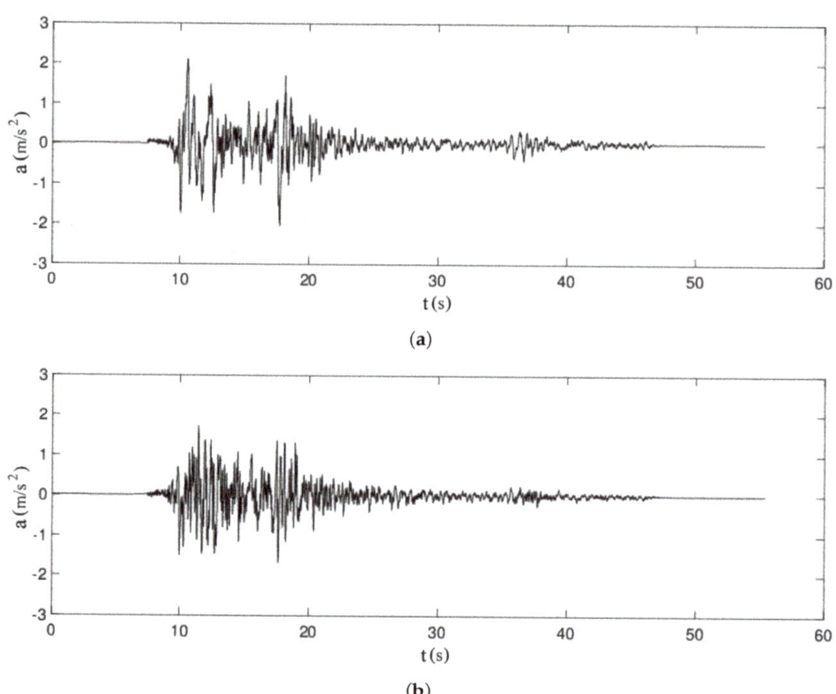

Figure 8. Processed acceleration time-histories of the Montenegro 1979 earthquake at the Albatros station: (a) east-west direction. (b) north-south direction [80].

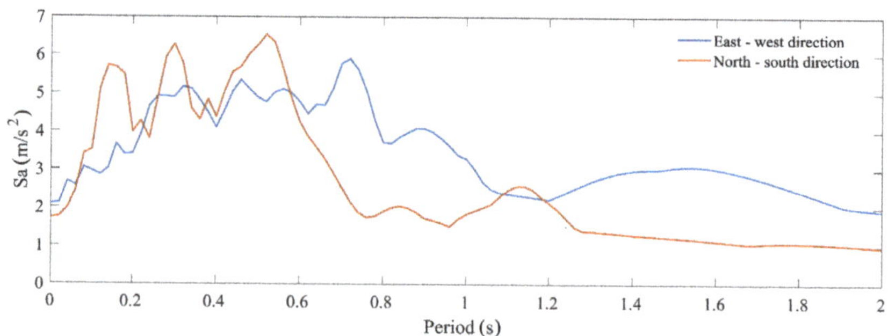

Figure 9. Acceleration response spectra of the Montenegro 1979 earthquake at the Albatros station for 5% damping ratio [80].

4. Numerical Results for the Case-Study Building

We started by validating the numerical model of the unretrofitted building against the experimental data from the uni-directional shake table-test with regard to displacement demands and damage mechanisms. We analysed the four configurations of the case study building, the original unretrofitted and three retrofitted configurations, with regard to the differences in floor diaphragm and wall-to-diaphragm connections. These analyses were performed using the bi-directional seismic excitation of the Montenegro 1979 record. The run time for running a single model IDA analysis, consisting of 12 dynamic analyses, was about 1 hour on an Intel Pentium i7 with 16 GB of RAM and a Nvidia Quadro P2000.

4.1. Model Validation

For each of the five runs, the experimental data set contained accelerations recorded at 22 positions on the building and its foundation. To derive the displacements, we double-integrated the signals and applied a band-pass filter before each integration. Because it was difficult to choose the single best set of corner frequencies for the band-pass filter, we chose several sets that all produced results that are in agreement with the video recording of the test [81]. The corner frequencies of the band-pass filter were judged reasonable if the maximum relative x-displacement between accelerometers F and T (Figure 5) was between 4–10 cm, which was the maximum sliding displacement estimated from the video of the test [81]. Based on this check, we chose two sets of corner frequencies (wide band: 0.5–450 Hz, narrow band: 1–40 Hz) and assume that the so-obtained displacements are the bounds of likely displacements. The numerical prediction was compared to the experimental results in terms of the average displacement of three measurements: (i) the 2nd storey, (ii) the stiff (east) facade of the 2nd storey, and (iii) the soft (west) facade of the 2nd storey. All displacements are in the x-direction.

The only model parameter that was calibrated was the damping ratio. To do this, we tested damping ratios between 1 and 5%, and the best fit was obtained for a damping ratio of 2.5%. As outlined in Section 2, we chose a secant-stiffness proportional damping model. Figure 10 compares the maximum displacement values per run. The predicted maximum values of the average displacement of the 2nd storey (Δ_{max}) lay within the bounds of the derived experimental values for almost all levels of excitation; the match is therefore very good. The model tended to underestimate the displacements of the stiff facade ($\Delta_{max,stiff}$), while the displacements of the soft facade ($\Delta_{max,soft}$) were well predicted, except for the last run, for which the model overestimated the displacement demand.

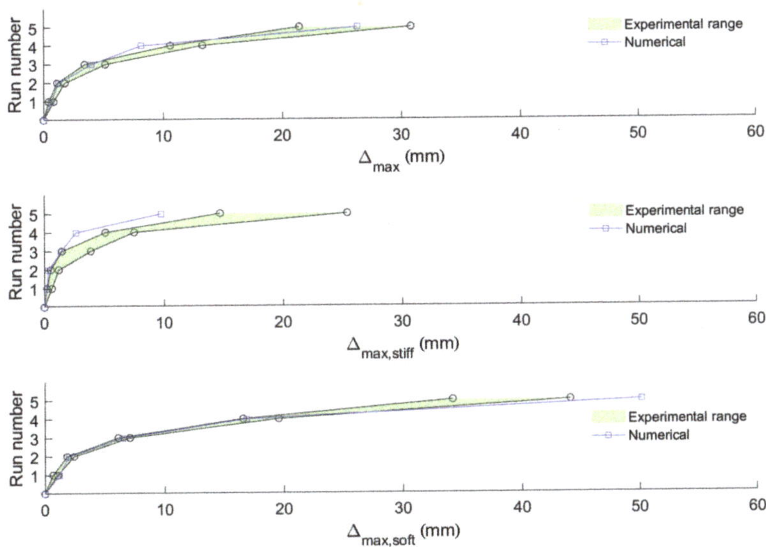

Figure 10. Comparison of experimental and numerical maximum displacements at each run.

The out-of-plane displacements of the gables were not measured during the test, but they can be observed via recorded videos [81–83] and the mechanism that developed during the final run (nominal PGA of 0.4 g) is sketched in [48]: (i) The north facade containing the opening developed an out-of-plane mechanism that involved the 2nd storey and gable wall of the facade as well as the adjacent pier and spandrel of the 2nd storey of the west facade. This part of the building rotated around the bottom of the 2nd storey (line A-B in Figure 5); (ii) The out-of-plane mechanism of the south facade involved only the gable wall, i.e., the gable rotated around its base (line G-H in Figure 5). The numerical model replicated the dominant out-of-plane behaviour of both out-of-plane facades, as shown in Figure 11.

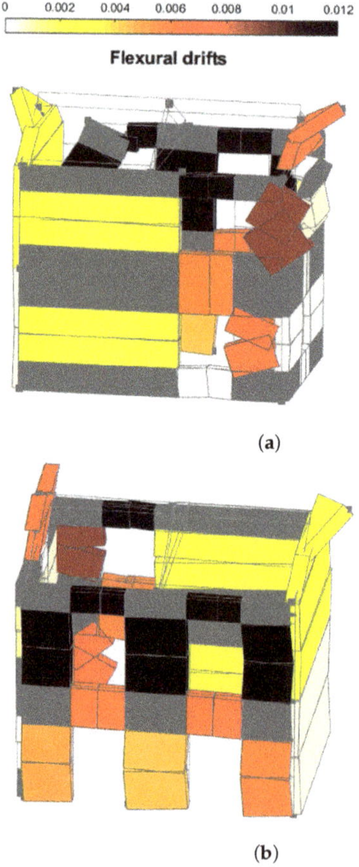

Figure 11. Out-of-plane behaviour (magnification factor ×10): (**a**) North facade-out-of-plane displacement involving the 2nd storey piers and the gable. (**b**) South facade-out-of-plane displacement involving only the gable.

4.2. Modelling Retrofitting Interventions

Once the model of the unretrofitted configuration was validated, it was used to model the following unretrofitted and three retrofitted scenarios:

- Unretrofitted (diaphragm and wall-to-diaphragm connection unretrofitted): This corresponds to the configuration of Building 1.
- Diaphragm retrofitted: The diaphragm stiffness was increased to reflect the effect of an additional layer of planks (Table 2).
- Wall-to-diaphragm connections retrofitted: The wall-to-diaphragm connections were modelled as infinitely stiff and strong using equal DOF constraints.

- Diaphragm and wall-to-diaphragm connections retrofitted: The two individual retrofitting conditions were combined.

Each of the models was subjected to an IDA (Incremental Dynamic Analysis) using the two horizontal components of the Montenegro 1979 record [80] (see Section 3), with the east-west component in the x-direction and the north-south component in the y-direction. The record was scaled—for the east-west direction, the starting PGA of 0.0525 g was increased in increments of 0.0525 g until a PGA of 0.63 g was reached (the effective PGA applied in the final run of the shake table test [Table 3]). For the north-south direction, the starting PGA of 0.043 g was increased in increments of 0.043 g until a PGA of 0.516 g was reached. Figure 12 shows the IDA curves for the x- and y-direction. The IDA curves are plotted as PGA vs. the absolute maximum value of the mean 2nd storey displacement.

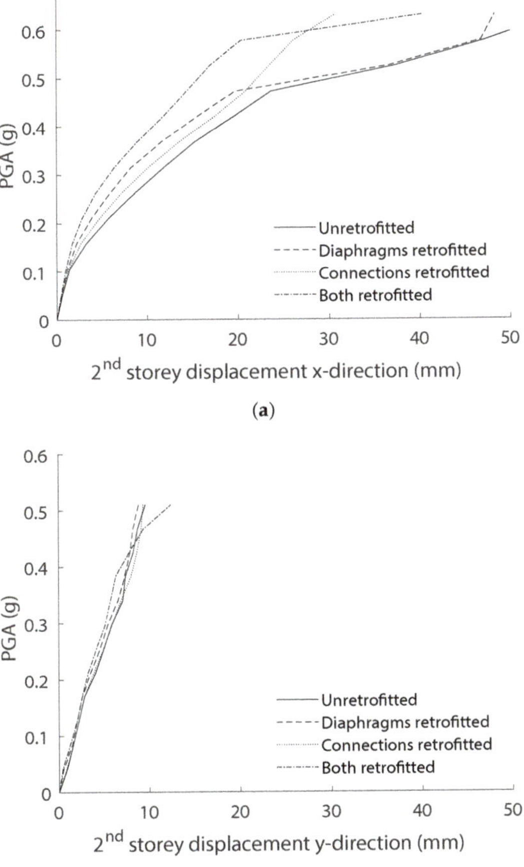

Figure 12. IDA curves for the four configurations in terms of PGA vs. the absolute maximum values of average 2nd storey displacement: (**a**) in the x-direction (**b**) in the y-direction.

The IDA curves for the x-direction show that for PGA values below a threshold value (here 0.4725 g), retrofitting the diaphragm reduced the displacements, as the sliding displacements of the diaphragm-wall connections were still small. Therefore, the increased shear stiffness of the retrofitted diaphragm proves effective. However, when the PGA rose above the threshold value, the capacity of the wall-to-diaphragm connection became the weak link, and significant sliding displacements occurred between the floor and wall. Then, in the two models with unretrofitted wall-to-diaphragm connections, sliding occurred at those connections, which drastically increased the mean 2nd storey displacement to produce a kink in the PGA-displacement curve. This means that the force transferred by the diaphragm is limited by the wall-to-diaphragm connections, and therefore, the beneficial effect of retrofitting the diaphragm is reduced. In fact, for PGAs larger than 0.45 g, the mean 2nd storey displacement of the original unretrofitted configuration and the configuration with only the diaphragm retrofitted were almost equal. Conversely, models with retrofitted connections make use of the full stiffness of the diaphragm.

The difference in the y-direction was significantly lower. This was partly due to the lack of significant out-of-plane behaviour in the y-direction for any of the modelling approaches in comparison with the out-of-plane displacements of the gables in the x-direction. The reduced impact of the shear stiffness on the redistribution of the loads between the in-plane walls also lowered the overall difference in the y-direction.

The effect of the retrofitting solutions was also visible when observing the deformed shapes, which are shown in Figure 13 for a PGA of 0.58 g. The unretrofitted and diaphragm-only retrofitted models showed a significant out-of-plane displacement of the gable walls. The in-plane deformation of the soft facade (west facade) was slightly lower for the diaphragm retrofitted model than for the unretrofitted model, but a larger effect was prevented by the limited force capacity of the wall-to-diaphragm connections. Otherwise, the two models with retrofitted connections did not show any significant out-of-plane displacements, which are successfully prevented by the rigid connections. The rigid connections fully exploit the beneficial effect of the increased shear stiffness of the retrofitted diaphragm, leading to lower displacements and drifts when compared to the model with only the connections retrofitted.

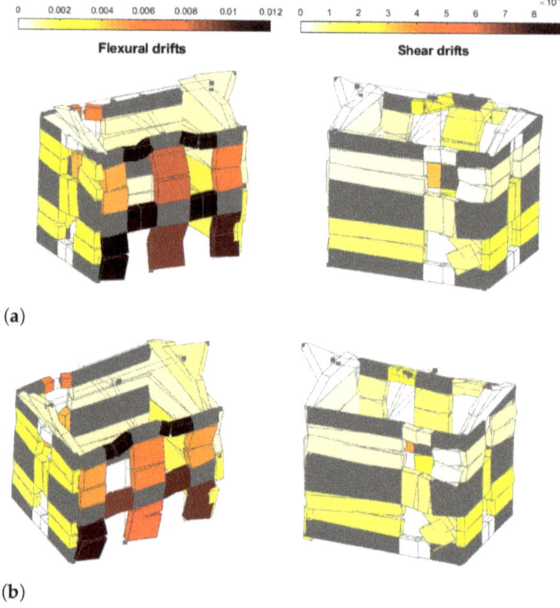

Figure 13. *Cont.*

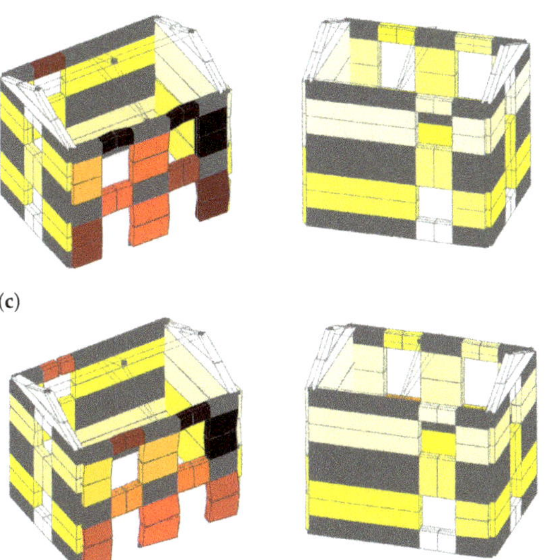

(**c**)

(**d**)

Figure 13. Deformed shapes and maximum element drifts in flexure and shear (magnification factor ×10): (**a**) Unretrofitted. (**b**) Diaphragms retrofitted. (**c**) Connections retrofitted. (**d**) Both retrofitted.

4.3. Force Demand on the Wall-to-Diaphragm Connections

The analyses showed that retrofitting wall-to-diaphragm connections significantly impacts the behaviour of the case-study building, shifting the failure mode from out-of-plane to in-plane. However, to ensure the behaviour as predicted by the model, the necessary pre-condition is that the wall-to-diaphragm connection must be reinforced by anchors able to sustain the force demand. The EqualDOF command, which was initially used to model the wall-to-diaphragm connections retrofitted by anchors (Section 2), did not provide the force transmitted by this connection. Therefore, to measure this force demand in another set of analyses, this connection was also modelled using very stiff elastic elements. The impact on the results was negligible, and the maximum tensile force demands on the wall-to-diaphragm connections for the example of PGA of 0.63 g are given in Table 4.

Table 4. Maximum tensile force demands on the anchors of the first storey, second storey, and gable of the model with connections retrofitted for the PGA of 0.63 g.

Position	Unit	Maximum Tensile Force
First floor	(kN/m)	20.58
Second floor	(kN/m)	7.56
Gable	(kN)	37.09

For the first storey, the largest tensile force demand of 20.58 kN/m was recorded for the anchors. This value was obtained by dividing the sum of the recorded tensile forces in the anchors with the length of the facade to obtain a value more practical for the design of the retrofitting intervention. For the second storey, the largest tensile force demand anchoring the roof to the wall was 7.56 kN/m. For the gable, the largest tensile force demand of 37.09 kN was recorded and reported here as a concentrated force, as a single anchor pair is assumed to anchor the ridge beam to the gable wall.

Ciocci et al. [75] tested the pull-out capacity of injection anchors in masonry of a similar typology to the case-study building. The anchor configurations each consisted of two horizontal anchors that failed at the cone. The test showed a direct relationship between the pull-out capacity and the overburden stress. For a very low overburden stress, such as the one acting on the roof-wall connection, the mean capacity of such a pair of anchors was approximately 30 kN; meaning that even for a very low overburden stress, the tested anchor design would achieve the assumed behaviour in the case-study building with a reasonable number of anchors per meter. Due to the possible high local concentration of force in the anchor pair, special attention should be paid to the anchorage of the ridge beam into the gable wall.

5. Conclusions

In this paper, we validated an equivalent frame model for unreinforced masonry buildings with unstrengthened timber floors and diaphragm-wall connections. This equivalent frame model completes the suite of models covering the full range of diaphragm and connection details found in unstrengthened and strengthened configurations of historical stone masonry buildings. Building configurations with strengthened floor diaphragms and wall-to-diaphragm connections had already been successfully modelled and validated by Penna et al. [68]. Modelling the unstrengthened configuration required three modifications: First, the new macro-element by Vanin et al. [38] captured in-plane and out-of-plane failure modes of piers and spandrels. Second, the wall-to-wall and wall-to-diaphragm connections were modelled with infinite stiffness but limited strength using nonlinear springs. Third, a new secant-damping model prevented the overdamping of rocking motions that is frequently observed for initial-stiffness proportional damping [78]. These changes allowed us to capture next-to-inelastic, in-plane deformations and large, nonlinear, out-of-plane displacements. The modelling approach was validated against the results of the Pavia Building 1 shake test [47].

In a second step, we investigated the influence of strengthening interventions on the global behaviour and local failure modes of the building under earthquake loading. The following modelling assumptions were made with regard to diaphragms and diaphragm-wall connections in their unstrengthened and strengthened configuration:

- In its unstrengthened configuration, the diaphragm consists of timber beams and a single layer of planks nailed to the timber beams. The diaphragm is modelled as an orthotropic elastic membrane. The properties of this membrane are determined according to Brignola et al. [70].
- The diaphragm is retrofitted by adding a layer of planks at a right angle to the first layer of planks. The increase in stiffness of the diaphragm is again calculated using the formulae provided in [70], with a slight modification to account for the deformation of the additional set of nails (Equation (3)).
- The wall-to-diaphragm connection in its unstrengthened configuration transfers loads only via friction. Representative friction coefficients were determined by Almeida et al. [73]. In finite element models of configurations with unstrengthened wall-to-diaphragm connections, the connection was modelled as rigid until the friction force was attained and sliding occurred.
- It was assumed that the wall-to-diaphragm connection was retrofitted by injection anchors that are relatively stiff until the peak force is attained. For this reason, they were modelled as infinitely rigid with infinite force capacity. Using the force capacities attained, it was computed how many anchors would be necessary to transfer the forces between the diaphragm and wall, which were recorded for the numerical model. Reasonable numbers were attained.

The effect of the two retrofit interventions (strengthening of floor diaphragms and of wall-to-diaphragm connections) was investigated by simulating virtual retrofit measures for Pavia Building 1. For this case-study building, strengthening the diaphragm alone only had an effect up to a threshold value of the PGA, where the capacity of the frictional wall-to-diaphragm connection was exceeded. For larger PGA levels, the seismic response of the building could only be improved if the wall-to-diaphragm connections were strengthened. Retrofitted wall-to-diaphragm connections increased the capacity because the failure mode changed from out-of-plane to in-plane. When the wall-to-diaphragm connections were retrofitted, also retrofitting the diaphragm improved the PGA values beyond the threshold value because this prevented sliding of the wall-to-diaphragm connections. Here, retrofitting the diaphragm led to smaller average peak displacements in comparison to the model where only the connections, but not the diaphragm, were retrofitted.

These observations lead to two conclusions: (i) When designing a retrofitting intervention, increasing the shear stiffness of a diaphragm can produce a more favourable response for lower PGA levels but will only negligibly affect the limit states closer to collapse when the wall-to-diaphragm connection becomes the weak link; and (ii) When modelling unreinforced masonry buildings using the equivalent frame approach, it is necessary to explicitly model the wall-to-diaphragm connection, as simplification to a perfect connection might lead to an unrealistic box-type behaviour. This is particularly important if the equivalent frame model should capture both global in-plane as well as local out-of-plane failure modes.

In addition to further validations of this modelling approach against large-scale experimental results, future work should address the nonlinear response of retrofitted wall-to-diaphragm connections and the non-linear response of diaphragms.

Author Contributions: Conceptualization, K.B. and F.V.; methodology, I.T.; software, F.V.; validation, I.T. and I.B.; formal analysis, I.T.; investigation, I.T.; resources, K.B.; data curation, I.T.; writing—original draft preparation, I.T.; writing—review and editing, K.B. and F.V.; visualization, I.T.; supervision, K.B.; project administration, K.B.; funding acquisition, K.B. All authors have read and agreed to the published version of the manuscript.

Funding: The project was supported by the Swiss National Science Foundation through grant 200021_175903/1: Equivalent frame models for the in-plane and out-of-plane response of unreinforced masonry building.

Institutional Review Board Statement: Not applicable.

Informed Consent Statement: Not applicable.

Data Availability Statement: The OpenSEES models used for producing the results presented in this paper as well as the data with results of the calibration procedure and IDAs are shared openly through the repository DOI:10.5281/zenodo.4659149.

Acknowledgments: We would like to express our gratitude to Ilaria Senaldi (EUCENTRE, Pavia, Italy) and Andrea Penna (University of Pavia, Italy) for providing the data of the experimental campaign as well as the equivalent frame models of Building 2 and 3.

Conflicts of Interest: The authors declare no conflict of interest.

Abbreviations

The following abbreviations are used in this manuscript:

DOF	Degree of freedom
PGA	Peak ground acceleration
IDA	Incremental dynamic analysis

References

1. Tomazevic, M. *Earthquake-Resistant Design of Masonry Buildings*; Imperial College Press: London, UK, 1999; Volume 1.
2. D'Ayala, D.; Speranza, E. Definition of collapse mechanisms and seismic vulnerability of historic masonry buildings. *Earthq. Spectra* **2003**, *19*, 479–509. [CrossRef]
3. Penna, A.; Morandi, P.; Rota, M.; Manzini, C.F.; da Porto, F.; Magenes, G. Performance of masonry buildings during the Emilia 2012 earthquake. *Bull. Earthq. Eng.* **2014**, *12*, 2255–2273. [CrossRef]
4. Carocci, C.F. Small centres damaged by 2009 L'Aquila earthquake: On site analyses of historical masonry aggregates. *Bull. Earthq. Eng.* **2012**, *10*, 45–71. [CrossRef]
5. Da Porto, F.; Munari, M.; Prota, A.; Modena, C. Analysis and repair of clustered buildings: Case study of a block in the historic city centre of L'Aquila (Central Italy). *Constr. Build. Mater.* **2013**, *38*, 1221–1237. [CrossRef]

6. Lourenço, P.B. Computations on historic masonry structures. *Prog. Struct. Eng. Mater.* **2002**, *4*, 301–319. [CrossRef]

7. Roca, P.; Cervera, M.; Gariup, G. Structural analysis of masonry historical constructions. Classical and advanced approaches. *Arch. Comput. Methods Eng.* **2010**, *17*, 299–325. [CrossRef]

8. Bui, T.T.; Limam, A. *Elventh International Conference on Computational Structures Technology*; Civil-Comp Press: Stirlingshire, UK, 2012; p. 119. [CrossRef]

9. Pulatsu, B.; Erdogmus, E.; Lourenço, P.B.; Lemos, J.V.; Tuncay, K. Simulation of the in-plane structural behavior of unreinforced masonry walls and buildings using DEM. In *Structures*; Elsevier: Amsterdam, The Netherlands, 2020; Volume 27, pp. 2274–2287.

10. DeJong, M.J.; Belletti, B.; Hendriks, M.A.; Rots, J.G. Shell elements for sequentially linear analysis: Lateral failure of masonry structures. *Eng. Struct.* **2009**, *31*, 1382–1392. [CrossRef]

11. Zhang, S.; Mousavi, S.M.T.; Richart, N.; Molinar, J.F.; Beyer, K. Micro-mechanical finite element modeling of diagonal compression test for historical stone masonry structure. *Int. J. Solids Struct.* **2017**, *112*, 122–132. [CrossRef]

12. Lourenco, P.B.; Silva, L.C. Computational applications in masonry structures: From the meso-scale to the super-large/super-complex. *Int. J. Multiscale Comput. Eng.* **2020**, *18*. [CrossRef]

13. Bracchi, S.; Rota, M.; Penna, A.; Magenes, G. Consideration of modelling uncertainties in the seismic assessment of masonry buildings by equivalent-frame approach. *Bull. Earthq. Eng.* **2015**, *13*, 3423–3448. [CrossRef]

14. Quagliarini, E.; Maracchini, G.; Clementi, F. Uses and limits of the Equivalent Frame Model on existing unreinforced masonry buildings for assessing their seismic risk: A review. *J. Build. Eng.* **2017**, *10*, 166–182. [CrossRef]

15. Parisi, F.; Augenti, N. Seismic capacity of irregular unreinforced masonry walls with openings. *Earthq. Eng. Struct. Dyn.* **2013**, *42*, 101–121. [CrossRef]

16. Siano, R.; Sepe, V.; Camata, G.; Spacone, E.; Roca, P.; Pelà, L. Analysis of the performance in the linear field of equivalent-frame models for regular and irregular masonry walls. *Eng. Struct.* **2017**, *145*, 190–210. [CrossRef]

17. Siano, R.; Roca, P.; Camata, G.; Pelà, L.; Sepe, V.; Spacone, E.; Petracca, M. Numerical investigation of non-linear equivalent-frame models for regular masonry walls. *Eng. Struct.* **2018**, *173*, 512–529. [CrossRef]

18. Berti, M.; Salvatori, L.; Orlando, M.; Spinelli, P. Unreinforced masonry walls with irregular opening layouts: Reliability of equivalent-frame modelling for seismic vulnerability assessment. *Bull. Earthq. Eng.* **2017**, *15*, 1213–1239. [CrossRef]

19. D'Altri, A.M.; Sarhosis, V.; Milani, G.; Rots, J.; Cattari, S.; Lagomarsino, S.; Sacco, E.; Tralli, A.; Castellazzi, G.; de Miranda, S. *Modeling Strategies for the Computational Analysis of Unreinforced Masonry Structures: Review and Classification*; Springer: Dordrecht, The Netherlands, 2020; Volume 27, pp. 1153–1185. [CrossRef]

20. Magenes, G.; Fontana, A. Simplified non-linear seismic analysis of masonry buildings. In *Proceedings of the British Masonry Society No. 8*; British Masonry Society: London, UK, 1998; pp. 190–195.

21. Kappos, A.J.; Penelis, G.G.; Drakopoulos, C.G. Evaluation of simplified models for lateral load analysis of unreinforced masonry buildings. *J. Struct. Eng.* **2002**, *128*, 890–897. [CrossRef]

22. Roca, P.; Molins, C.; Marí, A.R. Strength capacity of masonry wall structures by the equivalent frame method. *J. Struct. Eng.* **2005**, *131*, 1601–1610. [CrossRef]

23. Pasticier, L.; Amadio, C.; Fragiacomo, M. Non-linear seismic analysis and vulnerability evaluation of a masonry building by means of the SAP2000 V. 10 code. *Earthq. Eng. Struct. Dyn.* **2008**, *37*, 467–485. [CrossRef]

24. Belmouden, Y.; Lestuzzi, P. An equivalent frame model for seismic analysis of masonry and reinforced concrete buildings. *Constr. Build. Mater.* **2009**, *23*, 40–53. [CrossRef]

25. Rizzano, G.; Sabatino, R. An equivalent frame model for the seismic analysis of masnory structures. In Proceedings of the 8th Congresso de Sismologia e Engenharia Sismica, Aveiro, Portugal, 20–23 October 2010.

26. Caliò, I.; Marletta, M.; Pantò, B. A new discrete element model for the evaluation of the seismic behaviour of unreinforced masonry buildings. *Eng. Struct.* **2012**, *40*, 327–338. [CrossRef]

27. Lagomarsino, S.; Penna, A.; Galasco, A.; Cattari, S. TREMURI program: An equivalent frame model for the nonlinear seismic analysis of masonry buildings. *Eng. Struct.* **2013**, *56*, 1787–1799. [CrossRef]

28. Penna, A.; Lagomarsino, S.; Galasco, A. A nonlinear macroelement model for the seismic analysis of masonry buildings. *Earthq. Eng. Struct. Dyn.* **2014**, *43*, 159–179. [CrossRef]

29. Addessi, D.; Mastrandrea, A.; Sacco, E. An equilibrated macro-element for nonlinear analysis of masonry structures. *Eng. Struct.* **2014**, *70*, 82–93. [CrossRef]

30. Raka, E.; Spacone, E.; Sepe, V.; Camata, G. Advanced frame element for seismic analysis of masonry structures: Model formulation and validation. *Earthq. Eng. Struct. Dyn.* **2015**, *44*, 2489–2506. [CrossRef]

31. Peruch, M.; Spacone, E.; Camata, G. Nonlinear analysis of masonry structures using fiber-section line elements. *Earthq. Eng. Struct. Dyn.* **2019**, *48*, 1345–1364. [CrossRef]

32. Yousefi, B.; Soltani, M. An Equivalent Fiber Frame Model for Nonlinear Analysis of Masonry Structures. *Int. J. Archit. Herit.* **2021**, *15*, 644–668. [CrossRef]

33. Grande, E.; Imbimbo, M.; Sacco, E. Finite element analysis of masonry panels strengthened with FRPs. *Compos. Part B Eng.* **2013**, *45*, 1296–1309. [CrossRef]

34. Tondelli, M.; Rota, M.; Penna, A.; Magenes, G. Evaluation of uncertainties in the seismic assessment of existing masonry buildings. *J. Earthq. Eng.* **2012**, *16*, 36–64. [CrossRef]

35. Rota, M.; Penna, A.; Magenes, G. A framework for the seismic assessment of existing masonry buildings accounting for different sources of uncertainty. *Earthq. Eng. Struct. Dyn.* **2014**, *43*, 1045–1066. [CrossRef]

36. De Falco, A.; Guidetti, G.; Mori, M.; Sevieri, G. Model uncertainties in seismic analysis of existing masonry buildings: The Equivalent-Frame Model within the Structural Element Models approach. In Proceedings of the XVII Convengo Anidis, Pistoia, Italy, 17–21 September 2017; Pisa University Press: Pisa, Italy, 2017; pp. 63–73.

37. Tomić, I.; Vanin, F.; Beyer, K. Uncertainties in the Seismic Assessment of Historical Masonry Buildings. *Appl. Sci.* **2021**, *11*, 2280. [CrossRef]

38. Vanin, F.; Penna, A.; Beyer, K. A three-dimensional macroelement for modelling the in-plane and out-of-plane response of masonry walls. *Earthq. Eng. Struct. Dyn.* **2020**, *49*, 1365–1387.

39. Lagomarsino, S.; Cattari, S. PERPETUATE guidelines for seismic performance-based assessment of cultural heritage masonry structures. *Bull. Earthq. Eng.* **2015**, *13*, 13–47. [CrossRef]

40. Solarino, F.; Oliveira, D.V.; Giresini, L. Wall-to-horizontal diaphragm connections in historical buildings: A state-of-the-art review. *Eng. Struct.* **2019**, *199*, 109559. [CrossRef]

41. Tomaževič, M.; Weiss, P.; Velechovsky, T. The influence of rigidity of floors on the seismic behaviour of old stone-masonry buildings. *Eur. Earthq. Eng.* **1991**, *3*, 28–41.

42. Benedetti, D.; Carydis, P.; Pezzoli, P. Shaking table tests on 24 simple masonry buildings. *Earthq. Eng. Struct. Dyn.* **1998**, *27*, 67–90.

43. Dolce, M.; Ponzo, F.; Goretti, A.; Moroni, C.; Giordano, F.; De Canio, G.; Marnetto, R. 3d dynamic tests on 2/3 scale masonry buildings retrofitted with different systems. In Proceedings of the 14th World Conference on Earthquake Engineering, Beijing, China, 12–17 October 2008; Volume 1217.

44. Dolce, M.; Ponzo, F.C.; Di Croce, M.; Moroni, C.; Giordano, F.; Nigro, D.; Marnetto, R. Experimental assessment of the CAM and DIS-CAM systems for the seismic upgrading of monumental masonry buildings. In Proceedings of the PROHITECH, Rome, Italy, 21–24 June 2009; Volume 9.

45. Bothara, J.K.; Dhakal, R.P.; Mander, J.B. Seismic performance of an unreinforced masonry building: An experimental investigation. *Earthq. Eng. Struct. Dyn.* **2010**, *39*, 45–68. [CrossRef]

46. Mazzon, N.; Chavez, C.M.; Valluzzi, M.R.; Casarin, F.; Modena, C. Shaking Table Tests on Multi-Leaf Stone Masonry Structures: Analysis of Stiffness Decay. Structural Analysis of Historic Constructions. In *Advanced Materials Research*; Trans Tech Publications Ltd.: Bach, Switzerland, 2010; Volume 133, pp. 647–652. [CrossRef]

47. Magenes, G.; Penna, A.; Galasco, A. A full-scale shaking table test on a two-storey stone masonry building. In Proceedings of the 14th European Conference on Earthquake Engineering, Ohrid, North Macedonia, 30 August–3 September 2010.

48. Magenes, G.; Penna, A.; Senaldi, I.E.; Rota, M.; Galasco, A. Shaking table test of a strengthened full-scale stone masonry building with flexible diaphragms. *Int. J. Archit. Herit.* **2014**, *8*, 349–375. [CrossRef]

49. Senaldi, I.; Magenes, G.; Penna, A.; Galasco, A.; Rota, M. The effect of stiffened floor and roof diaphragms on the experimental seismic response of a full-scale unreinforced stone masonry building. *J. Earthq. Eng.* **2014**, *18*, 407–443. [CrossRef]

50. Costa, A.A.; Arêde, A.; Costa, A.C.; Penna, A.; Costa, A. Out-of-plane behaviour of a full scale stone masonry façade. Part 2: Shaking table tests. *Earthq. Eng. Struct. Dyn.* **2013**, *42*, 2097–2111. [CrossRef]

51. Vintzileou, E.; Mouzakis, C.; Adami, C.E.; Karapitta, L. Seismic behavior of three-leaf stone masonry buildings before and after interventions: Shaking table tests on a two-storey masonry model. *Bull. Earthq. Eng.* **2015**, *13*, 3107–3133. [CrossRef]

52. Pitilakis, D.; Iliou, K.; Karatzetzou, A. Shaking table tests on a stone masonry building: Modeling and identification of dynamic properties including soil-foundation-structure interaction. *Int. J. Archit. Herit.* **2018**, *12*, 1019–1037. [CrossRef]

53. Mouzakis, C.; Adami, C.E.; Karapitta, L.; Vintzileou, E. Seismic behaviour of timber-laced stone masonry buildings before and after interventions: Shaking table tests on a two-storey masonry model. *Bull. Earthq. Eng.* **2018**, *16*, 803–829. [CrossRef]

54. Kallioras, S.; Guerrini, G.; Tomassetti, U.; Marchesi, B.; Penna, A.; Graziotti, F.; Magenes, G. Experimental seismic performance of a full-scale unreinforced clay-masonry building with flexible timber diaphragms. *Eng. Struct.* **2018**, *161*, 231–249. [CrossRef]

55. Guerrini, G.; Senaldi, I.; Graziotti, F.; Magenes, G.; Beyer, K.; Penna, A. Shake-Table Test of a Strengthened Stone Masonry Building Aggregate with Flexible Diaphragms. *Int. J. Archit. Herit.* **2019**, *13*, 1078–1097.

56. Kim, S.C.; White, D.W. Nonlinear analysis of a one-story low-rise masonry building with a flexible diaphragm subjected to seismic excitation. *Eng. Struct.* **2004**, *26*, 2053–2067. [CrossRef]

57. Betti, M.; Galano, L.; Vignoli, A. Comparative analysis on the seismic behaviour of unreinforced masonry buildings with flexible diaphragms. *Eng. Struct.* **2014**, *61*, 195–208. [CrossRef]

58. Nakamura, Y.; Derakhshan, H.; Magenes, G.; Griffith, M.C. Influence of diaphragm flexibility on seismic response of unreinforced masonry buildings. *J. Earthq. Eng.* **2017**, *21*, 935–960. [CrossRef]

59. Nakamura, Y.; Derakhshan, H.; Griffith, M.C.; Magenes, G.; Sheikh, A.H. Applicability of nonlinear static procedures for low-rise unreinforced masonry buildings with flexible diaphragms. *Eng. Struct.* **2017**, *137*, 1–18. [CrossRef]

60. Gattesco, N.; Macorini, L.; Benussi, F. Retrofit of wooden floors for the seismic adjustment of historical buildings with high reversible techniques. Seismic Engineering in Italy. In Proceedings of the XII National Conference, Pisa, Italy, 25–27 October 2007; pp. 10–14.

61. Scotta, R.; Trutalli, D.; Marchi, L.; Pozza, L. Effects of in-plane strengthening of timber floors in the seismic response of existing masonry buildings. In Proceedings of the World Conference on Timber Engineering (WCTE), Vienna, Austria, 22–25 August 2016; pp. 22–25.

62. Scotta, R.; Trutalli, D.; Marchi, L.; Pozza, L.; Mirra, M. Seismic response of masonry buildings with alternative techniques for in-plane strengthening of timber floors. In Proceedings of the XII International Conference on Structural Repair and Rehabilitation (CINPAR), Porto, Portugal, 1 July 2017; pp. 26–29.

63. Mirra, M. Analisi Parametriche Dinamiche non Lineari Degli Effetti Dell' Irrigidimento di Solai in Legno in Edifici in Muratura Ordinaria. 2017. Available online: http://tesi.cab.unipd.it/59766/ (accessed on 12 April 2021).

64. Trutalli, D.; Marchi, L.; Scotta, R.; Pozza, L. Dynamic simulation of an irregular masonry building with different rehabilitation methods applied to timber floors. In Proceedings of the 6th ECCOMAS Thematic Conference (COMPDYN 2017), Rhodes Island, Greece, 12–14 June 2017; pp. 15–17.

65. McKenna, F.; Fenves, G.; Scott, M.; Jeremic, B. *Open System for Earthquake Engineering Simulation (OpenSees)*; University of California: Berkeley, CA, USA, 2000. Available online: http://opensees.berkeley.edu (accessed on 1 February 2021).

66. Ortega, J.; Vasconcelos, G.; Rodrigues, H.; Correia, M. Assessment of the influence of horizontal diaphragms on the seismic performance of vernacular buildings. *Bull. Earthq. Eng.* **2018**, *16*, 3871–3904. [CrossRef]

67. Vanin, F.; Penna, A.; Beyer, K. Equivalent-frame modeling of two shaking table tests of masonry buildings accounting for their out-of-plane response. *Front. Built Environ.* **2020**, *6*, 42. [CrossRef]

68. Penna, A.; Senaldi, I.E.; Galasco, A.; Magenes, G. Numerical simulation of shaking table tests on full-scale stone masonry buildings. *Int. J. Archit. Herit.* **2016**, *10*, 146–163. [CrossRef]

69. POLIMI. *Critical Review of Methodologies and Tools for Assessment of Failure Mechanisms and Interventions, Deliverable 3.3, Workpackage 3: Damage Based Selection of Technologies*; Technical Report; NIKER Project: Padova, Italy, 2010.

70. Brignola, A.; Podesta, S.; Pampanin, S. *In-Plane Stiffness of Wooden Floor*; Technical Report; Department of Civil and Natural Resources Engineering, University of Canterbury: Christchurch, New Zealand, 2008.

71. Brignola, A.; Pampanin, S.; Podestà, S. Experimental evaluation of the in-plane stiffness of timber diaphragms. *Earthq. Spectra* **2012**, *28*, 1687–1709. [CrossRef]

72. EC 1995-1-1-2004. *Eurocode 5—Design of Timber Structures—Part 1-1: General Rules and Rules for Buildings*; Technical Report; EN 1995-1-1: Bruxelles, Belgium, 2004.

73. Almeida, J.P.; Beyer, K.; Brunner, R.; Wenk, T. Characterization of mortar–timber and timber–timber cyclic friction in timber floor connections of masonry buildings. *Mater. Struct.* **2020**, *53*, 51. [CrossRef]

74. Moreira, S.M.T.; Oliveira, D.V.; Ramos, L.F.; Fernandes, R.; Guerreiro, J.; Lourenço, P.B. Experimental study on the seismic behavior of masonry wall-to-floor connections. In Proceedings of the 15th World Conference on Earthquake Engineering, Sociedade Portuguesa de Engenharia Sísmica (SPES), Lisbon, Portugal, 24–28 September 2012.

75. Ciocci, M.P.; Van Nimwegen, S.; Askari, A.; Vanin, F.; Lourenço, P.; Beyer, K. Experimental investigation on the behaviour of injection anchors in rubble stone masonry. 2021, in preparation.

76. Muñoz, R.; Lourenço, P.B.; Moreira, S. Experimental results on mechanical behaviour of metal anchors in historic stone masonry. *Constr. Build. Mater.* **2018**, *163*, 643–655. [CrossRef]

77. Contrafatto, L.; Cosenza, R. Prediction of the pull-out strength of chemical anchors in natural stone. *Frat. Integrità Strutt.* **2014**, *8*, 196–208. [CrossRef]

78. Vanin, F.; Beyer, K. Equivalent damping ratio for a macroelement modelling the out-of-plane response of masonry walls. 2021, in preparation.

79. Housner, G.W. The behavior of inverted pendulum structures during earthquakes. *Bull. Seismol. Soc. Am.* **1963**, *53*, 403–417.

80. Luzi, L.; Lanzano, G.; Felicetta, C.; D'Amico, M.C.; Russo, E.; Sgobba, S.; Pacor, F.; ORFEUS Working Group 5. *Engineering Strong Motion Database (ESM) (Version 2.0)*; Technical Report; Istituto Nazionale di Geofisica e Vulcanologia (INGV): Rome, Italy, 2020; [CrossRef]

81. EUCENTRE. PROVINO 1 0.40g 20090801 Telecamera 01. 2009. Available online: https://youtu.be/SspfD-nHCso?list=PL52352538BE37F38E (accessed on 2 April 2021).

82. EUCENTRE. PROVINO 1 0.40g 20090801 Telecamera 03. 2009. Available online: https://youtu.be/egYnpB1FlK0?list=PL52352538BE37F38E (accessed on 2 April 2021).

83. EUCENTRE. PROVINO 1 0.40g 20090801 Vista Multipla. 2009. Available online: https://youtu.be/7BFuXzrAbG0?list=PL52352538BE37F38E (accessed on 2 April 2021).

Article

Integration of Modelling Approaches for the Seismic Assessment of Complex URM Buildings: The Podestà Palace in Mantua, Italy

Sergio Lagomarsino *, Stefania Degli Abbati, Daria Ottonelli and Serena Cattari

Department of Civil, Chemical and Environmental Engineering, University of Genoa, 16145 Genoa, Italy; stefania.degliabbati@unige.it (S.D.A.); daria.ottonelli@unige.it (D.O.); serena.cattari@unige.it (S.C.)
* Correspondence: sergio.lagomarsino@unige.it

Abstract: This study investigated seismic assessments of the Podestà Palace in Mantua (Italy). This masonry palace has a complex geometrical configuration that resulted from the addition of various units stratified over centuries. This feature makes seismic assessment challenging from a modelling perspective due to the interaction among units. Here, an integrated use of three modelling strategies characterised by a different computational effort and degree of accuracy was employed: (i) the Structural Element Model, according to the Equivalent Frame Approach, to study the global response of the whole structure and to estimate the mutual dynamic interactions among units; (ii) the Macro-Block Model, to assess the out-of-plane response of facades prone to the activation of local mechanisms; and (iii) the Finite Element Model, to deepen the seismic response of some critical parts, highlighted by a global analysis but also roughly described by the Equivalent Frame Model. This integrated approach consists in the use of results achieved from one modelling approach as input for another. For example, the floor spectra estimated by (i) were used to define the seismic input in (ii); for assessing the most critical portions, more accurate models were addressed (as in case (iii)). The comprehensive interpretation of the seismic behaviour obtained by these models also allowed us to address more rationally possible strengthening solutions, such as the in-plane stiffening of vaults (particularly spread in the building), aimed to guarantee a better redistribution of seismic actions in such a complex building.

Keywords: seismic assessment; masonry; numerical models

Citation: Lagomarsino, S.; Degli Abbati, S.; Ottonelli, D.; Cattari, S. Integration of Modelling Approaches for the Seismic Assessment of Complex URM Buildings: The Podestà Palace in Mantua, Italy. *Buildings* **2021**, *11*, 269. https://doi.org/10.3390/buildings11070269

Academic Editor: Antonio Formisano

Received: 31 March 2021
Accepted: 21 June 2021
Published: 24 June 2021

Publisher's Note: MDPI stays neutral with regard to jurisdictional claims in published maps and institutional affiliations.

1. Introduction

The seismic protection of historical unreinforced masonry (URM) buildings represents a challenging issue because of their vulnerability and complex seismic behaviour. Their size and their complex configurations may combine the activation of both in-plane and out-of-plane (OOP) mechanisms, as testified by past earthquakes [1–6]. The spontaneous transformations, often stratified during their lifetime, may determine a significant in-plan and in-elevation geometric complexity. In many cases, the sequence of these additions led to structural units jointed together by means of connecting bodies (often built with timber floors and set up on arches or vaults), whose features govern the interactions among the different parts. As a consequence, the seismic assessment of such complex structures poses several critical issues.

The first one deals with the definition of the best modelling choice, which has to balance a reasonable computational effort with the need to guarantee a reliable assessment that is also able to catch the aforementioned interaction effects. Different modelling strategies can be adopted for masonry structures [1,7–10]. In particular, according to the classification proposed by [10], models can be distinguished in four classes, depending on the modelling scale (masonry material or structural elements) and the type of discretization (continuous or discrete):

- CCLM (Continuous Constitutive Law Models): finite element (FE) modelling with phenomenological or micromechanical homogenised constitutive laws;

- SEM (Structural Elements Models): equivalent frame modelling, based on the discretization in terms of piers and spandrels (where the nonlinear response is a priori concentrated) and other linear and nonlinear elements (e.g., aimed to simulate the presence of RC/steel/timber beams or tie rods);
- DIM (Discrete Interface Models): discrete modelling of blocks and interfaces;
- MBM (Macro-Blocks Models): modelling of rigid bodies, whose geometry represents the main input. As better described below, MBM typically implements limit analysis-based solutions that can be based on either static or kinematic theorems [11].

The choice of the best modelling strategy to be employed depending on the class of architectural assets under examination can be addressed by considering the recurring expected damage and seismic behaviour, as discussed in [12] (Figure 1). Moreover, an appropriate knowledge phase constitutes the preliminary but essential requisite to address this choice, which, on such complex assets, presupposes the integration of various tools, such as historical analysis, in-depth visual structural in-situ surveys, and experimental tests, whose invasiveness needs to be properly calibrated. Emblematic examples of this are illustrated in [13,14].

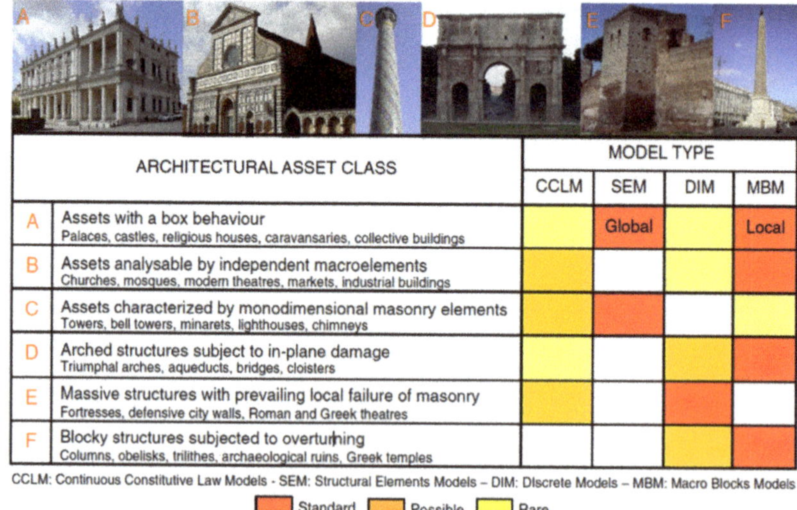

Figure 1. Classification of architectural assets and related possible modelling strategies (from [10]).

The second issue deals with the choice of analysis approach. Whatever modelling strategy has been adopted, it has to be able to accurately reproduce the nonlinear response of masonry. Indeed, masonry is characterised by a very low tensile strength that causes the onset of cracking phenomena for low stress levels. In general, two kinds of nonlinear analysis procedures may be used [9]:

- Incremental analysis procedures, which investigate, step by step, the evolution of the equilibrium conditions of a structure subjected to certain actions. This class includes both nonlinear static analyses and nonlinear dynamic (time-history) analyses.
- Limit-analysis-based solutions, which can be employed under the static or kinematic theorems and which are based on the assumptions of Heyman [11], who firstly applied the limit theorems of plasticity to masonry structures.

Finally, the last issue deals with the appropriate procedure to perform the seismic assessment and the verification of such complex URM structures. While standards (i.e., [15–17]) essentially focus on ordinary buildings, the assessment of cultural heritage assets is treated in some guideline documents ([18–20]), which are not specifically addressed to seismic vulnerability but consider all possible causes of damage and deterioration, with the aim of making a diagnosis and designing a strengthening intervention. In these documents, the use of qualitative approaches is usually favoured. Conversely, the Italian Guidelines for the seismic assessment of cultural heritage ([21]) clearly stated the additional need for a quantitative approach outlining general principles and proposing simplified approaches for a preliminary estimate. Despite this, specific recommendations on models or procedures for a detailed assessment are missing in this document. At present, various possibilities were explored in literature, although a standardised and unanimously recognised approach is not yet available. Casolo and Sanjust (2009) [22] applied a multi-step strategy that combines traditional CCLM models with a new rigid body spring model (RBSM) to assess the seismic vulnerability of a castle in Syracuse (Italy). Garofano and Lestuzzi (2016) [23] used a 3D applied element method (AEM) model [24] to assess the response of a historical URM building in Switzerland; the authors then identified other possible local failure mechanisms that were studied by means of kinematic limit analysis. Rossi et al. (2015) [25] applied the integrated use of three different modelling strategies (CCLM, MBM, and SEM) to analyse the Great Mosque of Algiers through nonlinear (static and kinematic) analyses, discussing also how the use of such models may change the current state of the building to a strengthened state. Degli Abbati et al. (2019) [26] proposed a numerical five-step procedure based on the use of nonlinear static analyses with load patterns calibrated from the results of a modal analysis; the procedure was firstly applied to a medieval fortress made by various towers interacting with the main body of the structure. Malcata et al. [27] compared and verified the combined use of the CCLM and SEM models on the Bonet building in Sintra.

Within this multifaceted context, the paper focuses on the seismic assessment of the Podestà Palace in Mantua (Italy), struck by the 2012 Emilia earthquake [2,28–31].

The palace was built and transformed over decades leading to its present configuration, characterised by the addition of units interconnected through deformable bodies (Section 2).

In order to interpret its seismic behaviour, we applied a strategy based on the use of three different modelling approaches, employed in an integrated way in order to limit the computational effort in performing nonlinear static and kinematic analyses. In particular, the approach adopted includes:

- Nonlinear kinematic analyses and an MBM modelling strategy, to study the response of facades most vulnerable to the activation of local out-of-plane mechanisms (Section 3). In particular, the north-east façade overlooking the Broletto square was analysed.
- Modal analysis and nonlinear static analyses executed on a 3D SEM model of the whole palace, to interpret the seismic response of this complex building. This model allowed us to explicitly account for the interactions between units and quantify the seismic safety index (Section 4). The global model clearly highlighted some vulnerable units that were then studied with more refined approaches. Moreover, the results of this 3D model were useful to refine the estimate of the seismic input (i.e., floor spectra) employed to verify the out-of-plane mechanisms.
- Nonlinear static analyses performed on the units identified by the SEM model and modelled with the CCLM approach (Sections 5 and 6).

Examining the seismic behaviour of the palace in the present state also allowed us to outline some possible strengthening solutions, parametrically investigated and specifically conceived to balance the aims of safety and preservation (Section 6).

2. Case-Study Description: The Podestà Palace in Mantua (Italy)

The examined case study is one of the most important buildings in the city of Mantua. It was built from the end of the twelfth century to the mid-fifteenth century to host the municipal government, and it has since transformed many times during its lifetime. At present, the palace is an aggregation of three main buildings (Figure 2), namely, the Podestà

Palace, the Masseria, and the Ragione Palace. The latter two are connected by two smaller bodies, the Arengario and the so-called Podestà's vaults. The Podestà Palace itself is composed of different units that date back to various periods and is located around a narrow courtyard. This fragmentation favoured the presence of many occlusion bodies, such as the thirteenth-century unit placed between the Torre delle Ore and the Palatium Novum, where the northeast façade overlooking the Broletto square is placed (Figure 2).

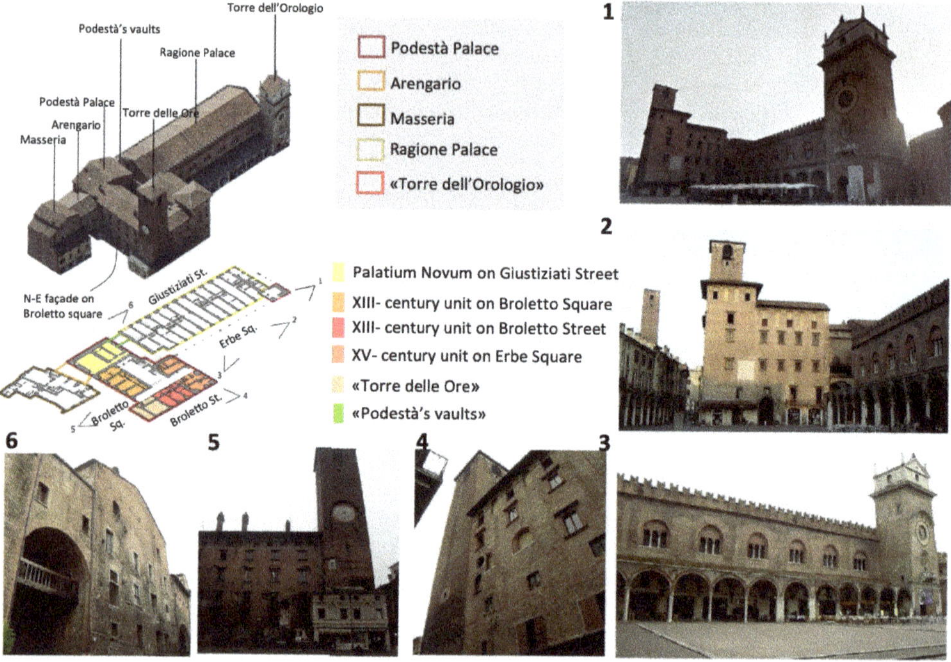

Figure 2. Overview of the different units of the palace.

When the Emilia earthquake occurred in May 2012, the palace was undergoing an extended restoration program, mainly addressed to reusing its spaces as a museum and as offices. Within this program, detailed in-situ surveys and in-situ non-destructive and minor-destructive tests were carried out (e.g., single and double flat jacks, Darmstadt tests, and endoscopic tests; Figure 3); they were then made available for the research aims herein presented. These data, integrated by the historical analysis, allowed us to define the structural features and the mechanical properties of materials: all essential aspects for the definition of a reliable numerical model. Moreover, flues, recesses, infill openings, and non-continuous walls were identified, useful for addressing specific solutions in the equivalent frame idealisation of walls. These irregularities are widespread in the buildings due to the structural and functional changes that often occurred over the centuries.

The building is composed of a solid brick and lime mortar masonry. The wall thickness ranges from 0.30 to 0.60 m, while it is higher in the Torre delle Ore (about 2 m thick). The diaphragms are flexible, with alternating timber floors or brick vaults (Figure 4a). Barrel vaults are prevailing at the lower levels, except for two cross vaults in the fifteenth-century unit. Roofs are realised with timber elements whose thrust is not always effectively absorbed.

Although the epicentre of the Emilia earthquake 2012 was rather far from Mantua, the palace exhibited some non-negligible damage. It mainly consisted of shear cracks in some walls of the Masseria, cracks in the vaults of the fifteenth-century unit, and cracks in the Arengario arches. Moreover, the earthquake determined a clear detachment of the façade overlooking the Broletto square. Such detachment was also detected by a laser scanner survey performed after the seismic event. The worsening in the detachment due to the earthquake was proven by comparing these data with those of another laser scanner survey dating back to 2007 (Figure 4b). In particular, the right part of the façade (coloured in red in Figure 4b) was subjected to a more significant deformation, indicating its possible incipient overturning. This diagnosis is also consistent with other evidence: cracks at the pavement level and detachments of barrel vaults from the main façade at the ground floor were not present, but they gradually increased as the upper floors were examined [32].

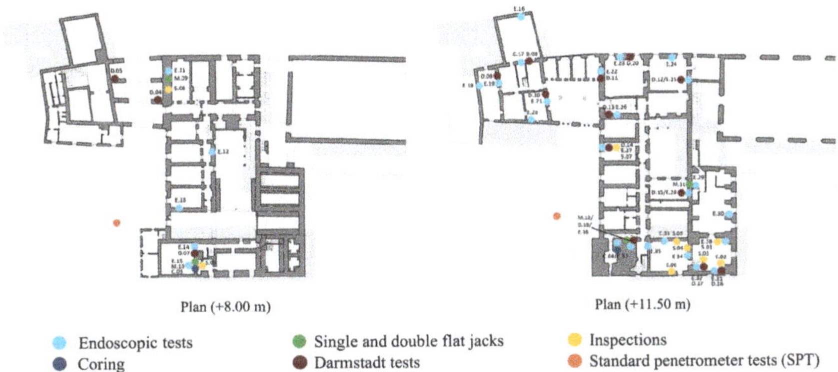

Plan (+8.00 m) Plan (+11.50 m)

- 🔵 Endoscopic tests 🟢 Single and double flat jacks 🟡 Inspections
- 🔵 Coring 🔴 Darmstadt tests 🟠 Standard penetrometer tests (SPT)

Figure 3. Location in plan of some available in-situ tests carried out at the 3rd and 4th levels.

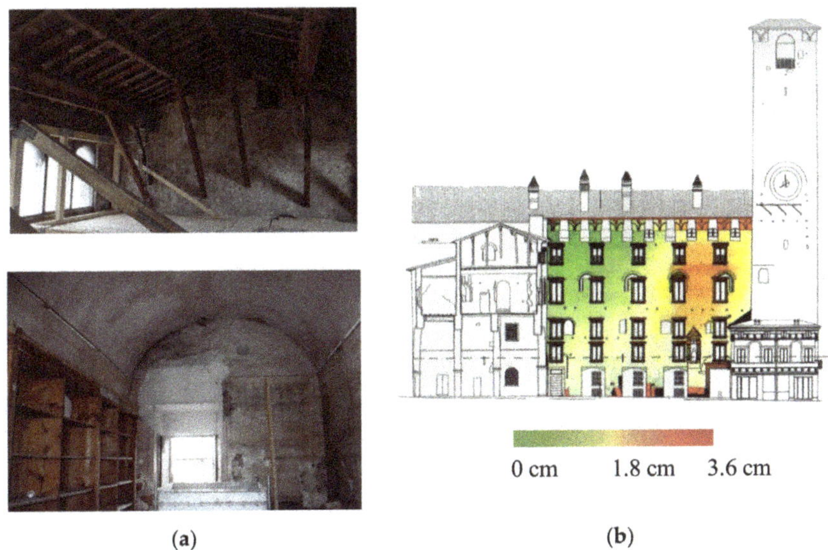

(a) (b)

Figure 4. (a) Floors types (timber diaphragms and vaults); (b) out-of-plane deformations on the façade overlooking the Broletto square from laser scanner measurements made before and after the seismic event.

3. Out-of-Plane Seismic Assessment of the Façade Overlooking the Broletto Square

3.1. Adopted Criteria for the Out-of-Plane Modelling by the Macro-Block Model Approach

Non-linear kinematic analyses on a macro-block model (MBM) of macroelements identified as those prone to the out-of-plane response were performed with the software MB-Perpetuate [33], which was originally developed in the framework of the Perpetuate

131

Project [10]. Among the various methods based on the limit analysis available in the literature and codes, in which the state of the-art is exhaustively described [34–36], in this paper, the OOP seismic assessment was carried out according to the criteria proposed by [37,38]. The aim of the assessment is to compute the maximum peak ground acceleration compatible with the fulfilment of given Limit States ($a_{g,OOP,LS}$). In the following, more details are provided for the evaluation of the capacity curve (i), the definition of the assumed limit states (ii), and the criteria adopted for the comparison with the seismic input (iii).

Regarding the first aspect (i), Figure 5a shows a capacity curve where four Damage Levels (DLs) are identified. The latter are then associated with corresponding LSs for the aim of the seismic verification according to codes. As can be seen, a bi-linear curve is assumed that differs from the ideal rigid block behaviour in order to account for the actual response of URM walls testified by experimental campaigns [38] and numerical investigations [39]. The curve is defined by the first elastic branch defined by two periods, namely, T_e (initial elastic period) and T_s (secant period obtained by the incremental kinematic analyses).

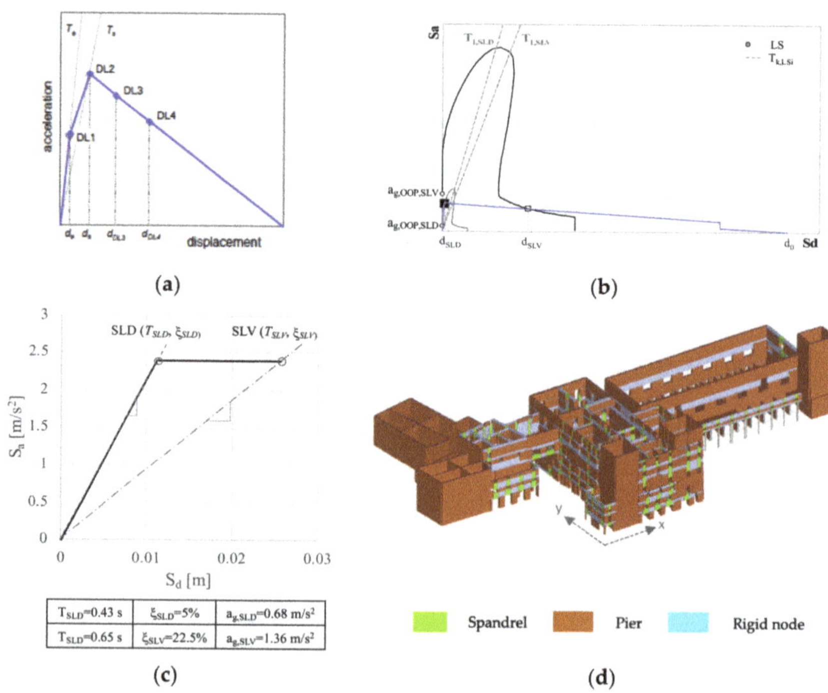

Figure 5. (a) Capacity curve of the local mechanism and LS definition; (b) floor spectra evaluation and seismic assessment—figures adapted from [32]; (c) value of T_{LSi} and ξ_{LSi} considered for the floor spectra definition as defined on the bi-linear curve of the global model; (d) 3D view of the global SEM model (see Section 4 for more details).

The seismic assessment was then performed by establishing the following correspondence between DL and LS (ii):

- DL2, which corresponds to the activation of the mechanism: it was assimilated to the Damage Limit State (SLD).
- DL3, which corresponds to a fraction of the displacement d_0, d_0 being the displacement associated with the value of the load multiplier α equal to zero. Although different proposals are available in the literature and codes (e.g., [17,36,40]), where DL3 was

assumed equal to $0.25d_0$, according to [37]. DL3 was then assimilated to the Life Safety Limit State (SLV).

The notation of LSs (SLD and SLV) is consistent with the Italian Technical Code [16,17].

Finally, passing to the third issue (iii), for local mechanisms that involve upper parts of the building (as in the examined case), the seismic demand is usually expressed in terms of floor spectra, which aim to account for the filtering effect provided by the main structure and the consequent amplification phenomena (Figure 5b). Among the available expressions proposed in the literature (e.g., [41,42]), the analytical formulation proposed by [43] was used. This expression gives the floor spectrum as a function of the acceleration response spectrum of the ground motion, the dynamic properties of the main structure (i.e., the periods T_{LSi} and the corresponding modal shapes), the equivalent damping of the main structure (ξ_{LSi}), and that of the portion involved in the local mechanism. This also allows us to consider the effect of higher modes that have to be properly combined. The suffix "LS" in the properties of the main structure recalls that they need to account for their evolution with the progress of the structure in a nonlinear range. The reliability of this analytical approach was recently proven in [44], thanks to the recordings acquired on some permanently monitored URM buildings.

The results of the global 3D model, discussed in Section 4.1, have been revealed to be particularly useful in defining the dynamic properties of the building and their evolution in the nonlinear range. In the examined case, the floor spectrum accounts only for the contribution of the first mode, while the structural period T_{LSi} and the corresponding damping ratio ξ_{LSi} were defined from the nonlinear analyses performed on the global model (Section 4). The floor spectrum was computed for each LS by adopting an equivalent secant period of the structure (T_{LSi}), derived from the pushover curves obtained by pushing the building in the X direction (the one that activates the out-of-plane response of the façade) and considering a pseudo-triangular load pattern (Figure 5d). The associated equivalent damping (ξ_{LSi}) was instead computed by applying some common expressions proposed in the literature [45]. Figure 5c shows the global pushover curve converted into an equivalent bi-linear one, defined according to [17]. The reference LSs are identified on this bilinear curve, together with the values of T_{LSi} and ξ_{LSi} considered for the floor spectrum definition. The corresponding values of the peak ground acceleration compatible with the fulfilment of the same LSs at the global scale ($a_{g,LS}$) are also specified (see Section 4 for more details). The values of $a_{g,OOP,LS}$ and $a_{g,LS}$ must be consistent with each other, and in general an iterative procedure is required [32], since the properties of the main structure corresponding to $a_{g,LS}$ constitute the input data for the computation of the floor spectrum necessary to evaluate $a_{g,OOP,LS}$.

3.2. Macroelements and Mechanisms Examined

Two macroelements were considered (Figure 6a): Macroelement 1 involves the entire façade overlooking the Broletto square (Figure 6a); Macroelement 2 involves a smaller part of this façade, as identified in Figure 6a2. The latter portion was defined starting from the observed seismic damage and the results of the laser-scanner analyses (Figure 4b). In both cases, a complete lack of connection among the façade, the orthogonal internal masonry walls, and the horizontal diaphragms was assumed. This assumption is coherent with the evidence from the in-situ survey. Furthermore, for each macroelement, two different mechanisms were evaluated (Figure 6c), by alternatively considering the retaining action guaranteed by the timber trusses of the roof (Figure 6b), as "not completely" effective (Type 1) or "fully" effective (Type 2). This aspect was included in the assessment as an epistemic uncertainty, since it was not possible to exhaustively verify it from the in-situ inspections.

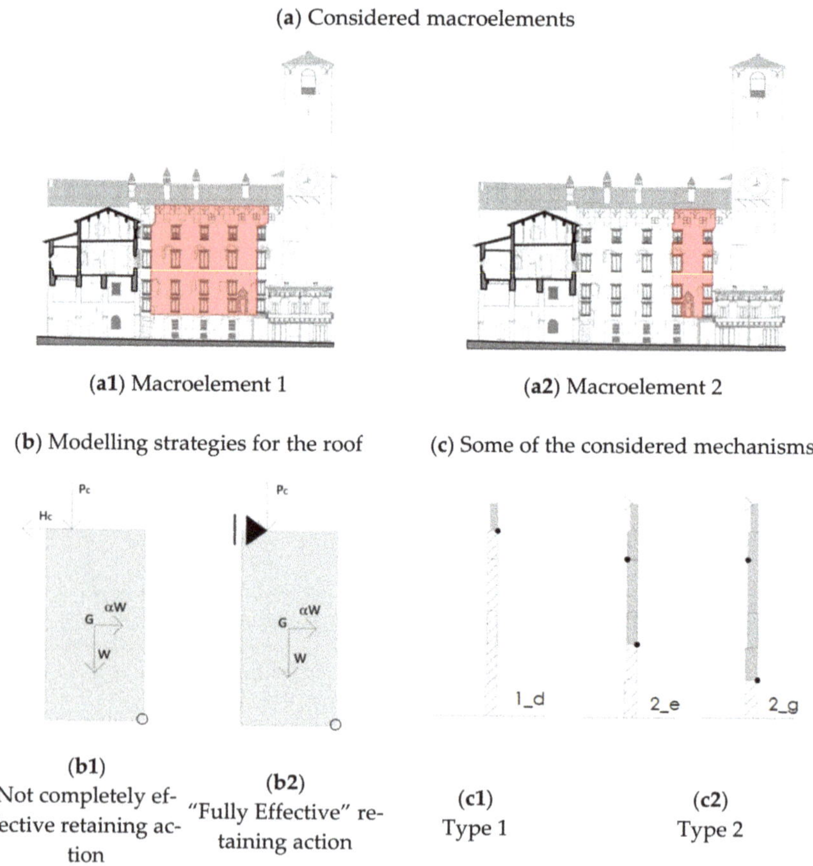

Figure 6. (**a**) Macroelements considered in the analyses; (**b**) different assumptions and corresponding modelling strategies for the roof action; (**c**) corresponding considered mechanisms (for Macroelement 2). Please refer to [32] for the abacus of all the considered mechanisms.

More specifically, the considered mechanisms are summarised as follows:

- Mechanisms of Type 1: These involve the overturning of a single block, with a single hinge that was parametrically considered at different heights of the façade (indicated by the suffixes "_a", "_b", etc.). The retaining effect of the trusses was modelled as an external stabilizing action, applied in their application point (about 20 cm). The interaction with the trusses presupposes two components: a vertical one (P_C) equal to the loads of the entire roof and an external horizontal action equal to $H_C = \mu P_C$ (where μ is the friction coefficient assumed equal to 0.2). In particular, in the analyses, the latter was considered effective until the truss extraction.
- Mechanisms of Type 2: In this case, the roof was modelled as a constraint able to prevent the overturning of the top of the façade. In this way, the development of mechanisms involving two blocks is likely to occur, with one hinge at the base and the other placed in an intermediate point between the base and the level of the roof. Again, parametric analyses were performed investigating the sensitivity of results to various hinge positions (indicated by the suffix "_e", "_g", etc.).

These two types of mechanisms were considered for both the identified macroelements. Figure 6c shows by way of example the mechanisms analysed for Macroelement 2 (associated with the hinge position "_d", "_e", and "g"). A more complete set of mechanisms is presented in [32]. It should be specified that, in all the analyses, the effect of the limited compressive strength of masonry was considered by properly moving back the plastic hinge across the section; the entity of that distance was calibrated as a function of the actual compressive stress acting on the various levels.

3.3. Results of Out-of-Plane Seismic Assessment

For the aim of seismic verification, the maximum peak ground acceleration compatible with the fulfilment of a given LS ($a_{g,OOP,LS}$) was compared with the one associated with the design spectrum ($a_{g,site,LS}$). Table 1 illustrates the results obtained for Macroelement 2, which was the most vulnerable one and the one actually activated, as also highlighted by the laser-scanner survey. In the table, for the most representative mechanisms, the results are illustrated in terms of the peak ground acceleration corresponding to the fulfilment of SLD ($a_{g,OOP,SLD}$) and SLV ($a_{g,OOP,SLV}$), the corresponding design peak ground acceleration of Mantua ($a_{g,site,SLD} = 0.5$ m/s^2; $a_{g,site,SLV} = 1.109$ m/s^2), and the resulting safety factors ($\alpha_{SLD} = a_{g,OOP,SLD}/a_{g,site,SLD}$ and $\alpha_{SLV} = a_{g,OOP,SLV}/a_{g,site,SLV}$). Values lower than 1 correspond to a not-satisfied verification. It should be noticed that, when the acceleration corresponding to the fulfilment of a given DL at the scale of the local mechanism was higher than the corresponding one at the scale of the whole building, the most precautionary condition was considered as a reference. Indeed, in the case of the SLV, the most punitive condition was always that at a global scale: this implies that the whole building reaches the SLV before the same condition occurs for the façade subjected to the out-of-plane response.

Table 1. Results for Macroelement 2.

	$a_{g,OOP,SLD}$ [m/s^2]	$a_{g,OOP,SLV}$ [m/s^2]	α_{SLD}	α_{SLV}	$\alpha_{Mantua,2012}$
Mech1_d	0.55	1.36 *	1.099	1.227	1.869
Mech2_e	0.14	1.36 *	0.280	1.227	0.476
Mech2_g	0.15	1.36 *	0.300	1.227	0.510

* In these cases, the verification at a global scale was more punitive. Thus, the value refers to this condition.

Furthermore, the OOP response of the selected mechanisms were also compared with the value of the acceleration estimated for the seismic event which hit Mantua in 2012 (equal to 0.29 m/s^2, as deduced from the shaking map—$\alpha_{Mantua,2012}$ in the table). In this latter case, the values of the coefficients $\alpha_{Mantua,2012}$ (computed by assuming as reference the $a_{g,OOP,SLD}$ values) are generally lower than 1. This agrees with the actual activation of the mechanism and confirms the reliability of the model employed.

These results highlighted the vulnerability of the façade to the OOP response and the necessity to design a proper strengthening intervention to prevent the mechanism activation. The simplest but at the same time effective proposal consists in improving the connection of the façade with the roof and with the transversal walls by means of steel tie rods. The optimization of this strengthening intervention was evaluated by varying the number and location of the tie rods, as exhaustively discussed in [32].

4. Seismic Assessment of the Global Response in the Current State of the Palace

4.1. Adopted Criteria for the Global Modelling by the SEM Approach

In order to explicitly consider the interactions between the different units of the palace, a SEM model of the whole complex building was set up. The model includes the main bodies object of the seismic assessment (i.e., the Podestà Palace and the Masseria), the adjacent ones (i.e., the urban aggregate close to the Masseria and the Ragione Palace), and the connection bodies (the Arengario and the Podestà's vaults). The 3D model was realised using the 3Muri software [46], based on the solver developed by [47]; the research version of this software was then adopted to perform the nonlinear analyses. As already mentioned, the software package works according to the equivalent frame (EF) model approach (Figure 7a).

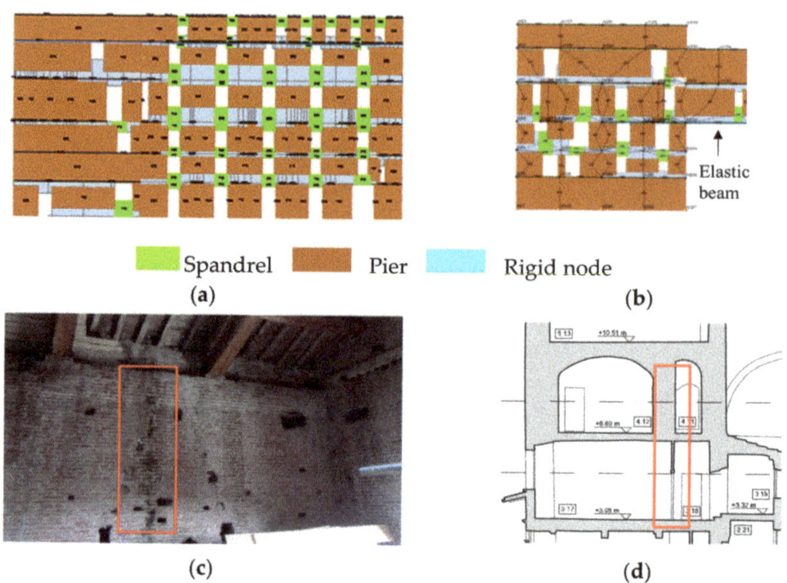

Figure 7. (**a**) Equivalent frame idealization of the façade overlooking the Broletto square; (**b**) example of modelling strategies assumed to guarantee a proper actions' repartition to the structural elements under the non-continuous walls; (**c**,**d**) examples of vulnerability-driving factors: a flue and a non-continuous wall, respectively.

According to the EF approach, only the in-plane response of the URM walls is considered, and each wall is discretised by a set of masonry panels (piers and spandrels), where the non-linear response is concentrated, connected by a rigid area (nodes). The wall idealisation into an equivalent frame is a tricky step. It affects both the elastic field, since it alters the actual deformability of the wall due to the simplification of introducing rigid nodes, and the nonlinear phase of the response, since the regions where the cracks and nonlinearity are likely to develop are assumed a priori. Despite these simplifications, this approach is one of the most spread both in engineering practice and at the research level thanks to its computational efficiency in performing nonlinear analyses (e.g., in [48]) and its reasonable accuracy, as proven by various numerical simulations in the literature (e.g., in [49,50]). In the examined case, performing a complete model of the whole aggregate with other approaches, such as FEM, would have been almost unfeasible. Of course, when applied to monumental buildings, particular attention has to be paid to its correct use starting from the aforementioned EF idealisations of walls that may present various sources of irregularities (in the opening layout, in the heterogeneity of materials, in the discontinuities, etc.). An overview on the issue, including the repercussions of irregularities in the performance-based assessment, is presented in [51]. Regarding the rules for EF idealisation specifically, although different criteria were proposed in the literature (e.g., [1,52–54]), no agreed rules have been unanimously yet recognised. Moreover, it was shown that, in general, the dispersion of the results increases with a pronounced irregularity in the opening patterns (e.g., [55,56]). Recent research ([55,57]), based on both numerical investigations and the analysis of the in-plane severe damage of real buildings, has highlighted that the criteria proposed in [1] give quite reliable results: thus, they were here assumed as a reference. Moreover, in general, it was verified that, in the model, the rigid nodes were not too big to limit the alterations induced to the wall stiffness. By way of example, Figure 7a shows the EF idealization of the façade overlooking the Broletto square.

Apart from that, the setup of the EF model requires great care and accuracy in transferring as much information acquired during the knowledge phase as possible into the model (as discussed for example also in [27,50]). Some critical situations (Figure 7c,d) that may potentially reduce the strength and stiffness of structural elements (such as the presence of flues, recesses, infill openings, and non-continuous walls) were taken into account with specific expedients:

- The flues and internal cavities were modelled as equivalent openings.
- The recesses were modelled alternatively: directly as openings (with a height equal to the one detected during the in-situ inspections), if their thickness was higher than half thickness of the wall, or as URM panels with a thickness and height equal to those measured in loco.
- All the buffered openings were modelled as URM panels with half the thickness of the original wall.
- Elastic beams were introduced under all the non-continuous walls in order to allow a proper transferring of the actions at ground level through the adjacent elements (Figure 7b).

Starting from the 2D modelling of walls, the complete 3D model was obtained by also introducing diaphragm elements, modelled as orthotropic membrane finite elements. The equivalent elastic moduli were calibrated according to [40,58,59] for vaults and timber floors, respectively. The ability of the software to model flexible and stiff diaphragms as well is crucial for the building under examination, for which the adoption of rigid floors would be inadequate. A view of the global 3D model is presented in Figure 5d.

Table 2 presents the mechanical parameters assumed to describe the response of URM panels; the latter correspond to the mean values of the range proposed in the Italian Technical Code [17] for a masonry built with solid bricks and lime mortar. The adoption of these values was supported by the results of the available diagnostic techniques performed on the building (Figure 3). Moreover, they are also consistent with the results of in-plane cyclic tests on masonry panels available in the literature (e.g., [60,61]).

Table 2. Mechanical parameters assumed in the SEM model.

	$E^{(1)}$ [MPa]	$G^{(1)}$ [MPa]	w [kN/m^2]	f_m [N/cm^2]	τ_0 [N/cm^2]
Solid brick and lime mortar	1500	500	18	320	7.6

$^{(1)}$ Un-cracked values; w: average specific weight; f_m: compressive strength; τ_0: shear strength. The strength values have to be divided by the CF, assumed equal to 1.2.

The strength values were reduced by applying a confidence factor (CF) equal to 1.2. This value corresponds to that proposed in Codes [15,16] assuming the attainment of a level of knowledge KL2 and, as known, aims to conventionally account for the residual uncertainties in the final assessment. In fact, despite the quite accurate knowledge acquired on structural details, the KL2 derives from the limited in-situ investigations on the materials, reduced in number to limit the invasiveness.

Both a linear modal and a nonlinear static analysis were performed on the model.

In the nonlinear field, the piecewise-linear constitutive laws proposed by [1,48] and implemented in Tremuri software were adopted to describe the response of masonry panels. This constitutive law (Figure 8) simulates the nonlinear response until very severe damage levels (from 1 to 5) are reached through progressive strength degradation (defined in terms of residual strength β_{Ei}) in correspondence with the assigned values of drift δ_{Ei}; the latter are differentiated as a function of the different possible failure modes (e.g., rocking, diagonal cracking, bed joint sliding, or mixed modes) and of the element type (if spandrel or pier). The ultimate strength of URM panels is computed according to the strength criteria commonly proposed in the literature (e.g., discussed in [62,63]) and codes (e.g., in [15,16,40]) for regular masonry, as that under examination. Table 3 summarises the main parameters assumed in the nonlinear analyses for piers and spandrels. Drift thresholds are compatible with those indicated by recent databases that have collected the results of several experimental results [60]. In the case of spandrels, the assumed residual strength is justified by the presence of an effective architrave; the value is compatible with those proposed in [40].

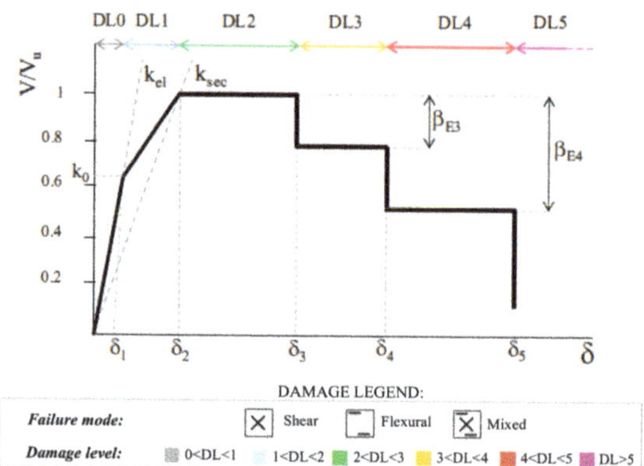

Figure 8. Piecewise-linear constitutive law and hysteretic response of the model.

Table 3. Thresholds used for piers and spandrels. For piers, the first and second values were assumed for a prevailing flexural behaviour and shear behaviour, respectively.

	δ_{E3} [−]	δ_{E4} [−]	δ_{E5} [−]	β_{E3} [%]	β_{E4} [%]
Piers	0.006–0.003	0.01–0.005	0.015–0.007	0–30	15–60
Spandrels	0.002	0.006	0.02	50	50

4.2. Evidence from the Modal Analysis on the Interaction among Units

First of all, the modal analysis was performed to interpret the dynamic response of the whole palace. The results of the building in the original configuration (hereafter referred to as "flexible floors") were compared with a stiffer one (hereafter referred to as "stiffer floors"), in order to evaluate the effects of a stiffening of the membranes on the global response. This stiffening aims to simulate the possible replacement of the original filler of the vaults (made by heavy incoherent materials and debris) with a new one, lighter but with non-negligible mechanical properties (see also Section 6). Thus, the simulation of this intervention also produces a decrease in the overall mass of the model.

Table 4 shows the main results of the modal analysis in terms of period (*T*), frequency (*n*), and percentage of participation mass (%M_x, %M_y, and %M_z). The results were illustrated by selecting the first 10 modes; the most significant ones in terms of participation mass are marked in grey. Although Mode 3 had >15% mass in the Y direction, it mostly activated the dynamic response of the Ragione Palace, which was not the main focus of the analyses; for this reason, Mode 3 is not highlighted in the table.

In general, it is possible to observe that the different units mutually interact in a very different way, considering the X or the Y directions. For example, in the Y direction, the Podestà Palace almost behaves as two distinguished bodies, i.e., the thirteenth-century unit overlooking the Broletto square and the fifteenth-century unit overlooking Erbe square. They are connected through highly flexible elements. Furthermore, the building overlooking the Giustiziati street strongly interacts with the Arengario and the Ragione Palace. Finally, the area overlooking the Broletto street is mostly affected by the interaction with the tower.

Figure 9 compares the most significant modal shapes in the X and Y directions, focusing on the area of the Podestà Palace and considering the two aforementioned hypotheses.

Table 4. Results of the modal analysis configurations with flexible and stiffer floors.

Flexible Floors (M_{TOT} = 30,996,265 kg)										
Mode	**1**	**2**	**3**	**4**	**5**	**6**	**7**	**8**	**9**	**10**
T [s]	0.61	0.52	0.51	0.51	0.51	0.48	0.47	0.43	0.41	0.41
n [Hz]	1.63	1.92	1.96	1.96	1.98	2.08	2.13	2.34	2.42	2.46
$\%M_x$	0.22	7.89	0.56	16.11	4.24	11.75	11.11	13.93	1.14	5.02
$\%M_y$	0	0.19	15.39	9.10	1.78	24.37	2.17	0.03	7.96	0
$\%M_z$	0	0	0	0	0	0	0	0	0	0
Stiffer Floors (M_{TOT} = 28,951,395 kg)										
Mode	**1**	**2**	**3**	**4**	**5**	**6**	**7**	**8**	**9**	**10**
T [s]	0.61	0.52	0.51	0.49	0.47	0.46	0.42	0.41	0.40	0.39
n [Hz]	1.63	1.92	1.96	2.05	2.15	2.18	2.39	2.42	2.47	2.58
$\%M_x$	0.17	4.44	0.65	25.47	12.07	0.01	8.26	1.02	0.74	14.25
$\%M_y$	0	0.19	8.18	4.60	12.14	24.24	4.28	0.20	0	2.08
$\%M_z$	0	0	0	0	0	0	0	0	0	0

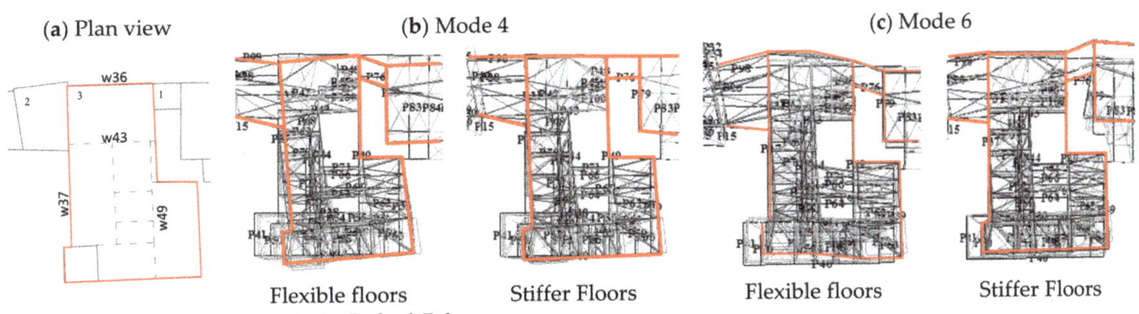

(a) Plan view   **(b)** Mode 4   **(c)** Mode 6

Flexible floors   Stiffer Floors   Flexible floors   Stiffer Floors

1—Podestà's vaults; 2—Arengario; 3—Podestà Palace

Figure 9. (**a**) Identification of mainly interacting units; comparison of modal shapes for Modes 4 (**b**) and 6 (**c**) in the configurations with flexible or stiffer floors.

The areas subjected to the most significant distortions are the unit facing the Broletto square, due to the deformability of the vaults (see Mode 4), and the connection bodies (namely, 1 and 2 in Figure 9), due to the very different stiffnesses that characterised the linked units (Mode 6). Moreover, it is worth observing that these distortions are reduced in the case of a slight stiffening of the diaphragms. In fact, this produces more regular behaviour, guaranteeing a better redistribution of the actions.

4.3. Results of Nonlinear Static Analyses

This section presents the results of the global nonlinear analyses performed on the palace in the present state (namely, flexible floors). In particular, the following steps were executed:

1. Nonlinear static analyses were performed on the global EF model.
2. Pushover curves of each single unit were extracted by plotting the shear at the base of each body (V) as a function of the mean displacement of the nodes placed at the last floor in the same area (d). Different control nodes were defined for each unit in order to accurately describe its evolution of the nonlinear response (Figure 10a). In particular, Nodes 1 and 2 were chosen for the pushover analyses performed on the Podestà Palace in the X and Y directions, respectively, while Node 3 was defined for the Masseria.
3. The corresponding capacity curves ($V^* - d^*$) of the equivalent Single Degree of Freedom (SDOF) system were defined, by following the general principles of [64], based on the evaluation of the participation coefficient Γ and the mass M^* of each unit (having extracted from the 3D model the data related to each of them). Thus, each

capacity curve was obtained by dividing the displacement d by Γ ($d^* = d/\Gamma$) and the base shear by the product ΓM^* ($V^* = V/(\Gamma M^*)$), where Γ and M^* were related to the unit under investigation.

4. Finally, for the seismic verification, the capacity curve of each unit was compared with the seismic demand.

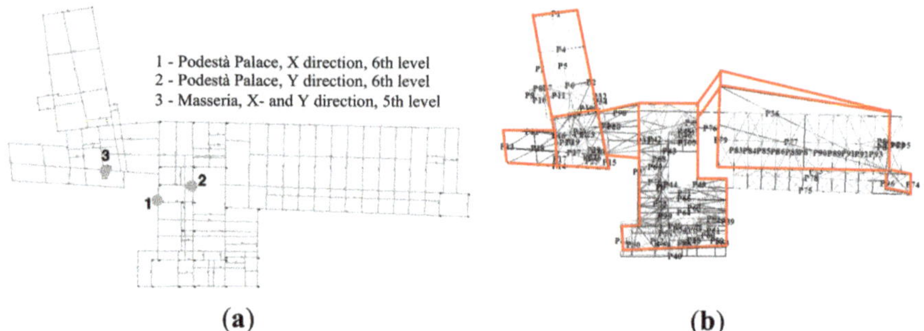

(**a**)   (**b**)

Figure 10. (**a**) Control nodes' definition; (**b**) deformed shape in plan of the whole complex in the +Y-direction (scale factor equal to 60).

The choice of the load pattern in such a complex case represents a tricky point, since the modal analysis highlighted several modes characterised by a moderate participant mass.

The use of a single load pattern kept invariant during the analysis may appear questionable, but other possibilities, such as the multimodal ([65,66]) or adaptive [67] pushover analyses, are, compared to that under investigation, equally poorly validated in the context of complex URM monumental buildings. Of course, nonlinear dynamic analyses (NLDA) would have represented a valuable alternative to more accurately investigate the effect of higher modes and describe the inertial forces activated. However, the execution of NLDA, besides requiring a significant increase in the computational effort, was beyond the scope of this research, mainly addressed at this stage more at integrating the use of different modelling approaches than at deepening the reliability of traditional nonlinear static analyses of URM complex aggregates. Thus, following an engineering-practice approach, the analyses were herein performed by adopting, for each examined direction (+X, −X, +Y, and −Y), two different load patterns (LPs): proportional to masses (hereafter referred to as "uniform") and proportional to the product mass per height (hereafter referred to as "pseudo-triangular"). Afterward, in order to ensure that the effects associated with higher modes were not underestimated, a flat rate increase of the participant mass was introduced, directly acting on the conversion of the SDOF. In particular, in those cases in which the activated mass of the equivalent SDOF system (e^*) resulted in lower than 75%, the values of $a_{g,LS}$ were increased by a corrective factor equal to $e^*/0.75$. Future development of the research will be addressed to apply the procedure proposed in [26], which involves the application of a set of LPs calibrated on the basis of results of the modal analysis and aimed to activate the inertial forces in different portions of the buildings.

Figure 10b shows the deformed shape of the whole monument obtained in the +Y direction by applying the uniform LP distribution (which was the most punitive in terms of verification for the Podestà Palace). The results refer to the analysis step corresponding to a 20% decay of the base shear, assumed as representative of the Life Safety limit state (SLV). In particular, the deformed shape highlights a seismic behaviour that is ruled by the presence of the connecting bodies (Arengario and Podestà's vaults).

Figure 11 illustrates instead the damage pattern of some walls of the Podestà Palace (enumerated as in Figure 9a) in the +X and +Y directions in correspondence with the maximum shear V_{max} or of 0.8 V_{max}.

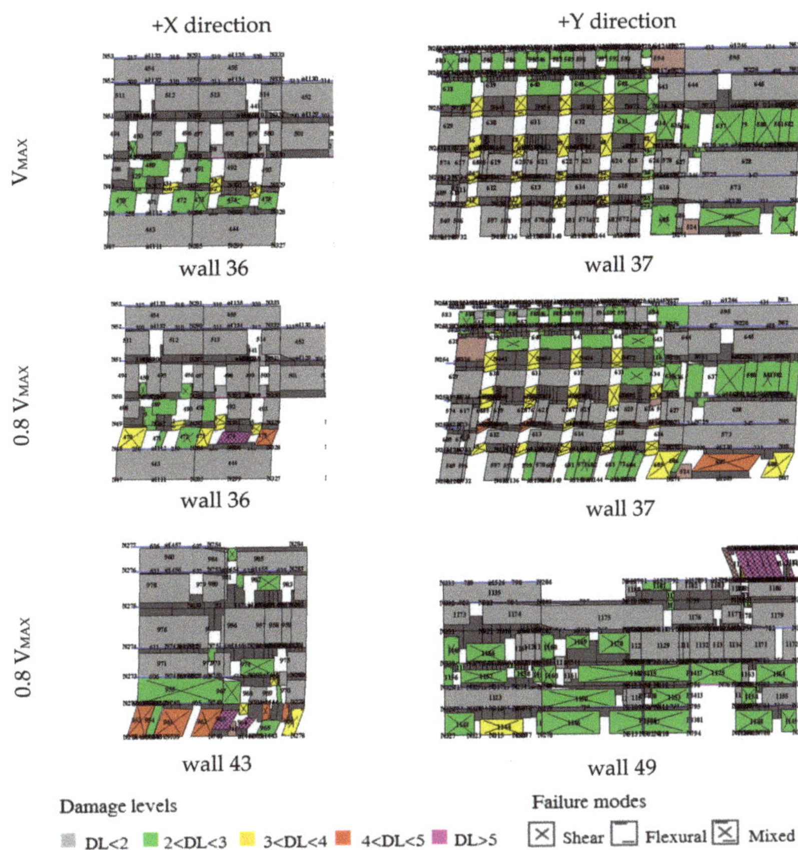

Figure 11. Damage pattern of some walls in the Podestà Palace (uniform LP distribution).

As is possible to see, for both directions, the recurring damage mechanism is soft story behaviour, placed in particular at Levels 2 and 3. Furthermore, the damage pattern highlights the vulnerability of the standing-out part of the fifteenth-century unit overlooking the Erbe square and the Broletto street. In the Masseria, instead, the damage is more pronounced in the X direction (mainly in the piers), but more spread out at the different levels. The most severe damage level is concentrated at the top levels, which are less confined by the adjacent bodies.

Finally, Table 5 illustrates the main parameters assumed in the seismic verification: participant mass PM; the corrective coefficient $e^*/0.75$ applied to $a_{g,LS}$; Γ and M^* computed according to [15,64]; the safety index (α_{SLV}), calculated as $a_{g,SLV}/a_{g,Site,SLV}$. The seismic demand refers to the response spectrum for the site of Mantua, evaluated assuming soil C, a nominal life V_N of 50 years, and a class of use IV (important and strategic structures). The value of $a_{g,SLV}$ was computed by applying the N2 method [64] through the expressions proposed in the Italian Technical Code [16], by imposing the equivalence between the maximum displacement demand of the anelastic system d^*_{max} and the maximum displacement capacity d_u^*.

The results of Table 5 are illustrated only for the most punitive analyses, which are performed by applying a uniform (U) load pattern for the Podestà Palace and a pseudo-triangular (PT) load pattern for the Masseria. In particular, it is interesting to observe that, for Masseria, the most punitive directions are +X and −Y, for which this unit is less influenced by the presence of the restraint provided by the adjacent buildings. As can be seen, the ratio α_{SLV} is higher than 1, except for in the analysis in the +Y direction for the Podestà Palace (although close to 1).

Table 5. Summary of the seismic verification results.

Dir. & Analysis		*PM* [%]	*e*/0.75*	Γ	*M** [kg]	$a_{g,SLV}$ [m/s^2]	α_{SLV}
Podestà	+X (U)	0.6616	0.88217	1.000	10,844,165.8	1.304	1.177
Palace	+Y (U)	0.6630	0.88402	0.892	1,135,266.4	1.068	0.963
Masseria	+X (PT)	0.4660	0.62129	1.249	1,197,424.3	1.426	1.286
	−Y (PT)	0.4500	0.59997	1.074	1,351,024.4	1.332	1.202

5. Analysis of the Arengario Connecting Body

This section presents the results of the analyses performed on the Arengario, the unit that links Masseria with the main building of the Podestà Palace. This connecting body is founded on four large arches set up on pillars with their own foundations. The arches support the two facades and an internal wall inside and decrease the span of the floor of the highest room (Figure 12a,b). The spans between one arch and the other have different dimensions: the first one (on the Broletto square) is about 4.5 m, the intermediate one is 3.5 m, and the last one (overlooking the Giustiziati street) is a bit higher. Furthermore, the arch spans are slightly different, and the entire three spans have a trapezoidal shape. The four large brick arches (Figure 12c) support a timber floor (Figure 12d) and two floors above, covered by another timber floor and the timber roof, respectively.

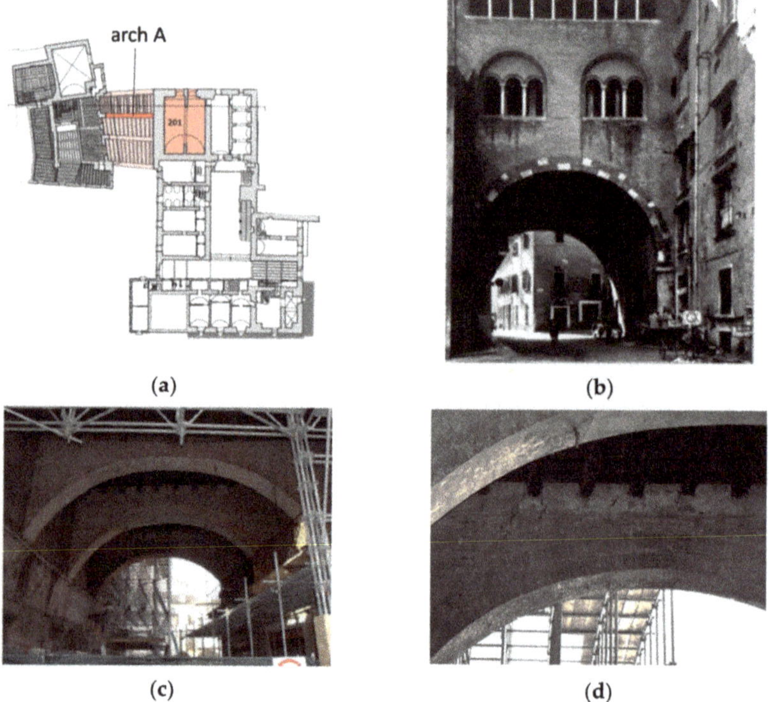

(a)

(b)

(c)

(d)

Figure 12. (**a**) Plan of the Arengario; (**b**) historical picture, taken from the Broletto square in the 1900s; (**c**) present picture of the arch system; (**d**) cracks in the arches and detail of the timber floor above.

The analysis of the global response and the results of the modal analysis (Section 4.2) clearly showed that the Arengario was one of the most vulnerable unit of the palace and possibly subjected to significant distortions. The damage occurred after the Emilia earthquake in 2012 and showed a significant damage pattern, made of non-negligible cracks and deformations; although they probably partially already existed before the seismic event, these cracks were worse (Figure 12d). Moreover, the analysis of the constructive details highlighted that the thrusts of the four arches were not adequately counterbalanced, due to the lack of internal walls in the Masseria and in the Podestà Palace (Figure 12a).

The following sections present the results of the nonlinear analyses performed in ANSYS [68] on an FEM model representative of this unit. In particular, the central arch (referred to as "arch A" and marked in red in Figure 12a) was considered as representative of the entire arch system of the Arengario. This arch was selected as the most vulnerable, since it is characterised by the complete lack of a contrast element in both the two adjacent bodies.

5.1. Geometrical and Numerical Model

The 3D geometrical model was developed in AutoCad 3D and includes the arch, its filler, the supporting pillars, and the supported wall (with doors and windows), up to the roof level. In order to address the mesh more efficiently, the model was discretised in volumes individuated on the basis of the presence of openings, different thicknesses, arch division in blocks, etc. Figure 13 presents the 3D geometrical (a,b) and structural (c) models. The structural model was realised using an eight-node 3D solid element (solid65 in the ANSYS library), defined by orthotropic material properties and able to simulate both the linear and nonlinear behaviour of materials. Consistently with what is already illustrated in Section 4.1, the mechanical properties of masonry were derived from the reference values of the Italian Technical Code [16,17] for a masonry made of solid brick and mortar joints. The following values were assumed: Young Modulus E_x = 1600 N/mm^2; Poisson Coefficient PR_{xy} = 0.2; density ρ = 1800 kg/m^3.

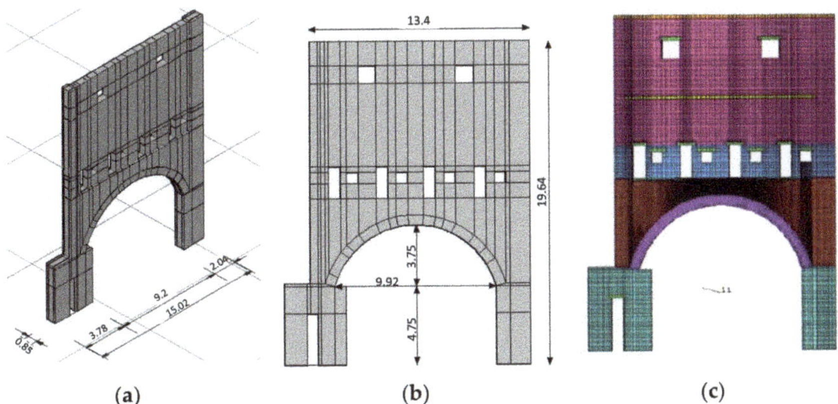

Figure 13. (**a,b**) 3D geometrical model of the Arengario; (**c**) FEM structural model.

Concerning the loads, they were modelled as volumes with proper values of density. Table 6 summarises the computed values and the corresponding mass associated with the volume added on the arch in the FEM model. Following the same logic, an additional load was then introduced on the biggest arch pillars, corresponding to the timber floor at the first level of the Masseria (equal to 120.18 kg). Finally, the Mery's method was applied to compute the thrust transmitted to the Arengario by the barrel vault placed just close to it (n. 201 in Figure 12a).

Table 6. Load analyses.

Element	Load [kN/m]	Equivalent Mass [kg]
First timber slab + internal walls	16.02	18,852.01
False timber ceiling	1.21	1422.26
Second timber slab	2.08	2456.58
Gable roof	1.94	2649.94
Pent roof	1.94	2284.09

5.2. Results of Nonlinear Static Analyses

Nonlinear static analyses were performed in ANSYS by applying, firstly, the gravity load and, afterwards, a horizontal acceleration equal to 0.15 g (in six progressive sub-steps, see Table 7) in the longitudinal direction. The latter aims to simulate the seismic demand computed from the design response spectrum at the SLV evaluated in correspondence with a period of 0.47 s; this reference value corresponds to that derived from the modal analysis performed on the Podestà palace (selecting the mode pertinent for the response of the Arengario).

Table 7. Acceleration values associated with the different analysis sets.

	Gravity Load	Seismic Analysis					
Set	1.1	2.1	2.2	2.3	2.4	2.5	2.6
a	-	0.025 g	0.05 g	0.075 g	0.1 g	0.125 g	0.15 g

Such an analysis was carried out more to interpret the seismic behaviour of this unit than to execute a rigorous seismic verification. The following stress states were analysed: the circumferential stress (S_y) along the arch, evaluated with respect to a cylindrical local coordinate system (Figure 14), and the vertical stress (S_z) and the main tensile stress (S_1), evaluated with respect to the global coordinates system (Figure 15).

From the analysis of the deformed shape and stress state of the arch (Figure 14), it is possible to observe how the arch tends to open itself as a result of the progressive separation of the pillars.

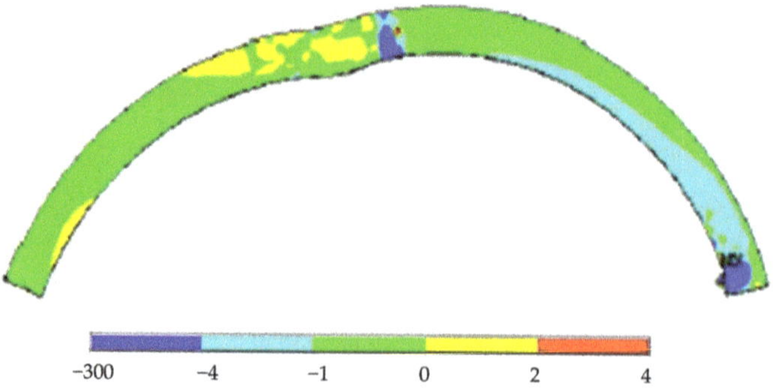

Figure 14. Stress state S_y in [N/mm²] at the end of Set 2.6 (0.15 g).

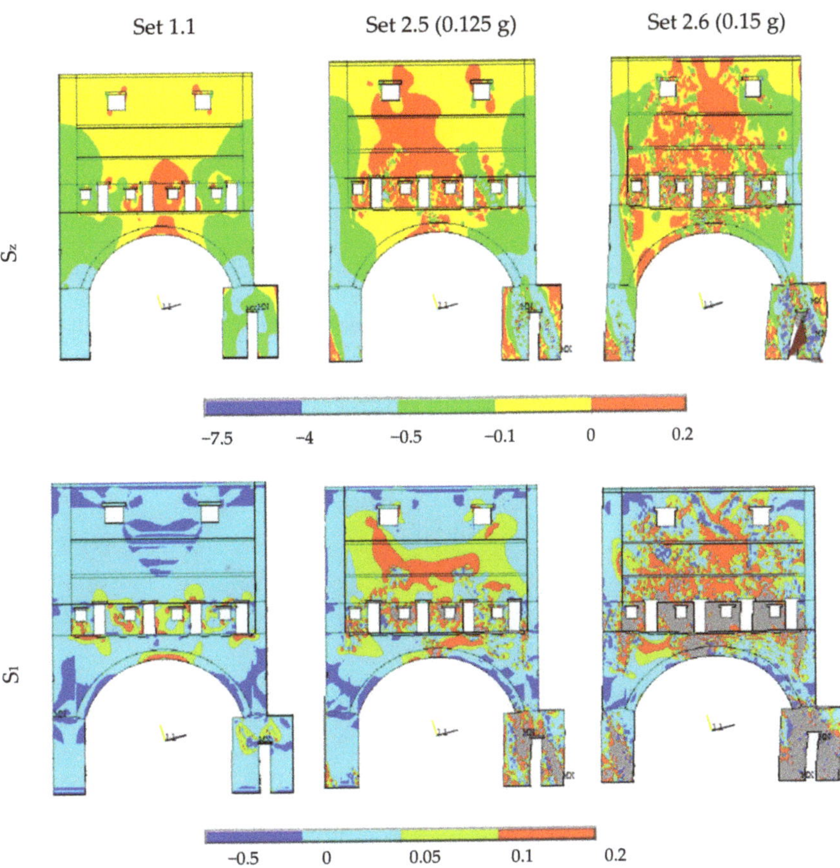

Figure 15. Stress state in [N/mm^2] at the end of different set of analysis.

Figure 15 shows that, even when starting from the application of the gravity load, the areas close to the pillars and to the arch keystone are subjected to the most significant stress state. In fact, while the pillars are characterised by a significant compression stress, the area near the arch keystone is subjected to tensile stress. When analysing S_z (Figure 15), it appears that the worse situation corresponds to the applied horizontal force equal to 0.15 g. In many areas close to the pillars and in the piers between the openings on the arch (in grey in the figure), the maximum detected stress is higher than the reference admissible one (assumed equal to 200 MPa for the masonry under examination). Finally, by considering the stress S_1, it is interesting to observe that this stress has a diagonal trend, starting from the pillars until the arch keystone. The pillar on the right is the most critical one, as it is characterised by the highest values of stress and displacements: this is probably due to the lack of contrast to the arch thrust in the adjacent Masseria unit.

6. Directions for the Retrofitting Strategies

The analyses illustrated in the previous sections mainly highlighted the following:

1. The usefulness of increasing the stiffness of some diaphragms, as pointed out by the global analysis through the SEM model (Section 6.1);
2. The need to improve the arch system of the Arengario connecting body, from both static and dynamic points of view (Section 6.2).

6.1. Possible Stiffening Strengthening Interventions of the Vaults

Regarding the first point (1), while different well-known strengthening intervention techniques can be adopted for timber floors to ensure a certain in-plane stiffening (as discussed, for example, in [59,69–71]), in the case of vaults, the issue is more complex. Though their contribution is often neglected, they may exhibit a non-negligible stiffness in the horizontal plane even from their original state, which, without specific interventions, can significantly decrease with the onset of cracking due to the seismic event, even for low levels of angular deformations ([58,72]). Especially for barrel vaults (which are quite common in the examined building), there is another problem: they transfer the loads only to the internal walls, while the arches are not connected to the external walls in the façade. This was also evident from the analysis of the damage pattern, which highlighted cracks spread in the vaults located at the interface with the orthogonal walls (Figure 16a,b).

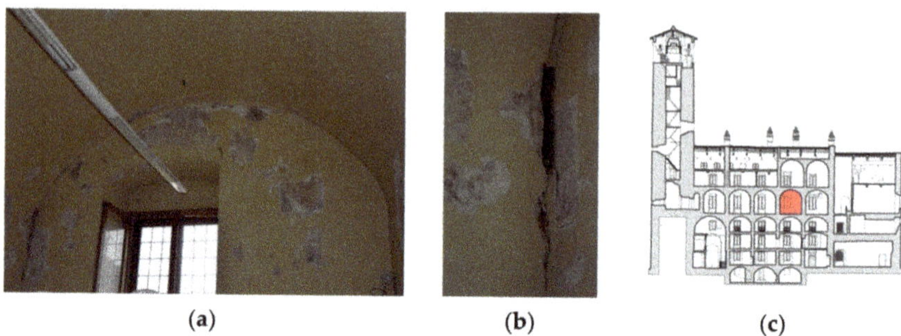

(a)   (b)   (c)

Figure 16. (**a,b**) Post-earthquake damage pattern; (**c**) identification of the examined vault.

In order to better understand the contribution provided by the vaulted system in the horizontal action repartition among walls, the prototype barrel vault identified in Figure 16c was analysed through an FEM model [68]. In particular, nonlinear static analyses were parametrically executed in order to investigate the sensitivity to different strengthening techniques aimed to gradually increase the stiffness of the membranes and the wall-to-wall connection.

The examined vault is placed at the fifth level of the thirteenth-century unit overlooking the Piazza Broletto (Figure 16c). It was selected as representative of many vaulted rooms in the Podestà Palace, analogous for its typology and dimensions. The brick vault covers a 4.16×6.82 m room (1.4 m height) and is characterised by a structural thickness of 30 cm. Figure 17 shows the 3D geometrical model realised in AutoCad 3D (a) and the FEM model developed in ANSYS (b). Besides the barrel vault itself, the model includes the two walls that support the vault, the filler, and the transversal walls parallel to the directrix of the arch.

Each configuration considered was aimed to simulate the effect of a different strengthening intervention:

- Model 1a is representative of the present state, characterised by a heavy and incoherent material as filler and by no wall-to-wall connection with the transversal walls, as the post-earthquake damage highlighted.
- Models 2a and 3a indicate the effects connected to an in-plane stiffening of the vault obtained, replacing the incoherent filler of Model 1a with a light, non-stiffening one (Model 2) or with a light, stiffening one (Model 3).
- Configurations b and c—tested in Models 1 and 3—indicate the possible benefits achieved by improving the wall-to-wall connection. In particular, two possible interventions were considered: a cuci-scuci intervention (b), realised only locally at the intersection of two orthogonal walls in order to improve the interlocking by a proper arrangement of blocks, and a more effective and continuous connection obtained through the realization of a reinforced masonry curb (c) (e.g., with two steel

plates—one for each side of the wall—transversally pinned and properly anchored in the façade).

- Configuration d—tested in Model 3—indicates the further increase of stiffening produced by the insertion of X cables (eventually equipped with dampers or fuses as described in [73]).

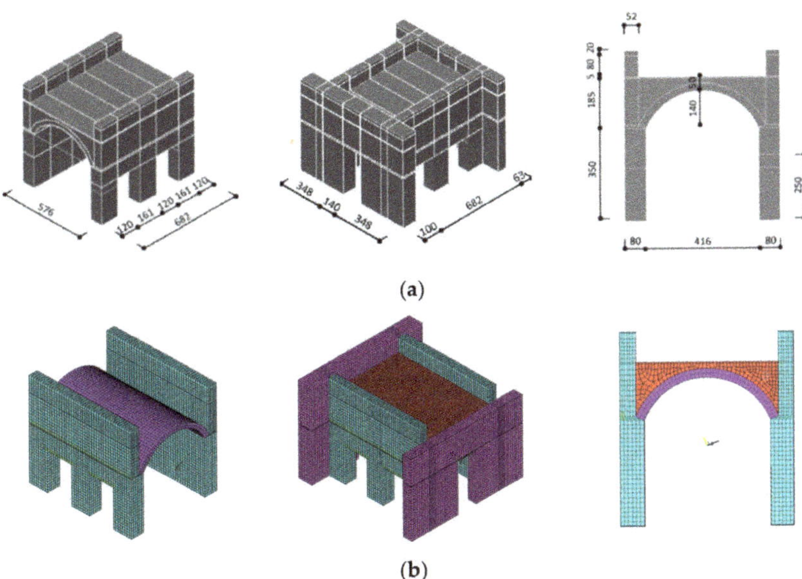

Figure 17. (**a**) Views and dimensions of the geometrical model; (**b**) FEM structural model.

Table 8 summarises all examined configurations, for each model.

Table 8. Examined configurations and percentage of seismic actions transferred to the transversal walls (last column).

Cfg	Model	Strengthening Intervention Acting on		%
		Vault Stiffening	**Wall-to-Wall Connection**	
Present State	1a	-	-	0
	1b	Incoherent material as filler	Local wall-to-wall connection by cuci-scuci at the intersecting walls	42
	1c	Incoherent material as filler	Continuous reinforced masonry curb	44
	2	Light, non-stiffening material as filler with RC slab fixed on the four sides	-	44
Design State	3a	Light, stiffening material	-	46
	3b	Light, stiffening material	Local wall-to-wall connection by cuci-scuci at the intersecting walls	57
	3c	Light, stiffening material	Continuous reinforced masonry curb	47
	3d	Light, stiffening material as filler and the use of X cables	-	19

The FEM model was realised by using an eight-node 3D solid element (solid65 in the ANSYS library) for the walls, the vaults, and the filler (in the present state) and the RC slab or the masonry curb (in the design state) and a link element (link180 in the ANSYS library) to model the X cables in the design state. The assumed orthotropic material properties are listed in Table 9.

Table 9. Materials properties used in the FEM model.

Element	Material	Young Modulus E_x [N/mm^2]	Poisson Coefficient PR_{xy}	Density ρ [kg/m^3]
Walls and vault	Brick masonry	1600	0.2	1800
Filler, present state	Incoherent material	negligible	0.2	1600
Filler, design state—Sol ii)	Light, non-stiffening material	negligible	0.2	500
Filler, design state—Sol iii)	Light, stiffening material	140	0.2	700
RC Slabs/architraves	RC	30,000	0.25	2500
Cables	Steel	210,000	0.3	7860

In order to conventionally simulate the state of the vault inserted into the actual building, concentrated masses were imposed in correspondence with the base of the two adjacent vaults not explicitly modelled (of an entity equal to half mass plus the filler); moreover, horizontal forces, equivalent to the thrust transmitted by the vaults and the filler of the adjacent cells, were applied.

After the application of the gravity loads, nonlinear pushover analyses were performed (Table 10) by applying incremental horizontal forces proportional to the masses separately in both longitudinal and transversal directions. In particular, the seismic analyses in the longitudinal direction aimed to evaluate the shear action activated in the vault due to the different deformability of the two longitudinal walls (in fact, one was a side wall, while the other was characterised by two openings as depicted in Figure 17). Instead, the seismic analyses in the transversal direction aimed to evaluate, for each design solution, the capacity of the vault to transfer the inertial actions to the transversal walls.

Table 10. Acceleration values associated with the different analysis sets.

	Gravity Load	Seismic Analysis							
Set	1.1	2.1	2.2	2.3	2.4	2.5	2.6	2.7	2.8
a_x–a_y	-	0.05 g	0.1 g	0.15 g	0.2 g	0.25 g	0.3 g	0.35 g	0.4 g

In the case of Models 2 and 3d, to simulate the actual stress and deformed states already present in the URM system before the realization of the intervention, analysis was performed by using specific tools available in ANSYS that allow one to activate some elements only in certain steps of the analysis. Thus, the slab and cables were activated only after the application of the gravity loads.

In the nonlinear field, a "smeared cracking" constitutive law was used. In particular, the following properties were assumed: a tensile strength (f_t) equal to 0.2 MPa, for Models 1 and 2, and to 0.1 MPa, for Model 3, and a compression strength (f_m) equal to 4 MPa, for Models 1 and 2, and to 6 MPa, for Model 3.

6.1.1. Response to the Gravity Loads

Figure 18 shows, for Models 1a, 2, and 3c, the vertical stresses S_Z in the walls. The figure shows that the highest values of compressive stresses are in the present state, due to the higher weight of the filler, especially in the wall with openings. The comparison between Model 2 (with the RC slab) and Model 3c (with the light and stiffening filler and the reinforced masonry curb) shows that, in the first case, the stresses are higher due to the major thrusts that occur. Furthermore, tensile stresses can be observed in correspondence with the link between the slab and the wall.

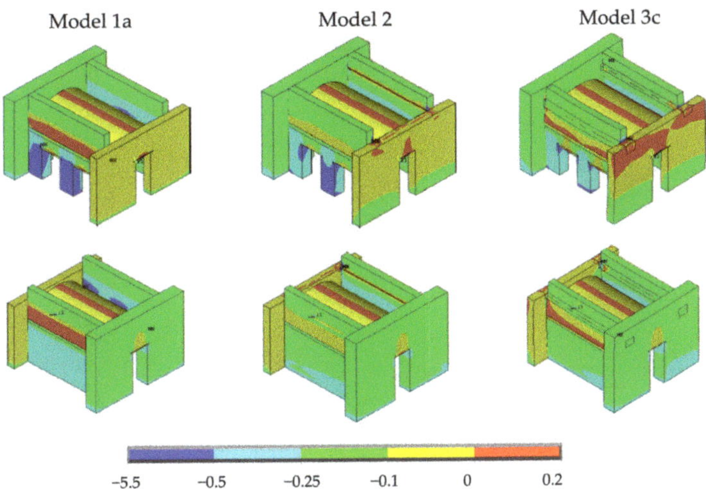

Figure 18. Vertical stresses in [N/mm²] in the walls (Set 1.1 of Table 9).

6.1.2. Seismic Response in the Transversal Direction

Depending on the level of wall-to-wall connection, the results of the three models (1, 2, and 3) differ in terms of the transfer of the inertial actions from the longitudinal to the transversal walls.

In Model 1a (present state), no redistribution is possible; in this case, the analysis shows the maximum horizontal acceleration capacity by relying only on the longitudinal walls that are out-of-plane-loaded. Conversely, in the other cases, different percentages of actions are transferred to the transversal walls for the different values of the horizontal seismic accelerations. The last column of Table 8 summarises, for each model, the percentage (%) of seismic actions transferred to transversal walls. In that way, it is possible to quantify the effects of the replacement of the original filler with a light or stiffening one (passing from Model 3c to 1c). Any substantial increase may be recognised. It is useful to specify that, in a precautionary way, the FEM model does not consider the contact between the transversal walls and the filler, which, in reality, may determine a confinement action with respect to wall deformation.

Figure 19 shows the deformed shape of the vault springer line in the different models:

- In the case of Model 1b, the vault has a significant deformability, with horizontal displacements in the centre of the springer line (2.7 mm), which is significantly higher than that of the edges (2 mm), where the wall is connected to the transversal walls.
- The insertion of a curb (Model 1c) slightly increases the transfer of the actions to the transversal walls and, as a consequence, the displacements (1.7 mm instead of 2 mm), but it does not affect the in-plane stiffness of the membrane.
- The use of the light and stiffening filler (Models 3b and 3c), instead, significantly decreases the vault deformation, whose springer line almost remains horizontal. The more significant transfer of the actions to the transversal walls determines a decrease in displacements, which at the edges of the wall are slightly higher than 1 mm. Analogous results are obtained through the insertion of the RC slab (Model 2).
- Finally, with the strengthening intervention with X cables and no wall-to-wall connections, the displacements are higher (3 mm), but the cables are able to limit vault deformation in the central part. The use of such cables, together with the improvement of the wall-to-wall connection, may thus represent an alternative solution when the light and stiffening filler cannot be realised (to limit the invasiveness at the extrados) or if it is characterised by an insufficient thickness.

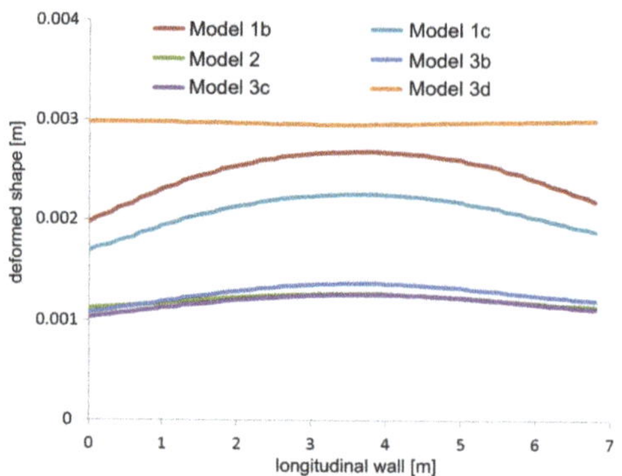

Figure 19. Deformed shape of the vault springer line in different models ($a_x = 0.135$ g).

Finally, Figure 20 shows the vertical stresses in the walls. It is clear that the stresses in the longitudinal walls are significantly higher in Models 1a and 1c than in the others. Obviously, in Model 1, the inertial action on the barrel vault has to be absorbed only by the walls that are out-of-plane-loaded, but the presence of the curb only is not sufficient to limit the stresses.

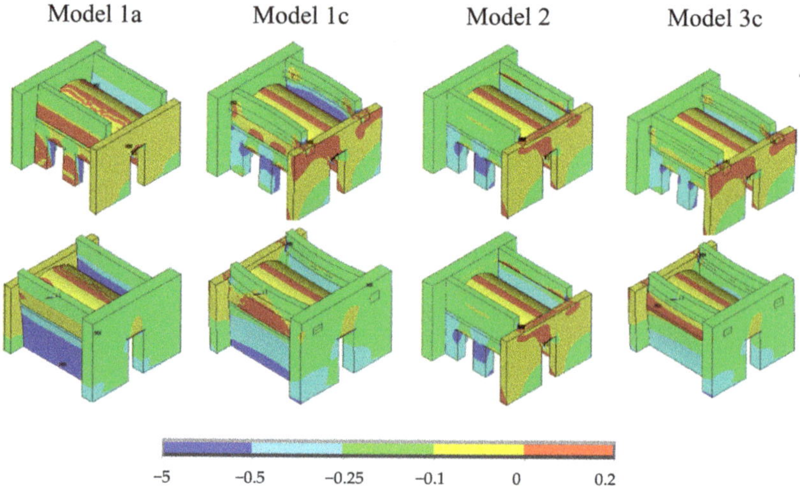

Figure 20. Vertical stresses in [N/mm^2] in the walls (Set 2.3 of Table 9).

6.2. Benefits from the Improvement of the Static and Dynamic Response of Arengario

Finally, to improve the arch system from both a static and dynamic point of view (Point 2 of Section 6), the intervention discussed in the following consists in the addition of an intrados tie rod system. Two different solutions were studied:

- Sol (i): a tie rod placed upon the arch springing level. This leaves a clear path for traffic flow.
- Sol (ii): a tie rod placed in the springer area, where the main thrusts of the arch are concentrated and where the major displacements were detected.

The tie rod was modelled in the FEM with a link element (link10 in the ANSYS library [68]), characterised by a steel material where $E = 210{,}000$ N/mm^2, $PR_{xy} = 0.3$, $\rho = 7860$ kg/m^3. Table 11 summarises the steps of analyses performed. Similarly to what was already described in the case of Models 2 and 3d, in order to more realistically simulate the insertion of the tie rods in a deformed configuration, the tie rod effect was activated only after the application of the gravity load of the URM elements.

Table 11. Acceleration values associated with the different analysis sets.

| | Gravity Load | | Seismic Analysis | | | | | |
	No Tie Rod	With Tie Rod						
Set	1.1	2.1	3.1	3.2	3.3	3.4	3.5	3.6
a	-	-	0.025 g	0.05 g	0.075 g	0.1 g	0.125 g	0.15 g

Figure 21 shows the stress state at the end of Set 3.6 in the two different hypotheses of the intervention. It is clear that tie rods placed at the lowest height determine a stress reduction. The effectiveness of Sol (ii) is also evident from Figure 22, which shows a comparison in terms V (shear at the base of the pillars) vs. Δl (relative displacements between the pillars).

Until $V = 300$ N (corresponding to Set 3.4), the trend is almost equal in the two solutions, but for increasing values of V, the relative displacement is higher in Sol. (i); this result demonstrates that here the tie rod is less effective.

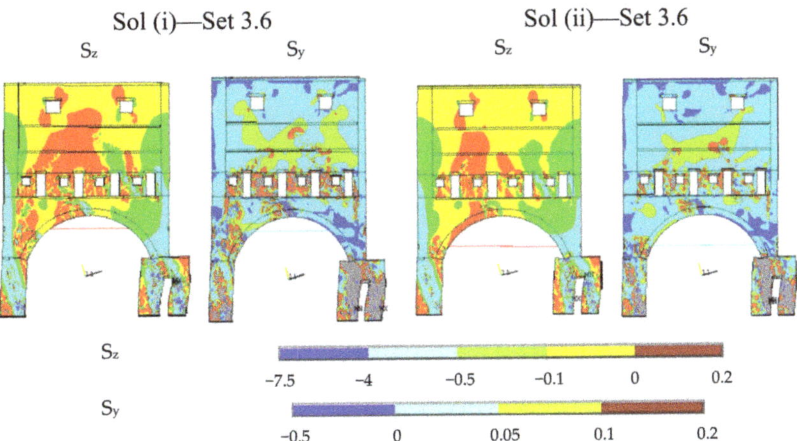

Figure 21. Stress states in [N/mm^2] in the two hypotheses of intervention for the Arengario.

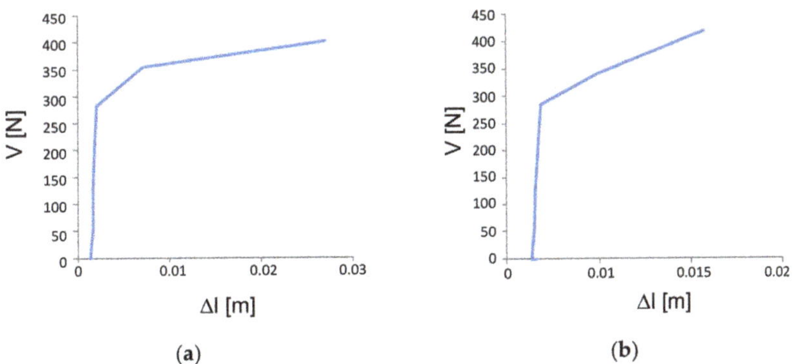

Figure 22. Relative displacement between pillars (Δl) vs. shear (V) obtained with the two hypotheses of interventions: (**a**) Sol (i); (**b**) Sol (ii).

7. Concluding Remarks

This paper investigates seismic assessments of the Podestà Palace in Mantua (Italy), an important palace formed by an aggregation of units stratified over centuries. Different numerical models were integrated to interpret its seismic response, strongly affected by the dynamic interaction among units.

In particular, nonlinear kinematic analyses were performed by referring to an MBM approach in order to assess the seismic response associated with the activation of local mechanisms. In particular, the OOP response of the façade overlooking the Broletto square was investigated. The damage that occurred after the 2012 Emilian earthquake highlighted the vulnerability of this macroelement, mainly due to the lack of connection with orthogonal walls or tie rods. This vulnerability was also confirmed by computations that allowed us to assess the reliability of the approach. An interesting aspect in the assessment of the OOP response was the use of data achieved from the in-plane global 3D model to refine the estimate of floor spectra, by accounting for the filtering effect provided by the main structure.

Then, modal and nonlinear static analyses were performed on the global EF approach of the palace that also included the aggregate and adjacent units. These analyses allowed us to explicitly consider the mutual dynamic interactions among units and to quantify the seismic safety index of the entire building. Results—with safety indexes higher than or close to 1—highlighted that the most significant criticalities of the palace consist in the response of specific critical portions, such as the connection body known as Arengario and the diaphragm system. Both are subjected to significant distortions, as also shown by the results of the modal analysis. For this reason, the latter were further studied with more refined models based on the FEM modelling strategy.

The paper outlines how main seismic behaviour involving a global or local response can be identified in a complex historical building and how a feasible but correct assessment procedure can be selected. Indeed, global modelling of the entire building, considering the interaction between in-plane and out-of-plane behaviour, would not be feasible and even insignificant, because of the limited slenderness of each single masonry pier.

On the basis of the structural deficiencies indicated, possible strengthening solutions were investigated with the aim of minimizing the impact on the construction. As far as the stiffening of barrel vaults is concerned, various strategies were explored that alternatively imply an intervention at the intrados or the extrados. In the case of the Arengario arch system, the best location of tie rods was proven by the nonlinear analyses carried out.

Author Contributions: S.D.A.: modelling and analysis, result post-processing, figure editing, and writing—original draft preparation; D.O.: modelling and analysis, result post-processing, figure editing, and writing—original draft preparation; S.C. and S.L.: conceptualization, supervision, software, result interpretation, and writing—review. All authors have read and agreed to the published version of the manuscript.

Funding: This research received no external funding.

Institutional Review Board Statement: Not applicable.

Informed Consent Statement: Not applicable.

Data Availability Statement: The data presented in this study are available on request from the corresponding author.

Acknowledgments: The authors would like to thank the building enterprise *CMSA* (*Cooperativa di Muratori, Sterratori ed Affini*) for the material made available to the building as well as Eng. Camilla Ciaravolo, Eng. Margherita Cecchinelli, and Eng. Ilaria Marassi for the help in setting up the models and performing the analyses.

Conflicts of Interest: The authors declare no conflict of interest.

References

1. Binda, L.; Modena, C.; Casarin, F.; Lorenzoni, F.; Cantini, L.; Munda, S. Emergency actions and investigations on cultural heritage after the L'Aquila earthquake: The case of the Spanish Fortress. *Bull. Earthq. Eng.* **2011**, *9*, 105–138. [CrossRef]
2. Cattari, S.; Degli Abbati, S.; Ferretti, D.; Lagomarsino, S.; Ottonelli, D.; Tralli, A. Damage assessment of fortresses after the 2012 Emilia earthquake (Italy). *Bull. Earthq. Eng.* **2014**, *12*, 2333–2365. [CrossRef]
3. Parisi, F.; Augenti, N. Earthquake damages to cultural heritage constructions and simplified assessment of artworks. *Eng. Fail. Anal.* **2013**, *34*, 735–760. [CrossRef]
4. Penna, A.; Calderini, C.; Sorrentino, L.; Carocci, C.F.; Cescatti, E.; Sisti, R.; Borri, A.; Modena, C.; Prota, A. Damage to churches in the 2016 central Italy earthquakes. *Bull. Earthq. Eng.* **2019**, *17*, 5763–5790. [CrossRef]
5. Sorrentino, L.; Cattari, S.; da Porto, F.; Magenes, G.; Penna, A. Seismic behaviour of ordinary masonry buildings during the 2015 central Italy earthquakes. *Bull. Earthq. Eng.* **2019**, *17*, 5583–5607. [CrossRef]
6. Vlachakis, G.; Vlachaki, E.; Lourenço, P.B. Learning from failure: Damage and failure of masonry structures, after the 2017 Lesvos earthquake (Greece). *Eng. Fail. Anal.* **2020**, *117*, 104803. [CrossRef]
7. Lourenço, P.B. Computations on historic masonry structures. *Prog. Struct. Eng. Mater.* **2002**, *4*, 301–319. [CrossRef]
8. Roca, P.; Cervera, M.; Gariup, G.; Pelà, L. Structural Analysis of Masonry Historical Constructions. Classical and Advanced Approaches. *Arch Comput. Methods Eng* **2010**, *17*, 299–325. [CrossRef]
9. D'Altri, A.M.; Sarhosis, V.; Milani, G.; Rots, J.; Cattari, S.; Lagomarsino, S.; Sacco, E.; Tralli, A.; Castellazzi, G.; de Miranda, S. Modeling strategies for the computational analysis of unreinforced masonry structures: Review and classification. *Arch. Comput. Methods Eng.* **2020**, *27*, 1153–1185. [CrossRef]
10. Lagomarsino, S.; Cattari, S. PERPETUATE guidelines for seismic performance-based assessment of cultural heritage masonry structures. *Bull. Earthq. Eng.* **2015**, *13*, 13–47. [CrossRef]
11. Heyman, J. The stone skeleton. *Int. J. Solids Struct.* **1966**, *2*, 249–279. [CrossRef]
12. Lagomarsino, S.; Abbas, N.; Calderini, C.; Cattari, S.; Rossi, M.; Ginanni Corradini, R.; Marghella, G.; Mattolin, F.; Piovanello, V. Classification of cultural heritage assets and seismic damage variables for the identification of performance levels. In Proceedings of the 12th International Conference on Structural Studies, Repairs and Maintenance of Heritage Architecture (STREMAH), Chianciano Terme, Italy, 5–7 September 2011; Volume 118, pp. 697–708.
13. Ponte, M.; Bento, R.; Vaz, S.D. A Multi-Disciplinary Approach to the Seismic Assessment of the National Palace of Sintra. *Int. J. Archit. Herit.* **2021**, *15*, 757–778. [CrossRef]
14. Cámara, M.; Romero, M.; Pachón, P.; Compán, V.; Lourenço, P.B. Integration of disciplines in the structural analysis of historical constructions. The Monastery of San Jerónimo de Buenavista (Seville-Spain). *Eng. Struct.* **2021**, *230*, 111663. [CrossRef]
15. EN 1998-3. *Eurocode 8: Design of Structures for Earthquake Resistance—Part 3: Assessment and Retrofitting of Buildings*; CEN (European Committee for Standardization): Brussels, Belgium, 2005.
16. NTC 2018. Italian Technical Code, Decreto Ministeriale 17/1/2018. Aggiornamento delle Norme Tecniche per le Costruzioni. Ministry of Infrastructures and Transportation, G.U. n.42 of 20/2/2018. (In Italian). Available online: https://www.studiopetrillo.com/ntc2018.html (accessed on 31 March 2021).
17. Ministry of Infrastructures and Transportation, C.S.Ll.PP. n.7 del 21/01/2019. Istruzioni per L'applicazione Dell'aggiornamento delle Norme Tecniche per le Costruzioni di cui al D.M. 17/01/2018 G.U. S.O. n.35 of 11/2/2019. (In Italian). Available online: https://www.gazzettaufficiale.it/eli/id/2019/02/11/19A00855/sg (accessed on 31 March 2021).
18. ICOMOS. *Recommendations for the Analysis, Conservation and Structural Restoration of Architectural Heritage*; International Scientific Committee for Analysis and Restoration of Structures and Architectural Heritage (ISCARSAH): Barcelona, Spain, 2005.
19. ISO 13822. *Bases for Design of Structures—Assessment of Existing Structures*, 2nd ed.; ISO International Standard: Geneva, Switzerland, 2010.
20. CIB 335. *Guide for the Structural Rehabilitation of Heritage Buildings*; CIB Commission W023—Wall Structures: Rotterdam, The Netherlands, 2010; ISBN 978-90-6363-066-9.
21. DPCM 9/2/2011, 2008. Linee Guida per la Valutazione e la Riduzione del Rischio Sismico del Patrimonio Culturale con Riferimento alle Norme Tecniche delle Costruzioni di cui al Decreto del Ministero delle Infrastrutture e dei Trasporti del 14 Gennaio 2008. (In Italian). Available online: http://www2.ing.unipi.it/~{}a005843/Consolidamento%202016-17/Normativa/LineeGuida_BBCC_2010_11_26_1.pdf (accessed on 31 March 2021).
22. Casolo, S.; Sanjust, C.A. Seismic analysis and strengthening design of a masonry monument by a rigid body spring model: The "Maniace Castle" of Syracuse. *Eng. Struct.* **2009**, *31*, 1447–1459. [CrossRef]
23. Garofano, A.; Lestuzzi, P. Seismic Assessment of a Historical Masonry Building in Switzerland: The "Ancien Hôpital De Sion". *Int. J. Archit. Herit.* **2016**, *10*, 975–992. [CrossRef]
24. Meguro, K.; Tagel-Din, H. Applied element method for structural analysis: Theory and application for linear materials. *JSCE Struct. Eng. Earthq. Eng.* **2000**, *17*, 21–35.
25. Rossi, M.; Cattari, S.; Lagomarsino, S. Performance-Based assessment of the Great Mosque of Algiers. *Bull. Earthq. Eng.* **2015**, *13*, 369–388. [CrossRef]
26. Degli Abbati, S.; D'Altri, A.M.; Ottonelli, D.; Castellazzi, G.; Cattari, S.; de Miranda, S.; Lagomarsino, S. Seismic assessment of interacting structural units in complex historic masonry constructions by nonlinear static analyses. *Comput. Struct.* **2019**, *213*, 51–71. [CrossRef]

27. Malcata, M.; Ponte, M.; Tiberti, S.; Bento, R.; Milani, G. Failure analysis of a Portuguese cultural heritage masterpiece: Bonet building in Sintra. *Eng. Fail. Anal.* **2020**, *115*, 104636. [CrossRef]

28. Cattari, S.; Degli Abbati, S.; Ferretti, D.; Lagomarsino, S.; Ottonelli, D.; Rossi, M.; Tralli, A. The seismic behaviour of ancient masonry buildings after the earthquake in Emilia (Italy) on May 20th and 29th, 2012. *Ing. Sismica* **2012**, *29*, 87–111.

29. Penna, A.; Morandi, P.; Rota, M.; Manzini, C.F.; da Porto, F.; Magenes, G. Performance of masonry buildings during the Emilia 2012 earthquake. *Bull. Earthq. Eng.* **2014**, *12*, 2255–2273. [CrossRef]

30. Bracchi, S.; da Porto, F.; Galasco, A.; Graziotti, F.; Liberatore, D.; Liberatore, L.; Magenes, G.; Mandirola, M.; Manzini, C.F.; Masiani, R.; et al. Comportamento degli edifici in muratura nella sequenza sismica del 2012 in Emilia. *Progett. Sismica* **2012**, *3*, 141–161.

31. Andreini, M.; de Falco, A.; Giresini, L.; Sassu, M. Structural damages in the cities of Reggiolo and Carpi after the earthquake on May 2012 in Emilia Romagna. *Bull. Earthq. Eng.* **2014**, *12*, 2445–2480. [CrossRef]

32. Degli Abbati, S.; Cattari, S.; Marassi, I.; Lagomarsino, S. Seismic out-of-plane assessment of Podestà Palace in Mantua. In Proceedings of the 4th MuRiCo, Ravenna, Italy, 9–11 September 2014.

33. Lagomarsino, S.; Ottonelli, D. A Macro-Block Program for the Seismic Assessment (MB-PERPETUATE). PERPETUATE (EC-FP7 Project), Deliverable D29. Available online: http://www.perpetuate.eu (accessed on 1 January 2012).

34. Abrams, D.P.; AlShawa, O.; Lourenço, P.B.; Sorrentino, L. Out-of-plane seismic response of unreinforced masonry walls: Conceptual discussion, research needs and modelling issues. *Int. J. Archit. Herit.* **2017**, *11*, 22–30.

35. Sorrentino, L.; D'Ayala, D.; de Felice, G.; Griffith, M.; Lagomarsino, S.; Magenes, G. Review of out-of-plane seismic assessment techniques applied to existing masonry buildings. *Int. J. Archit. Herit.* **2017**, *11*, 2–21. [CrossRef]

36. Degli Abbati, S.; Cattari, S.; Lagomarsino, S. Validation of displacement-based procedures for rocking assessment of cantilever masonry elements. *Structures* **2021**. [CrossRef]

37. Lagomarsino, S. Seismic assessment of rocking masonry structures. *Bull. Earthq. Eng.* **2015**, *13*, 97–128. [CrossRef]

38. Degli Abbati, S.; Lagomarsino, S. Out-of-plane static and dynamic response of masonry panels. *Eng. Struct.* **2017**, *150*, 803–820. [CrossRef]

39. De Felice, G. Out-of-plane seismic capacity of masonry depending on wall section morphology. *Int. J. Archit. Herit.* **2011**, *5*, 466–482. [CrossRef]

40. New Zealand Society for Earthquake Engineering; Structural Engineering Society New Zealand Inc. (SESOC); New Zealand Geotechnical Society Inc.; Ministry of Business, Innovation and Employment; Earthquake Commission. The Seismic Assessment of Existing Buildings (Technical Guidelines for Engineering Assessments), Part C—Detailed Seismic Assessment. Auckland, New Zealand. 2017. Available online: https://www.building.govt.nz/building-code-compliance/b-stability/b1-structure/seismic-assessment-existing-buildings/ (accessed on 31 March 2021).

41. Calvi, P.M.; Sullivan, T.J. Estimating floor spectra in multiple degree of freedom systems. *Earthq. Struct.* **2014**, *7*, 17–38. [CrossRef]

42. Vukobratovic, V.; Fajfar, P. Code-Oriented floor acceleration spectra for building structures. *Bull. Earthq. Eng.* **2017**, *15*, 3013–3026. [CrossRef]

43. Degli Abbati, S.; Cattari, S.; Lagomarsino, S. Theoretically-based and practice-oriented formulations for the floor spectra evaluation. *Earthq. Struct.* **2018**, *15*, 565–581.[CrossRef]

44. Cattari, S.; Degli Abbati, S.; Lagomarsino, S. Floor spectra validation through actual data from the 2016/2017 earthquake in Central Italy. In Proceedings of the 17th WCEE, Sendai, Japan, 27 September–2 October 2021.

45. Blandon, C.; Priestley, M. Equivalent viscous damping equations for direct displacement based design. *J. Earthq. Eng.* **2005**, *9*, 257–278. [CrossRef]

46. S.T.A. DATA 2012, 3Muri Program, Release 5.0.4. Available online: http://www.stadata.com (accessed on 31 March 2021).

47. Lagomarsino, S.; Penna, A.; Galasco, A.; Cattari, S. TREMURI program: An equivalent frame model for the nonlinear seismic analysis of masonry buildings. *Eng. Struct.* **2013**, *56*, 1787–1799. [CrossRef]

48. Cattari, S.; Camilletti, D.; Lagomarsino, S.; Bracchi, S.; Rota, M.; Penna, A. Masonry Italian Code-Conforming Buildings. Part 2: Nonlinear Modelling and Time-History Analysis. *J. Earthq. Eng.* **2018**, *22*, 2010–2040. [CrossRef]

49. Marino, S.; Cattari, S.; Lagomarsino, S.; Dizhur, D.; Ingham, J.M. Post-Earthquake Damage Simulation of Two Colonial Unreinforced Clay Brick Masonry Buildings Using the Equivalent Frame Approach. *Structures* **2019**, *19*, 212–226. [CrossRef]

50. Cattari, S.; Degli Abbati, S.; Alfano, S.; Brunelli, A.; Lorenzoni, F.; da Porto, F. Dynamic calibration and seismic validation of numerical models of URM buildings through permanent monitoring data. *Earth Eng. Struct. Dyn.* **2021**. [CrossRef]

51. Asıkoglu, A.; Vasconcelos, G.; Lourenço, P.B. Overview on the Nonlinear Static Procedures and Performance-Based Approach on Modern Unreinforced Masonry Buildings with Structural Irregularity. *Buildings* **2021**, *11*, 147. [CrossRef]

52. Augenti, N. Seismic behaviour of irregular masonry walls. In Proceedings of the 1st European Conference on Earthquake Engineering and Seismology, Geneva, Switzerland, 3–8 September 2006.

53. Dolce, M. Schematizzazione e modellazione degli edifici in muratura soggetti ad azioni simiche. *Ind. Costr.* **1991**, *25*, 44–57. (In Italian)

54. Moon, F.L.; Yi, T.; Leon, R.T.; Kahn, L.F. Recommendations for Seismic Evaluation and Retrofit of Low-Rise URM Structures. *J. Struct. Eng.* **2006**, *132*, 663–672. [CrossRef]

55. Cattari, S.; Camilletti, D.; D'Altri, A.M.; Lagomarsino, S. On the use of continuum Finite Element and Equivalent Frame models for the seismic assessment of masonry walls. *J. Build. Eng.* **2021**, *43*, 102519. [CrossRef]

56. Berti, M.; Salvatori, L.; Orlando, M.; Spinelli, P. Unreinforced masonry walls with irregular opening layouts: Reliability of equivalent-frame modelling for seismic vulnerability assessment. *Bull. Earth. Eng.* **2017**, *5*, 1213–1239. [CrossRef]

57. Ottonelli, D.; Manzini, C.F.; Marano, C.; Cordasco, E.A.; Cattari, S. A comparative study on a complex URM building. Part I: Sensitivity of the seismic response of different modelling options in the equivalent frame models. *Bull. Earth. Eng.* **2021**. [CrossRef]

58. Cattari, S.; Resemini, S.; Lagomarsino, S. Modelling of vaults as equivalent diaphragms in 3D seismic analysis of masonry buildings. In Proceedings of the 6th International Conference on Structural Analysis of Historical Constructions, Bath, UK, 2–4 July 2008.

59. Brignola, A.; Pampanin, S.; Podestà, S. Experimental evaluation of the in-plane stiffness of timber diaphragms. *Earthq. Spectra* **2012**, *28*, 1687–1909. [CrossRef]

60. Morandi, P.; Albanesi, L.; Graziotti, F.; Li Piani, T.; Penna, A.; Magenes, G. Development of a dataset on the in-plane experimental response of URM piers with bricks and blocks. *Constr. Build. Mater.* **2018**, *190*, 593–611. [CrossRef]

61. Krzan, M.; Gostic, S.; Cattari, S.; Bosiljkov, V. Acquiring reference parameters of masonry for the structural performance analysis of historical building. *Bull. Earth. Eng.* **2017**, *13*, 203–236. [CrossRef]

62. Magenes, G.; Calvi, G.M. In plane seismic response of brick masonry walls. *Earth Eng. Struct. Dyn.* **1997**, *26*, 1091–1112. [CrossRef]

63. Calderini, C.; Cattari, S.; Lagomarsino, S. In-Plane strength of unreinforced masonry piers. *Earth Eng. Struct. Dyn.* **2009**, *38*, 243–267. [CrossRef]

64. Fajfar, P. Capacity spectrum method based on inelastic demand spectra. *Earth Eng. Struct. Dyn.* **1999**, *28*, 979–993. [CrossRef]

65. Azizi-Bondarabadi, H.; Mendes, N.; Lourenco, P.B. Higher mode effects in pushover analysis of irregular masonry buildings. *J. Earthq. Eng.* **2019**, *25*, 1459–1493. [CrossRef]

66. Reyes, J.C.; Chopra, A.K. Three-Dimensional Modal Pushover Analysis of Unsymmetric-Plan Buildings Subjected to Two Components of Ground Motion. *Geotech. Geol. Earthq. Eng.* **2013**, *24*, 203–217.

67. Pinho, R.; Antoniou, S.; Casarotti, C.; López, M. Displacement-Based adaptive pushover for assessment of buildings and bridges. In *Advances in Earthquake Engineering for Urban Risk Reduction. Nato Science Series: IV: Earth and Environmental Sciences*; Wasti, S.T., Ozcebe, G., Eds.; Springer: Dordrecht, The Netherlands, 2006; Volume 66, pp. 79–94. [CrossRef]

68. ANSYS, Engineering Simulation Software, Release 13. 2013. Available online: http://www.ansys.com (accessed on 31 March 2021).

69. Parisi, M.A.; Piazza, M. Seismic strengthening and seismic improvement of timber structures. *Constr. Build. Mater.* **2015**, *97*, 55–66. [CrossRef]

70. Gubana, A. State-of-the-Art Report on high reversible timber to timber strengthening interventions on wooden floors. *Constr. Build. Mater.* **2015**, *97*, 25–33. [CrossRef]

71. Giongo, I.; Wilson, A.; Dizhur, D.; Derakhshan, H.; Tomasi, R.; Griffith, M.; Quenneville, P.; Ingham, J. Detailed seismic assessment and improvement procedure for vintage flexible timber diaphragms. *Bull. N. Zealand Soc. Earthq. Eng.* **2014**, *47*, 97–118. [CrossRef]

72. Rossi, M.; Calderini, C.; Lagomarsino, S. Experimental testing of the seismic in-plane displacement capacity of masonry cross vaults through a scale model. *Bull. Earthq. Eng.* **2016**, *14*, 261–281. [CrossRef]

73. Farzampour, A.; Eatherton, M.R. Parametric computational study on butterfly-shaped hysteretic dampers. *Front. Struct. Civ. Eng.* **2019**, *13*, 1214–1226. [CrossRef]

Article

Seismic Vulnerability Assessment of Portuguese Adobe Buildings

Samar Momin [1], Holger Lovon [1], Vitor Silva [1,2], Tiago Miguel Ferreira [3,*] and Romeu Vicente [1]

[1] RISCO, Research Centre for Risks and Sustainability in Construction, Civil Engineering Department, Campus Universitário de Santiago, University of Aveiro, 3810-193 Aveiro, Portugal; samar@ua.pt (S.M.); holger.lovonq@ua.pt (H.L.); vitorsilva@ufp.edu.pt (V.S.); romvic@ua.pt (R.V.)

[2] Universidade Fernando Pessoa, Praça de 9 de Abril 349, 4249-004 Porto, Portugal

[3] Department of Civil Engineering, Campus de Azurém, University of Minho, ISISE, 4800-058 Guimarães, Portugal

* Correspondence: tmferreira@civil.uminho.pt

Abstract: Adobe construction represents 5.3% of the total Portuguese building stock according to the latest National Housing Census. The distribution of these adobe buildings is scattered across the country, with higher density in the central region and in Algarve in the south, where the seismic hazard is highest. A large proportion of these buildings are still in use for residential and commercial purposes and are of historical significance, contributing to the cultural heritage of the country. Adobe buildings are known to exhibit low seismic resistance due to their brittle behavior, thus making them vulnerable to ground shaking and more prone to structural damage that can potentially cause human fatalities. Three buildings with one-story, two-stories, and two-stories plus an attic were numerically modeled using solid and contact elements. Calibration and validation of material properties were carried out following experimental results. A set of 30 ground motion records with bi-directional components were selected, and non-linear time-history analyses were performed until complete collapse occurred. Two novel engineering demand parameters (EDPs) were used, and damage thresholds were proposed. Finally, fragility and fatality vulnerability functions were derived. These functions can be used directly in seismic risk assessment studies.

Keywords: adobe buildings; seismic vulnerability assessment; fragility functions; physical damage estimation; fatality vulnerability functions; indoor fatalities; LS-DYNA; Portugal

Citation: Momin, S.; Lovon, H.; Silva, V.; Ferreira, T.M.; Vicente, R. Seismic Vulnerability Assessment of Portuguese Adobe Buildings. *Buildings* **2021**, *11*, 200. https://doi.org/10.3390/buildings11050200

Academic Editors: Rita Bento and Ana Simões

Received: 6 April 2021
Accepted: 7 May 2021
Published: 10 May 2021

1. Introduction

In this study, the seismic vulnerability assessment of three Portuguese adobe buildings was conducted to derive fragility and fatality vulnerability functions, which can be used to estimate losses due to earthquakes. Over the past centuries, Portugal has experienced numerous earthquakes; a lesser-known event from the 18th century, the 1722 Algarve earthquake, occurred 33 years before the well-known great 1755 Lisbon earthquake and tsunami, and caused widespread damage in Algarve. More recent events in Portugal such as the 1909 Mw 6.0 Benavente earthquake, the 1969 Mw 8.0 Algarve earthquake, and the 1980 Mw 6.8 Azores earthquake are examples of events where significant damage was observed. The 1909 Benavente earthquake, despite its recorded moderate magnitude (Mw 6.0), is known to be the largest crustal earthquake in the Iberian Peninsula [1]. It occurred around a period during which the Art Nouveau movement was gaining popularity in the central region, and adobe constructions were in vogue. According to a post-earthquake survey report of the 1909 Benavente earthquake [2], 879 buildings were damaged: 20% reported light damage, 40% reported moderate damage, and the remaining 40% were completely damaged. Furthermore, there were severe casualties that resulted in 46 deaths, out of which 30 were from the village of Benavente. The direct and indirect consequences of this event played an important role in influencing the local seismic building culture.

The majority of the past studies regarding the seismic vulnerability of adobe buildings were focused on the South American building stock. Tarque et al. [3] developed analytical fragility functions following the displacement-based earthquake loss assessment (DBELA) methodology [4] for commonly found one story buildings from Peru. Ahmad et al. [5] broadly proposed fragility functions for a range of commonly found buildings, including adobe structures in Pakistan. A study focused broadly on the South American building stock [6] developed fragility functions for unreinforced masonry with adobe blocks, using single-degree-of-freedom (SDoF) oscillators to model the response of the buildings. Recently, Martins and Silva [7] developed analytical fragility and vulnerability functions using censored cloud regression analysis, for one, two, and three-story adobe buildings. Finally, Sumerente et al. [8] developed fragility functions, combining in-plane and out-of-plane loading conditions, for typical one- and two-story adobe buildings in Peru.

According to Abeling and Ingham [9], the building volume loss is considered a better damage descriptor for estimating risk to occupants as compared to traditional damage states because it can be directly correlated with earthquake fatalities. Such data is essential to better estimate earthquake-related risk and losses, thus minimizing economic losses and mitigating fatalities. The fragility and vulnerability functions of the Portuguese adobe buildings with one-story, two-stories, and two-stories plus an attic were derived following the steps listed below.

1. Building surveys that contained well-documented evidence such as building schematics and descriptions of the buildings were examined. However, due to the scarcity of a large dataset specific to adobe buildings, three buildings with one archetype per building class of one-story, two-stories, and two-stories plus an attic were selected for the case study presented herein.
2. These selected buildings were numerically modeled using the advanced finite element method-based software LS-DYNA [10]. These models were tested by following experimental validation and a series of sensitivity analyses to check the influence of various parameters.
3. A set of 30 bi-directional ground motion records were selected based on the peak ground acceleration (PGA) as the intensity measure (IM), and used to perform several non-linear time-history analyses until the complete collapse of the buildings.
4. To derive the fragility functions, two novel engineering demand parameters (EDPs)—namely the crack propagation ratio (CPR) and volume loss ratio (VLR)—were selected, and damage thresholds were proposed such that they can be correlated with the damage classifications of the European Macroseismic Scale (EMS-98) [11].
5. The fragility functions were derived using the well-established cloud analysis [12] to quantify physical structural damage due to a given IM.
6. Furthermore, a fatality vulnerability function was derived to estimate indoor fatalities.

2. Characterization of Adobe Buildings in Portugal

According to the 2011 Population and Housing Census of Portugal (INE) [13], the total building stock of Portugal comprises 3,353,610 buildings, out of which the adobe building stock amounts to 178,422 buildings, i.e., about 5.32% of the whole Portuguese building stock (Figure 1).

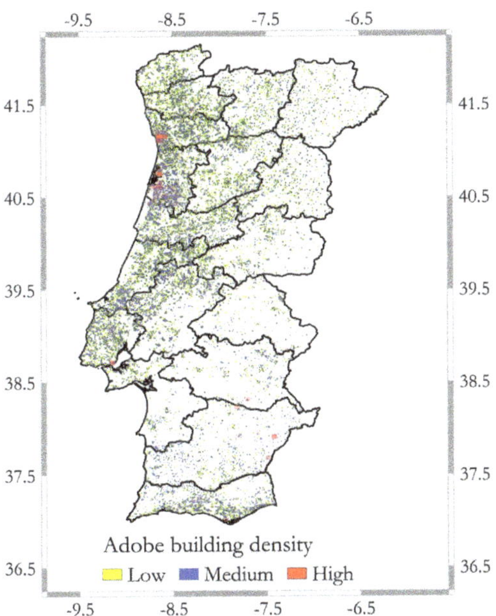

Figure 1. Geographical distribution of adobe buildings in Portugal, as reported in the INE 2011 [13].

In Portugal, earthen materials have been used to construct loadbearing walls in the form of adobe or rammed earth for the construction of buildings, especially in the central and southern regions of the country. Locally, earth-based building materials are divided into three different types of building techniques: rammed earth, known as "taipa"; wattle-and-daub, "tabique"; and adobe [14]. Currently, there is a lack of reliable vulnerability models for adobe buildings in Portugal. In this study, the advances in computational modeling capabilities were utilized for the development of sophisticated numerical models that can simulate the complete collapse of a full-scale building subjected to bi-directional loading and are capable of predicting crack propagation and volume loss. As previously mentioned, there are many adobe buildings in the central and southern part of the country, due to the existence of sandy soils and the presence of lime, which has been reported to be used as a stabilizing agent. It can be observed that the central region of Portugal has the highest density of adobe buildings, since it was the prevailing construction system adopted for the first half of the 20th century [15] and the legacy from the patrimony built in that period is still significant. As part of the endeavor towards the preservation of the inheritance and cultural heritage of the country, from the historical past to the present, adobe buildings irrespective of the grandeur or simplicity are extremely important. In fact, many patrimonial buildings of high historical, cultural, and architectural value [16] are still in a reasonable conservation state [17]. The factor that led to the upsurge in adobe as a construction material of choice, especially in the central region, is credited to the Art Nouveau movement that spread across Europe in various forms. This movement had a significant influence in some Portuguese cities [18], particularly in Aveiro, where it is estimated that 30–40% of the buildings are still adobe buildings [19]. The Southern region of Algarve, where the seismic hazard is relatively high [20], has a high density of adobe buildings, as shown in Figure 1.

Database of Adobe Buildings Information

An appropriate selection of the buildings to be modeled depends largely on the availability of building survey data containing documented evidence with building schematics, dimensions, and detailing with photographs that can be consulted to infer information about the exterior walls, interior walls, openings, lintel beams, roofing systems, attic, and features such as gable end walls.

Three building surveys pertaining to earth-based buildings in continental Portugal were examined in this study, namely, the Algarve survey report [21], the Alentejo survey report [22], and the survey of façade walls in Aveiro [19]. From these surveys, it was concluded that it was essential to segregate these earthen buildings according to the primary, secondary, and tertiary construction material that was used to build the external walls and then the internal walls, respectively. The surveys reported a vast amount of buildings constructed using taipa as both their primary and secondary construction material, along with some buildings that used adobe and tabique as the secondary and tertiary construction materials for internal walls. From these surveys, we selected a one-story building with gable end walls in which both the primary and secondary construction material used was adobe. The selection of buildings with two and more stories was carried out following studies [23] that were archived at the University of Aveiro's database, which documented detailed building drawings constructed with adobe as the primary construction material. It is noteworthy to mention that these buildings, located in the municipality of Ílhavo, are currently in use, as shown in Figure 2a,b.

(a)   (b)

Figure 2. Photographs of the main façades: (**a**) two-story (Building 2); (**b**) 2-story plus an attic (Building 3).

3. Case Study Buildings—Geometrical and Material Properties

The information obtained from the documented evidence discussed in Section 2 was studied in detail and incorporated in the LS-DYNA software environment to build the numerical models and perform the vulnerability assessment of the three selected case study buildings, namely Buildings 1–3.

3.1. Geometrical Properties

The selected buildings are unique buildings due to their geometrical properties such as exterior wall thickness, story or inter-story heights, total area, percentage of openings, and building features such as lintel beams and gable-end walls. Each case study building is described below.

3.1.1. Building 1

This one story building with gable-end walls, was located in the county Castro Marim in Algarve was utilized for residential purposes, but was abandoned at the time of the survey. Its floorplan is rectangular, and it is divided into four compartments. The primary and secondary construction material is adobe, and the external walls are 30 cm thick, remaining constant along the height of the building. The thickness of the walls was modeled by adopting a "two-parts" scheme in order to avoid computational instabilities, as shown in Figure 3c. The internal walls are 15 cm thick and divide the various compartments of the building. However, for simplifying the numerical models and reducing the computational cost, these internal walls were not modeled in this study. The main façade has a door and two windows on either side. Each visible multicoloured block is termed as a 'part', and these different parts constitute the prisms that are numbered from 1–7, as depicted in Figure 3a. The interlocking of blocks was modeled by arranging them in a staggered pattern, and the addition of constraints at the intersections simulate the interlocking of two orthogonal walls, as shown in Figure 3c. Since buildings are built using lintel beams, these elements were included above the openings in the numerical models. These are elastic members constituted by a set of blocks whose nodes are constrained at the edges in the x, y, and z directions, as shown in Figure 3a. These are structural components that are important to model, as they play an important role in preventing the onset of premature collapse mechanisms due to seismic loading. The gable-end walls compose both the left and right façades of the building and do not have any openings, as shown in Figure 3b. The roof is composed of simple wood beams and trusses, which support the ceramic roofing tiles. The beams of the roofing system were modeled using a discrete element with compression-only springs added in the two orthogonal directions with equal spacing, as shown in Figure 3a,c. The mechanical properties of the wooden beams of the roofing system were defined according to prescriptions from Eurocode 5 [24]. The dimensions of the building are summarized in Table 1, and the geometrical details and features that have been modeled in the LS-DYNA environment are shown in Figure 3a–c. Further details pertaining to numerical modeling are discussed in Section 4.

Table 1. Building characteristics and dimensions.

Building Characteristics		Building 1	Building 2	Building 3
Total no. of stories		1	2	2 + Attic
Length (m)	X-direction	6.30	10.80	8.20
	Y-direction	7.80	9.30	12.00
Area (m^2)		49.14	100.44	98.40
Height (m)	1st Story	2.85	3.30	2.40
	2nd Story	-	3.30	3.60
	Attic	-	-	3.00
Total height (m)		2.85	6.60	9.00
Total no. of external walls		4	4	4
External wall thickness (cm)		30	60	40
Gable-end walls		Yes	No	Yes
Lintel beams		Yes	Yes	Yes
Total no. of window openings		4	15	16
Total no. of door openings		2	3	4
Total percentage of openings (%)		10	20	15

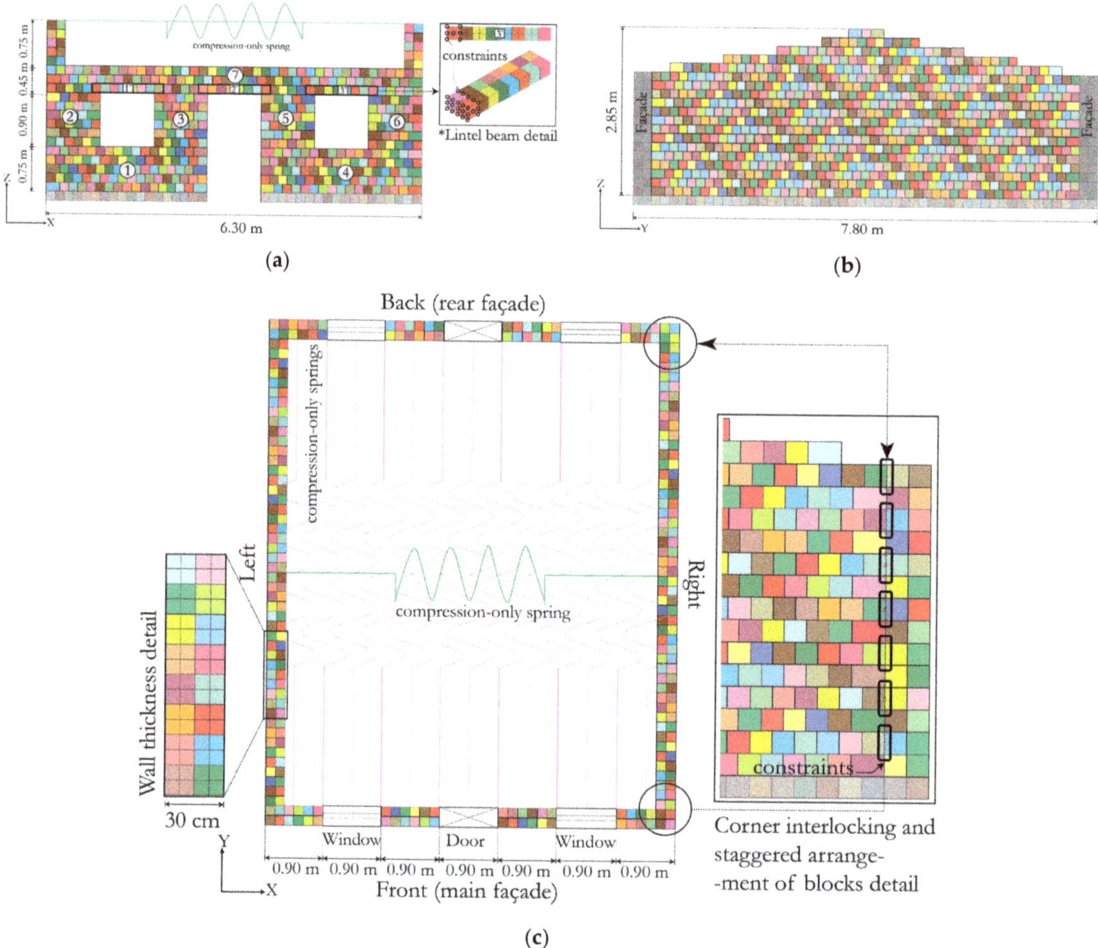

Figure 3. Dimensions of Building 1 and details of the modeling in the LS-DYNA environment: (**a**) elevation view showing the front (main façade), (inset right) close-up of a lintel beam; (**b**) elevation view showing the symmetric left and right gable end walls; (**c**) plan view of the building and position of compression-only springs: (left inset) wall thickness detail (right inset) close-up view of a corner intersection, constraints, and interlocking.

3.1.2. Building 2

The two-story residential building is located in the municipality of Ílhavo and is currently in use (see Figure 2a). Its floorplan is rectangular and composed of a ground and first floor with equal inter-story heights. The primary construction material is adobe, while the external walls are 60 cm thick as a constant throughout the entire height of the building. The interior walls are 15 cm thick and were built using a secondary construction material (tabique). These thin tabique internal walls have a much lower structural capacity when compared to the thick adobe external walls. For this reason, as well as to reduce the computational demand of the numerical models, these elements were not modeled in this study. Nonetheless, a recent study by Battaglia et al. (2020) [25] indicates that tabique walls can lead to a slight reduction in the structural capacity. Each of the façades has multiple openings, as presented in Figure 4a,b. The lintel beams were modeled above each of the openings, as previously described. The roof is composed of simple wood beams and trusses, which support the ceramic roofing tiles. The beams of the roofing system were modeled using a discrete element with compression-only springs in one direction, as shown in Figure 4c. The mechanical properties of the wooden beams of the roofing system were defined according to prescriptions from Eurocode 5 [24]. The dimensions of the building are summarized in Table 1, and the geometrical details are shown in Figure 4a–c. Further details pertaining to numerical modeling are discussed in Section 4.

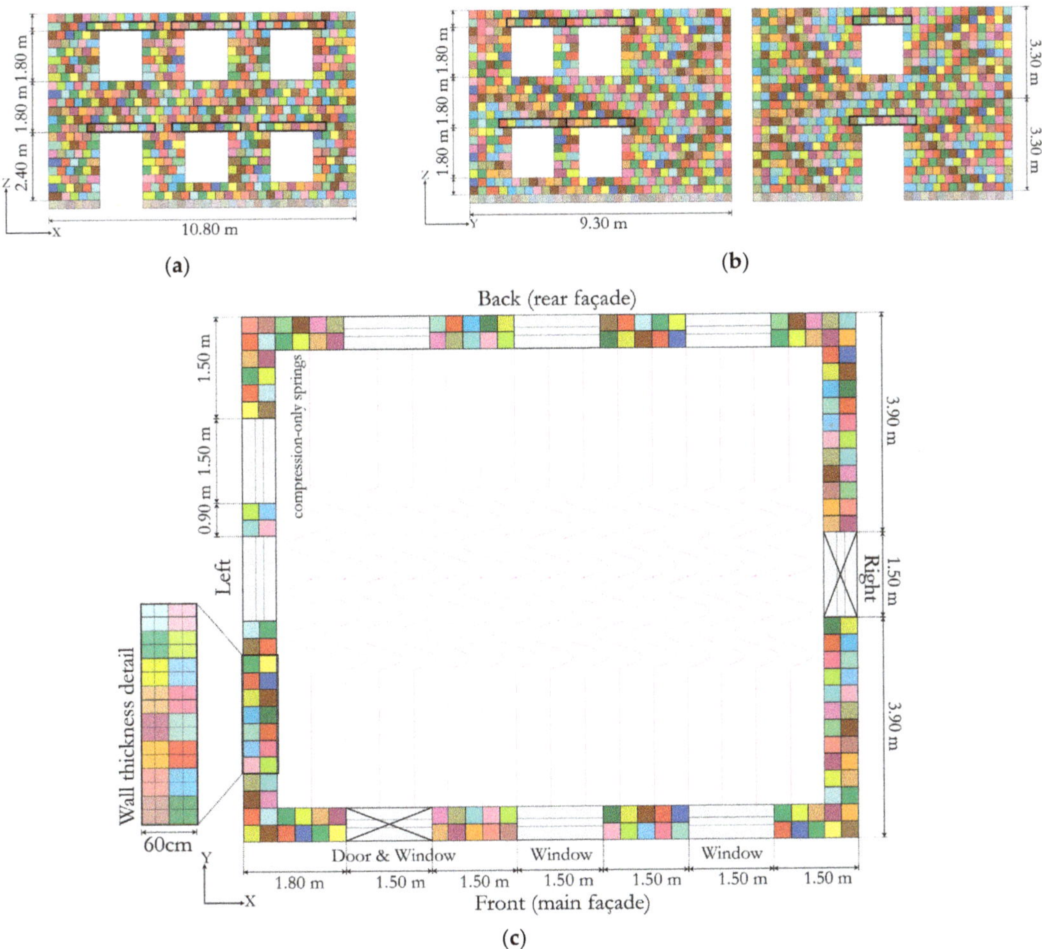

Figure 4. Dimensions of Building 2 and details of the modeling in the LS-DYNA environment: (**a**) elevation view showing the front (main façade); (**b**) elevation view showing the left and right walls; (**c**) plan view of the building and position of compression-only springs (left inset) wall thickness detail.

3.1.3. Building 3

The two-story plus attic with gable-end walls is a residential building also located in the municipality of Ílhavo. Similar to Building 2, this one is also currently in use (see Figure 2b). Its floorplan is rectangular and is composed of a basement which serves as a storage and a wine-cellar with a lower height than the other two stories, a habitable first floor with multiple compartments, and an attic at the top. The primary construction material is adobe. The external walls are 40 cm thick, which is constant throughout the entire height of the building. The interior walls are 15 cm thick and were built using a secondary construction material (*tabique*); these walls were not numerically modeled for the reasons described for Building 2. Each of the façades has multiple openings, as shown in Figure 5a,b. The lintel beams were modeled above each of the openings, as mentioned previously. In the basement level, there are windows that are fixed and unopenable, serving only for lighting, whereas the above story and attic have openable windows that are used for lighting and ventilation. The flooring system consists of wooden beams that are equally spaced at 40 cm and the roof that consists of wooden trusses. The beams of the roofing system were modeled using a discrete element with compression-only springs added in the two orthogonal directions with equal spacing as shown in Figure 5a,c. The mechanical properties of the wooden beams of the roofing system were defined according to prescriptions from Eurocode 5 [24]. The dimensions of the building are summarized in Table 1, and the geometrical details are shown in Figure 5a–c. Further details pertaining to numerical modeling are discussed in Section 4.

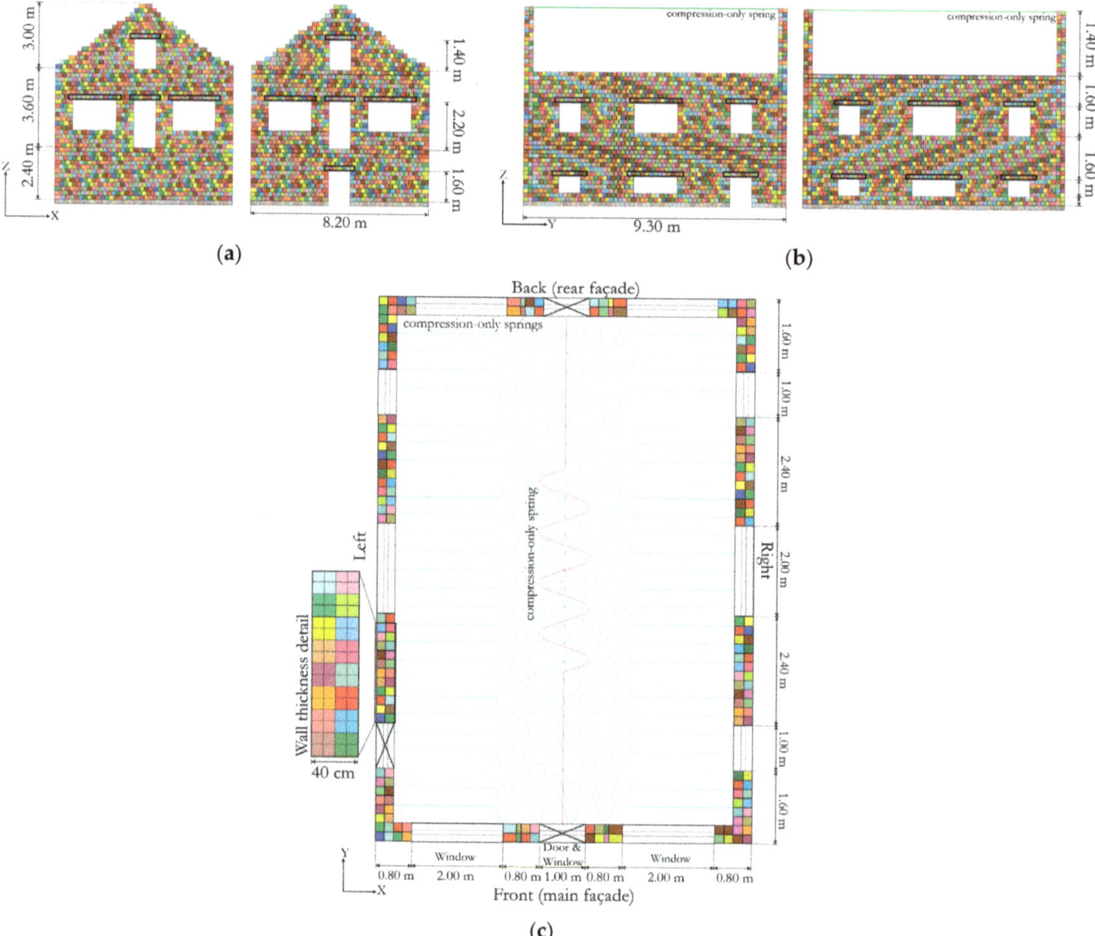

Figure 5. Dimensions of Building 3 and details of the modeling in the LS-DYNA environment: (**a**) elevation view showing the front (main façade) and back (rear façade) showing the gable end walls; (**b**) elevation view showing the left and right walls; (**c**) plan view of the building and position of compression-only springs (left inset) wall thickness detail.

3.2. Material Properties

The study of adobe mechanical properties is of paramount importance to improve the reliability of numerical models, which can then be used to derive fragility and vulnerability functions. Several studies and experimental campaigns have been conducted at the Civil Engineering Department of the University of Aveiro since 2005, providing crucial insights and information to sustain more robust numerical models. Adobe has also been extensively studied in Portugal, but mainly on the characterization of its mechanical properties [26,27]. The experimental campaigns on in-plane cyclic tests on full-scale double-T shaped adobe walls carried out by Silveira et al. [28] provided the value of Young's modulus (E) adopted herein. The material properties reported by Silveira et al. [28] were incorporated in a case study by Sarchi et al. [29] to create two finite element models, that were then used to develop fragility functions.

3.2.1. Testing the Modeling Platform

In the context of this study, it was pertinent to first test the modeling platform, herein LS-DYNA, then check if the results of the experimental tests were reproducible and thus applicable. A numerical model of the adobe double-T wall experiment reported in Silveira et al. [28] was recreated in the LS-DYNA environment as shown in Figure 6a, and a trial range of values of Young's modulus was applied and tested in the linear range. Each of the eigenmodes and their corresponding eigenfrequencies were checked, and it was found that at a Young modulus of 738 MPa, the corresponding frequency was 23 Hz, as shown in Figure 6b, henceforth providing a match to the experimental results. This value was corroborated by another study [30] that suggested that the Young modulus should be lower than 1 GPa for this kind of material.

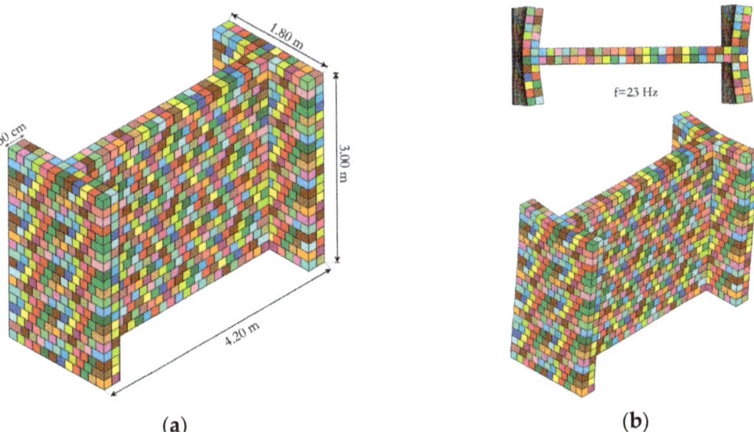

Figure 6. Recreated single adobe double-T wall in the LS-DYNA environment (**a**) dimensions and detailing; (**b**) Top and isometric view of the in-plane mode of vibration with its corresponding matched frequency f = 23 Hz.

3.2.2. Sensitivity Analysis

A sensitivity analysis was performed in which the various parameters involved in the numerical modeling such as tensile strength, normal energy, and shear energy release rates, and coefficients of friction were tested by using different proposals from the literature. It is a well-known fact from the scientific literature that adobe masonry exhibits a brittle behavior, owing to its low compressive, tensile, and shear strength [31–35]. For this reason, when subjected to cyclic action, adobe masonry usually presents a low resistance capacity. The tensile strength was assumed to be 0.05 MPa [31], and the recommendations from Lourenço and Pereira [36] were followed to define the initial range of parameters such as the normal energy release rate and shear energy release rate. According to AlShawa et al. [37], it is acceptable to assume equal shear and tensile parameters for masonry. The parameters that influenced the structural response the most were the normal and shear energy release rates. Thus, the final values of normal energy and shear energy release rate adopted in this study for buildings 1, 2, and 3 are 10 N/m, 30 N/m, and 20 N/m, respectively. From the literature, there was no clear recommendation for the coefficient of friction. Thus, a static coefficient of 0.4 and a dynamic coefficient of 0.3 was assumed. A summary of all the mechanical properties and input parameters adopted for the numerical models of the adobe buildings are given in Table 2.

Table 2. Mechanical properties and input parameters of the numerical models.

Element	Mechanical Properties	Value	Units
SOLID elements	Young's modulus	0.74	GPa
	Poison's ratio	0.30	-
	Density	1500	kg/m^3
Cohesive elements	Static coeff. of friction	0.4	-
	Dynamic coeff. of friction	0.3	-
	Scale factor for segment penalty stiffness	1.0	-
	Normal and shear failure stress	0.05	MPa
	Normal and shear energy release rate	10, 30, 20	N/m
	Normal (CN) and tangential stiffness	0.74	GPa
Springs	Timber elasticity modulus	7.00	GPa
	Timber elasticity modulus (5%)	4.70	GPa
	Design compressive strength	16.00	MPa
	Design bending strength	14.00	MPa

4. Framework for Vulnerability Assessment

The framework for vulnerability assessment of the adobe buildings is shown in Figure 7, and each of these steps is discussed in detail in the following subsections.

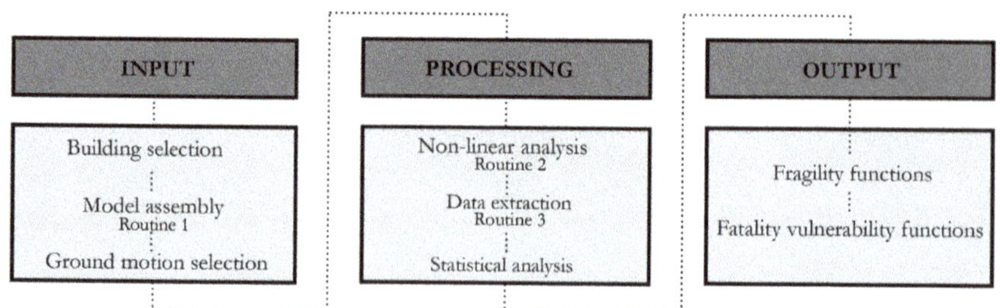

Figure 7. Framework for vulnerability assessment of adobe buildings.

4.1. Numerical Modeling

LS-DYNA, a general-purpose finite element software capable of simulating complex real-world problems, was used in this study. A shared memory parallel (SMP) double precision solver was utilized to improve the convergence of the non-linear analyses, and a combined finite-discrete element [38] strategy was adopted to simulate various stages of damage, such as cracking of walls and volume loss. The development of the numerical models in this study followed the recommendations from Karanikoloudis and Lourenço [31], and Lourenço and Pereira [36].

The pre-processing, processing, and post-processing of the numerical models can be broadly categorised into three steps, as shown in Figure 8. Firstly, the detailed building drawings are used to define the geometry in the software environment (LS-PrePost [39]) Secondly, the assembled models have to be processed and the data stored. Finally, the relevant data is extracted and further post-processed to obtain results such as the maximum displacements or fraction of cracked walls. To improve the computational performance and accuracy of these steps, automatic processes have been implemented in the MATLAB [40] routines (1–3). For each of the three steps mentioned above, a schematic of these routines is shown in Figure 8. Routine 1 was used for assembling the required geometry of the building based on a simplified input protocol. It efficiently generates the nodes and the solid elements and connects the segments to produce the complete building model. Routine 2 was used to execute the second step of this process, that is, running the static and dynamic analyses, which are performed in a stepwise manner. It generates a set of instructions that can be run from the system console that controls the LS-DYNA solver and can run multiple analyses simultaneously while efficiently segregating and storing the data generated. Routine 3 was used to execute the third step of this process: the extraction of the relevant data from the analyses, which, in the present study, mainly relates to crack propagation and volume loss.

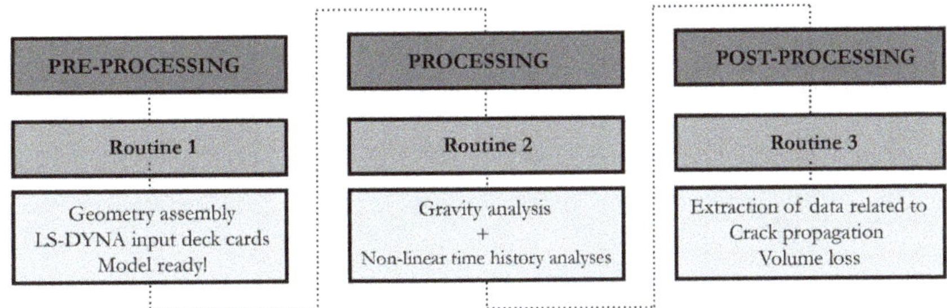

Figure 8. Procedure utilizing the automated algorithms in LS-DYNA for pre-processing, processing, and post-processing of the building models.

4.1.1. Solids Elements for Walls

To suit this study's objectives, it is pertinent to model the collapse mechanism of the buildings. This process allows quantifying the crack propagation and the volume loss of material. According to a classification given by D'Altri et al. [41], this modeling strategy can be considered a block-based approach, and a similar novel computational modeling strategy was proposed by Pulatsu et al. [42]. To this end, the building was discretized uniformly into blocks, as shown in Figure 9. In this study, the buildings were modeled utilizing three different mesh size discretization to reproduce the required wall thickness. The thickness of the walls is modeled adopting a "two-parts" scheme, in order to avoid computational instabilities. Thus, to reproduce a wall of 30 cm thickness, we used two blocks of 15 cm each. Furthermore, each block is constituted of two elements with a mesh size of 7.5 cm each, as shown in Figure 3c. The same approach has been applied to the other two buildings as shown in Figures 4C and 5C. Finally, the computational cost plays an important role, and larger mesh sizes help in reducing the cost significantly. A summary of the components of the models is given in Table 3.

Table 3. Numerical modeling characteristics.

	Building 1	Building 2	Building 3
Mesh size (cm)	7.5	15	10
Block size (cm)	15	30	20
No. of nodes	141,408	123,156	300,654
No. of solid elements	42,223	37,040	89,962
No. of parts	5162	4433	10,932
Total volume of blocks composing the walls (m^3)	17.80	124.90	89.90
Roof	Spring-based	Spring-based	Spring-based

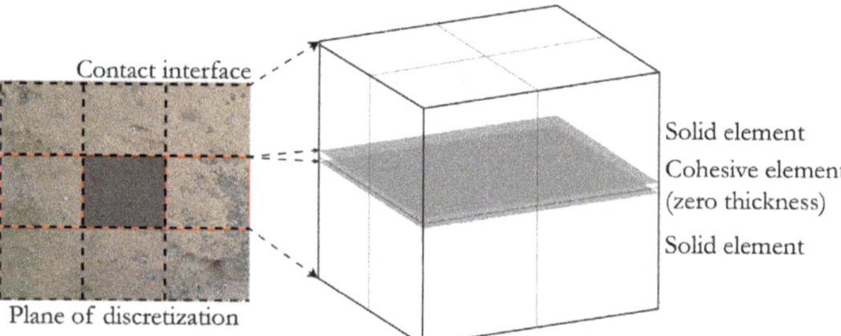

Figure 9. Procedure utilizing the automated algorithms in LS-DYNA for pre-processing, processing, and post-processing of the building models.

The elements used in our models were LS-DYNA's standard three-dimensional solid elements formed by eight nodes (hexahedron) ELFORM_TYPE_1, based on linear shape functions using one-point integration with constant stress [10]. The material properties attributed to these blocks is *MAT_001, an isotropic elastic material model [10]. This combination of solid elements and associated material properties proved to be efficient, enabling multiple non-linear time history analyses. However, while working with solid elements, it is possible to develop Hourglassing (HG). This is a pathology that can potentially cause numerical inconsistencies due to the development of nonphysical, zero energy modes of deformation that produce zero strain and no stress in the solids. To control this numerical error, an HG control (IHQ4) that follows a stiffness form of type 2 Flanagan-Belytschko has been utilized, with a Hourglass coefficient (QH) of 0.05 [37].

4.1.2. Segment Contact Interface Elements and Cohesive Mixed-Mode Material Model

The solid elements were connected using zero thickness master and slave interfaces. A penalty stiffness approach was used to model the separation, impact, and sliding with friction between blocks. A *CONTACT_AUTOMATIC_SURFACE_TO_SURFACE_TIEBREAK was used as these automatic contacts are non-oriented, and can detect penetration coming from either side of the master and slave segments. The selected contact algorithm is based on the *MAT_138 [10], which has a Discrete Crack Model, and it was assumed that the tensile behavior is governed by a bilinear traction separation law with linear softening, as shown in Figure 10a. This material model has a pre-defined failure criterion, and the cohesive elements were removed after meeting the criteria. No damage was assumed in the initial state of the buildings. The schematic of a master and slave node is shown in Figure 10b, and a summary of the mechanical properties and input parameters adopted for the numerical models are given in Table 2.

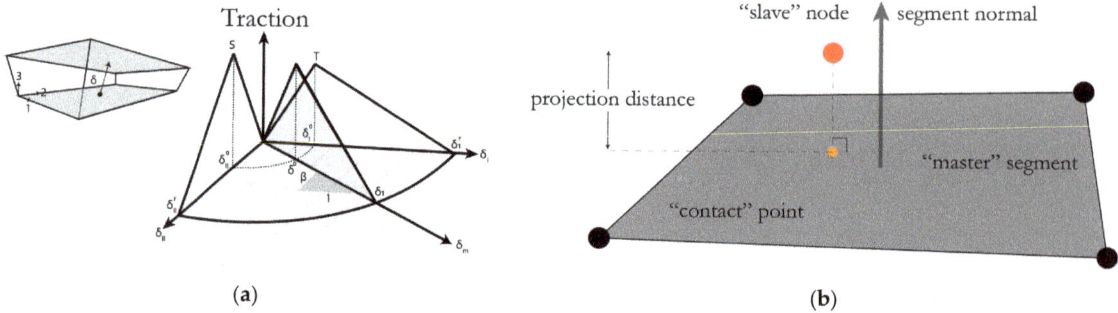

Figure 10. Material models in LS-DYNA: (**a**) Mixed-mode traction-separation law adapted from [10]; (**b**) assessment of projection distance for a free slave node adapted from [43].

4.1.3. Optimization

Due to the large deformations of the solid elements from their original position, an element may become so distorted that its volume becomes negative. This behavior increases processing time and eventually causes premature termination of the analysis. A function to remove such elements was incorporated using the *DEFINE_ELEMENT_DEATH_SOLID_SET [10] option. This functionality tracks solid blocks that fall below the base of the building and removes them from the calculation. This optimization significantly improved the computational performance and required storage, while maintaining a steady iterative timestep.

4.2. Eigenvalue Analysis

Eigenvalue analysis is fundamental towards understanding the fundamental period of vibration of the buildings. The modes of vibration corresponding to the X and Y direction of the building models with significant mass participations are shown in Figure 11a–f. The local modes with corresponding low values of mass participation are not shown for the sake of brevity.

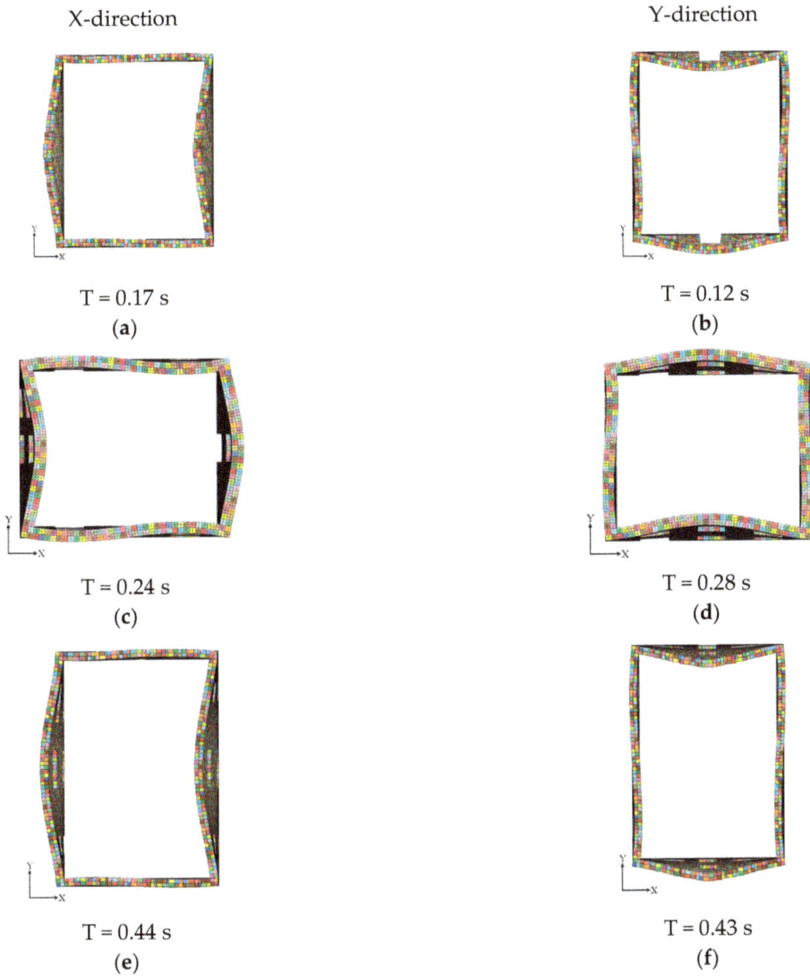

Figure 11. Fundamental periods of vibration and the corresponding modes for translation in X-direction and Y-direction: (**a,b**) building 1; (**c,d**) building 2; (**e,f**) building 3.

4.3. Hazard Demand

A set of 30 bi-directional ground motions records were selected based on the local tectonic regime in Portugal and a seismic hazard disaggregation study for the most populated regions [44]. The IM considered was PGA, and the records were segregated into five bins between 0.05 g and 1.05 g. Some of these records were scaled to cover the range of PGA required to induce collapse. A limit was set for the duration of the records, since it was not computationally feasible to use ground motion records that exceeded 60 s in duration. Figure 12 illustrates the acceleration spectra for the selected ground motion records along with the median spectrum and the 16th and 84th percentile spectra.

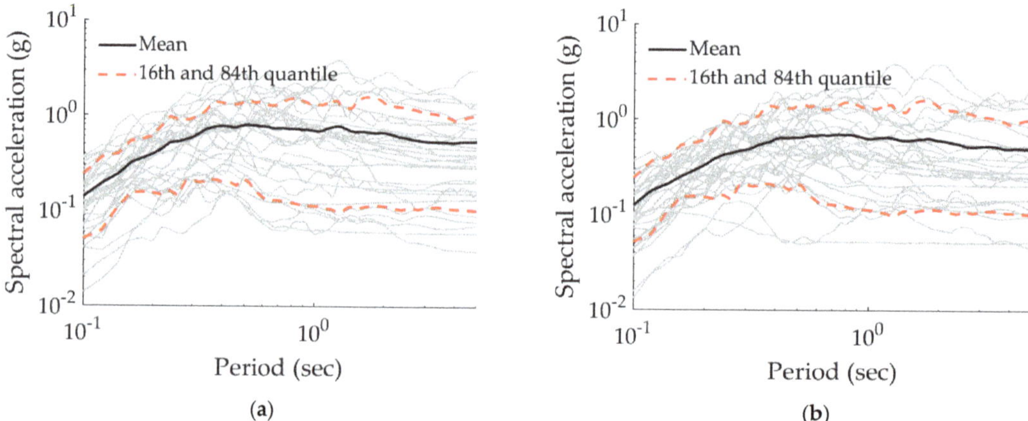

Figure 12. Response spectra of the bi-directional components of the selected ground motion records: (**a**) horizontal component in X-direction; (**b**) horizontal component in Y-direction.

4.4. Non-Linear Time History Analysis

Non-linear time history analyses were performed on all three building models using the set of selected ground motion records in the two orthogonal directions. A gravity analysis prior to a non-linear time history analysis ensures that the model is preloaded prior to applying dynamic loads. In this manner, it is possible to clearly distinguish the preloading phase from the transient phase [45]. The pre-loading was performed using a high damping in order to get a soft response (pseudo-static) to avoid any dynamic effects, whereas for the non-linear time histories, a damping of 3% was assumed [31].

In these analyses, it was expected that depending on the intensity of the ground motion records, the contact interfaces develop cracks due to the energy input. This effect causes the solid elements of the walls get displaced from their original position, and trigger a combined collapse mechanism. The failure modes of these buildings under seismic loading depend on the mechanical properties and on the overall geometric configuration of the building [17,46]. Examples of the end stage of the non-linear time history analyses according to the different assumed damage states are shown in Figure 13a–o.

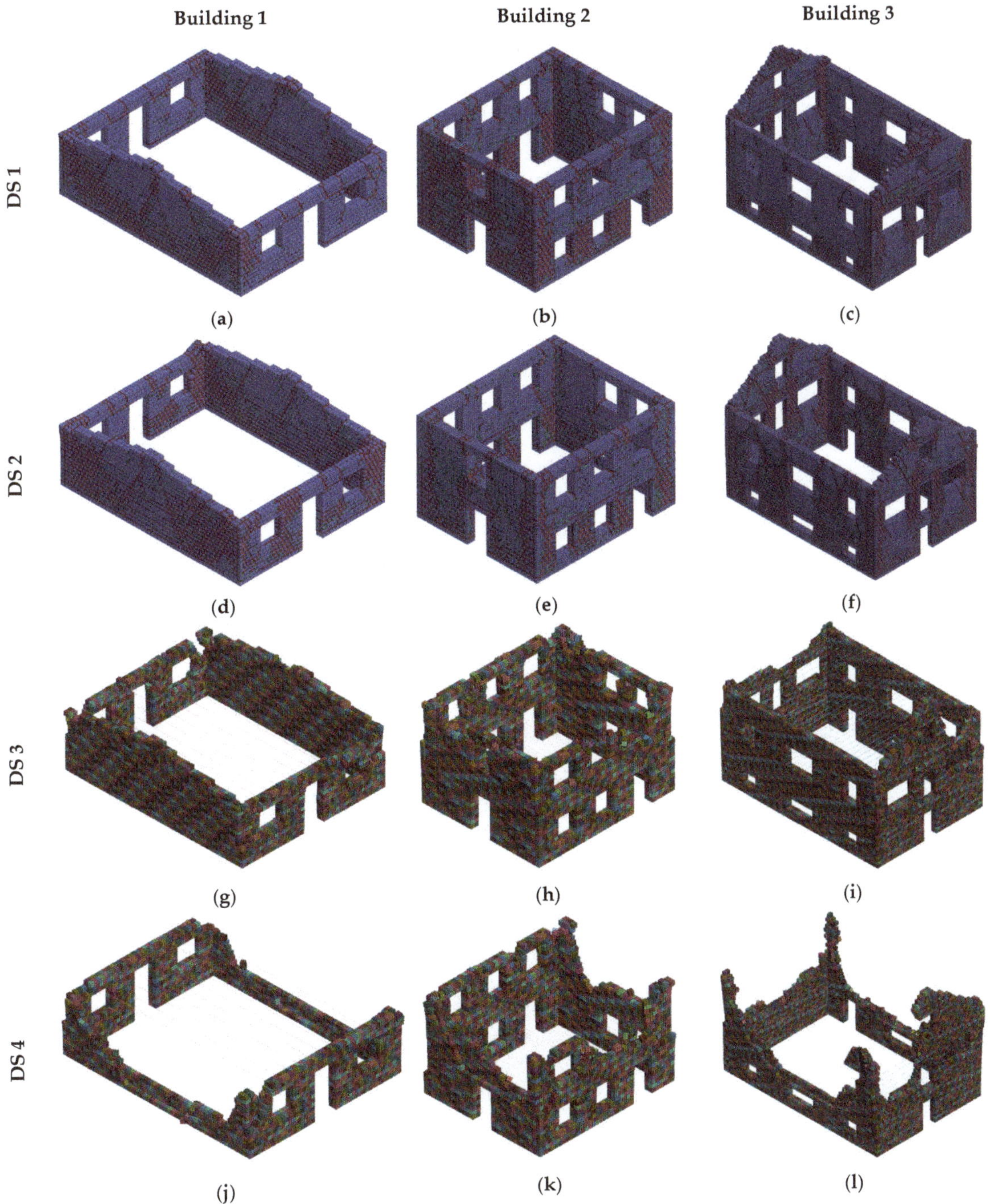

Figure 13. *Cont.*

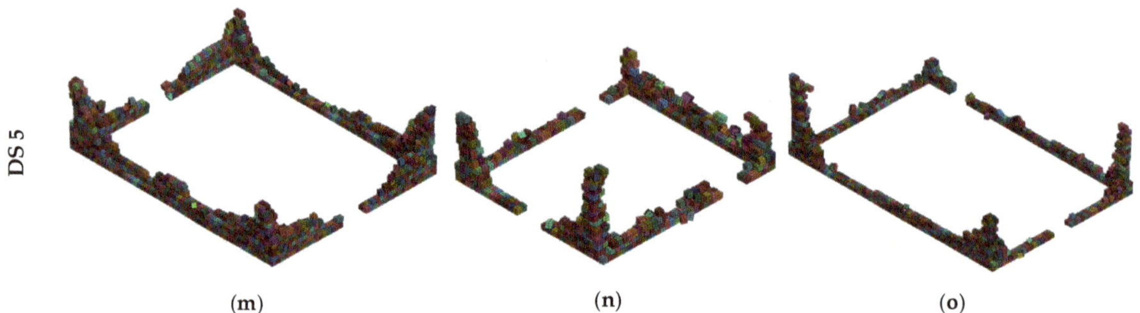

(m)   (n)   (o)

Figure 13. Crack formation and volume loss shown at the end stage of the non-linear time history analyses sustained by buildings 1–3 for the 5 damage states: (**a,d,g,j,m**) building 1; (**b,e,h,k,n**) building 2; (**c,f,i,l,o**) building 3.

4.5. EDPs and DSs

To compute the probability of exceeding a set of damage states, it is necessary to establish clear criteria relating each damage state with an engineering threshold (i.e., engineering demand parameter). We adopted the damage states from the European Macrosceismic Scale (EMS 98) [11], and defined a damage criteria based on two EDPs: (1) the crack propagation ratio, which refers to the cracks that develop on the walls prior to the onset of detachment of the blocks, and (2) the volume loss ratio, which refers to the post-failure movements of the blocks that have the potential to cause fatalities. The two novel EDPs are sub-categorized into the five damage states (DS1 to DS5) in Table 4. The thresholds for the novel EDPs were defined following the post-earthquake damage observations from So and Pomonis (2012) [47], in which typical volume loss for collapse of European masonry buildings were indicated. The data generated from the non-linear time history analyses pertaining to crack propagation and volume loss were extracted using the automated MATLAB routines and segregated for further analysis.

Table 4. EDPs and DSs and the corresponding damage thresholds.

EDPs	DSs	Damage Description	Threshold
Crack propagation ratio	DS 1	Negligible to slight	15%
	DS 2	Moderate	25%
Volume loss ratio	DS 3	Substantial to heavy	10%
	DS 4	Very heavy	25%
	DS 5	Destruction	40%

4.6. Cloud Analysis

Fragility assessment is an important step in seismic risk assessment [12]. Fragility functions define the conditional probability of exceeding a damage state for a given IM. Incremental Dynamic Analysis (IDA) [48] and Multiple Stripe Analysis (MSA) [49] are commonly used in fragility studies. However, these methods usually require a large computational effort due to the need to scale the ground motion records, and perform a large number of dynamic analysis. The issue of computational effort is particularly important in the current study, as the detailed numerical models are computationally-demanding. Thus, for the sake of simplicity, the Cloud analysis procedure was adopted [12]. This procedure has been used in recent structural fragility studies and it is capable of considering the record-to-record variability and other sources of uncertainty related to structural modeling. The steps followed for the fragility assessment are described below, and illustrated in Figure 14.

1. The results extracted from the non-linear time history analysis were plotted (*IM* vs. *EDP*) in a logarithmic scale, and a linear regression was performed to generate a trendline that best fits the data.
2. The following equation was used to calculate the expected value of the dependent variable (*EDP*) given an *IM*. A homogeneity of variance was assumed for the *IM-EDP* random variables; hence the model can be described as shown in Equation (1).

$$E[\log(EDP)|IM] = b \times \log(IM) + \log(a)$$
$$\sigma_{rec-rec} = \sqrt{\frac{\sum_{i=1}^{n}(\log(EDP_i) - E[\log(EDP)|IM_i])^2}{n-2}}$$
(1)

where $E[\log(EDP)|IM]$ stands for the expected logarithm of *EDP* given an *IM*, b and $\log(a)$ are the regression parameters, $\sigma_{rec-rec}$ is the record-to-record variability, n is the no. of records, and EDP_i corresponds to the *i*-EDP value obtained from the non-linear analysis for the corresponding IM_i.

3. The structural fragility functions obtained from the probabilistic model can be expressed using Equation (2).

$$P[EDP \geq ds_i|IM] = 1 - \Phi\left(\frac{EDP_{ds_i} - E[\log(EDP|IM)]}{\sigma_{rec-rec}}\right)$$
(2)

where Φ is the cumulative normal standard distribution and the EDP_{ds_i} is the damage threshold (e.g., 15%) corresponding to a given damage state (e.g., DS1).

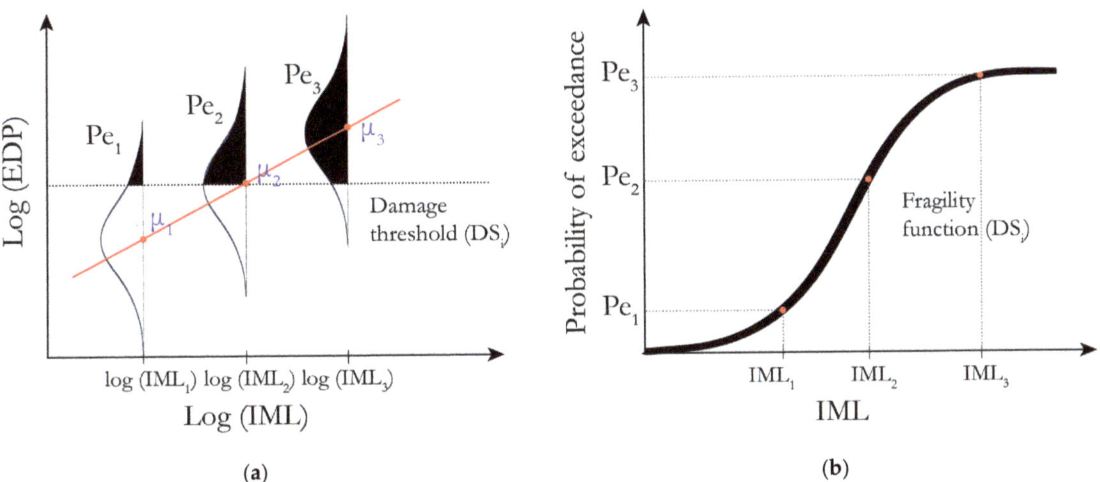

Figure 14. A schematic of the fragility assessment procedure: (**a**) probability of exceedance for three IMLs given a damage threshold; (**b**) Fragility function for a given damage state.

4.7. Fatality Vulnerability Functions

The method to estimate fatality rates in this study follows the methodology proposed by Abeling and Ingham [9], which uses a relationship between the volume loss of the building and an expected fatality ratio. The buildings volume loss is a damage descriptor that measures the ratio between the volume of debris (V_d) and the space capacity (V_c) measured up-to a height of 2 m from the floor level, as shown in Figure 15. The volume loss can be calculated using Equation (3).

$$Volume\ loss = \frac{V_d}{V_c}$$
(3)

A cloud of fatality ratio (FR)—*IM* data was produced by multiplying the volume loss obtained from each of the structural analysis by the corresponding FR proposed by Abeling and Ingham [9]. This cloud of points was then fitted to a lognormal cumulative density function, as defined in Equation (4)

$$Fatality\ ratio = 0.52 \times \Phi\left(\frac{\ln(IM) - \theta}{\beta}\right) \tag{4}$$

where 0.52 is the maximum fatality ratio corresponding to 100% volume loss as per [9], Φ is the cumulative normal standard distribution, θ is the logarithmic mean, and β is the logarithmic standard deviation, as presented in Table 5.

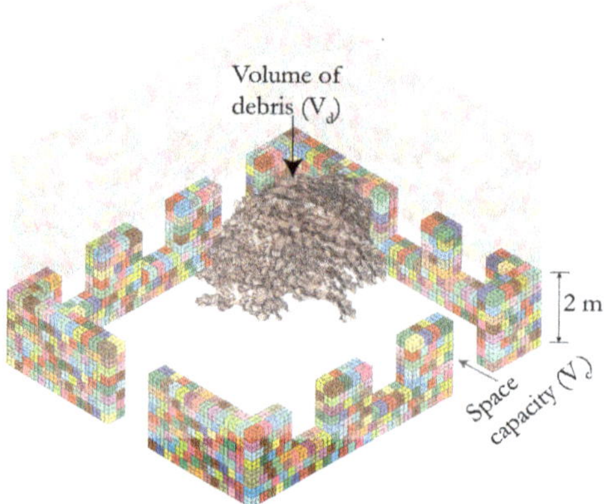

Figure 15. Illustration of volume loss using Building 2 (adapted from Okada [50]).

Table 5. Fatality vulnerability parameters for adobe buildings.

Building Class	θ *	β **
1-story	1.06	0.88
2-story	0.45	0.75
2-story plus attic	0.45	0.77

* θ—logarithmic mean, ** β—logarithmic standard deviation.

5. Results

The results of this study for the three adobe buildings are presented in terms of the derived fragility and fatality vulnerability functions, as shown in Figure 16a–f with comparisons to the recent study by Martins and Silva (2020) [7]. The corresponding fragility parameters (i.e., μ—logarithmic mean and σ—logarithmic standard deviation) are presented in Table 6 and the fatality vulnerability parameters are presented in Table 5. It can be inferred from the values of dispersion (i.e., σ) in Table 6, that there are two different values for the dispersion, one value corresponding to DS1-DS2 and another for DS3-DS5. The difference arises due to the two EDPs selected to predict the different phenomenon. It is intuitive to understand that a building can develop cracks after which volume loss occurs. The 1-story building is less vulnerable in the first damage states, due to the absence of a story above and the low height. It becomes vulnerable in the last damage states, since post-cracking, the walls lose their load bearing capacity, and these blocks are immediately displaced from their original position. Also, the elastic period of the buildings has an influence on DS1 and DS2, while DS3-DS5 are in the non-linear range. According to Ginell and Tolles [32] the post-elastic behavior of adobe is different from ductile building materials, since it is a brittle material characterized by low compressive, tensile, and shear strength. It is worth noting that all the buildings are unreinforced, and their load bearing capacities rests solely on the integrity of the walls.

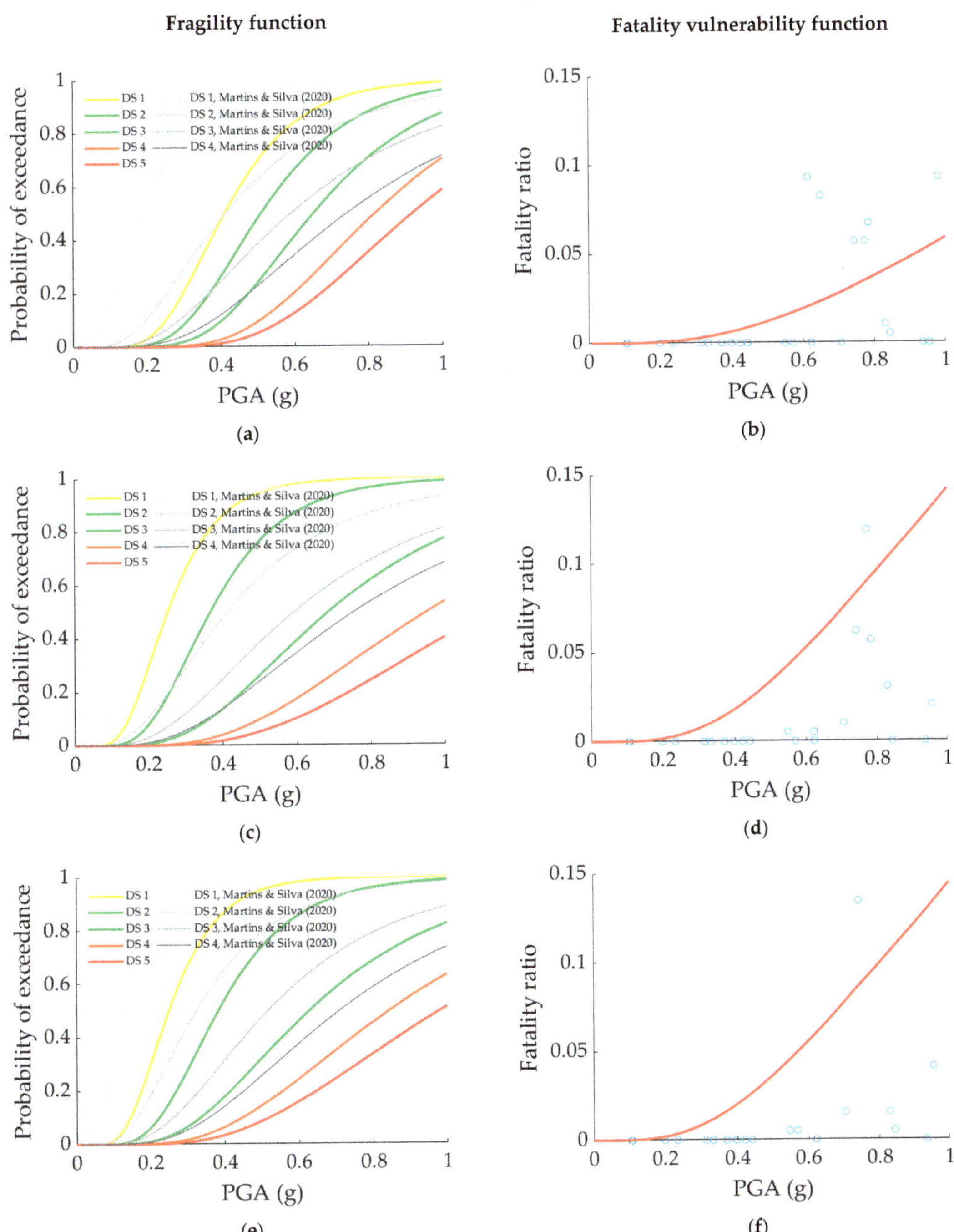

Figure 16. Fragility and fatality vulnerability function for each of the building classes, respectively (**a**,**b**) 1-story; (**c**,**d**) 2-story; (**e**,**f**) 2-story plus an attic.

Table 6. Fragility parameters for adobe buildings.

Building Class	IM	DS 1		DS 2		DS 3		DS 4		DS 5	
		μ *	σ **	μ	σ	μ	σ	μ	σ	μ	σ
1-story	PGA	−0.89	0.39	−0.67	0.39	−0.43	0.37	−0.20	0.37	−0.08	0.37
2-story	PGA	−1.38	0.43	−1.00	0.43	−0.38	0.50	−0.05	0.50	0.12	0.50
2-story plus attic	PGA	−1.39	0.42	−0.96	0.42	−0.46	0.49	−0.17	0.49	−0.02	0.49

* μ—logarithmic mean, ** σ—logarithmic standard deviation.

6. Conclusions

This study presented the development of fragility and vulnerability functions for typical adobe buildings in Portugal, one of the most vulnerable types of construction in the country. The selection of the buildings to numerically model was carried out based on survey data comprising building drawings, photographs, and information regarding the construction practice and material properties. The sensitivity of mechanical properties and modeling options was assessed, and calibration and validation of material properties were carried out following the results from a full-scale experimental in-plane cyclic tests on adobe walls. A set of 30 ground motion records with bi-directional components were selected, and non-linear time-history analyses were performed until collapse was reached. Two novel EDPs were adopted and damage thresholds were proposed such that they can be correlated with damage classifications in agreement with the European Macroseismic Scale (EMS-98) [11]. Fragility functions were then derived following the cloud analysis methodology [12] and compared to a recent study [7].

The fragility functions proposed herein can be used directly in the assessment of damage due to earthquake scenarios, or in probabilistic risk analysis to identify regions in the country where the probability of collapse of these structures is particularly high. Similarly, the vulnerability functions in terms of fatality rates can be employed to assess human losses, which is particularly important for the development of risk management measures.

Author Contributions: Conceptualization, V.S., R.V. and T.M.F.; methodology, S.M., H.L., V.S., R.V. and T.M.F.; software, H.L., S.M.; validation, S.M., H.L., V.S., T.M.F. and R.V.; investigation, S.M., H.L.; resources, R.V.; data curation, H.L., S.M.; writing—original draft preparation, S.M.; writing—review and editing, T.M.F., R.V., H.L. and V.S.; visualization, S.M., H.L.; supervision, V.S., R.V. and T.M.F.; project administration, R.V.; funding acquisition, R.V., V.S. and T.M.F. All authors have read and agreed to the published version of the manuscript.

Funding: This work was funded by the Foundation for Science and Technology (FCT)—Aveiro Research Centre for Risks and Sustainability in Construction (RISCO), Universidade de Aveiro, Portugal in the framework of research project PTDC/ECI-EST/31865/2017 MitRisk—Framework for seismic risk reduction resorting to cost-effective retrofitting solutions.

Institutional Review Board Statement: Not applicable.

Informed Consent Statement: Not applicable.

Data Availability Statement: The data presented in this study are available on request from the corresponding author.

Conflicts of Interest: The authors declare no conflict of interest. The funders had no role in the design of the study; in the collection, analyses, or interpretation of data; in the writing of the manuscript, or in the decision to publish the results.

References

1. Sousa, G.; Gallaecia, E.S. Historical Seismicity in Portugal. In *Seismic Retrofitting: Learning from Vernacular Architecture*; Taylor & Francis: Abingdon, UK, 2015.
2. Choffat, P. Le tremblement de terre du 23 Avril 1909 dans le Ribatejo. In *Revista de Obras Pu´blicas e Minas, Tomo XLIII*; Imprensa Nacional: Lisboa, Spain, 1912. (In French)

3. Tarque, N.; Crowley, H.; Pinho, R.; Varum, H. Displacement-Based Fragility Curves for Seismic Assessment of Adobe Buildings in Cusco, Peru. *Earthq. Spectra* **2012**, *28*, 759–794. [CrossRef]
4. Crowley, H.; Pinho, R.; Bommer, J.J. A Probabilistic Displacement-based Vulnerability Assessment Procedure for Earthquake Loss Estimation. *Bull. Earthq. Eng.* **2004**, *2*, 173–219. [CrossRef]
5. Ahmad, N.; Ali, Q.; Crowley, H.; Pinho, R. Earthquake loss estimation of residential buildings in Pakistan. *Nat. Hazards* **2014**, *73*, 1889–1955. [CrossRef]
6. Villar-Vega, M.; Silva, V.; Crowley, H.; Yepes, C.; Tarque, N.; Acevedo, A.B.; Hube, M.A.; Gustavo, C.D.; María, H.S. Development of a Fragility Model for the Residential Building Stock in South America. *Earthq. Spectra* **2017**, *33*, 581–604. [CrossRef]
7. Martins, L.; Silva, V. Development of a fragility and vulnerability model for global seismic risk analyses. *Bull. Earthq. Eng.* **2020**, *2020*, 1–27. [CrossRef]
8. Sumerente, G.; Lovon, H.; Tarque, N.; Chácara, C. Assessment of Combined In-Plane and Out-of-Plane Fragility Functions for Adobe Masonry Buildings in the Peruvian Andes. *Front. Built Environ.* **2020**, *6*, 52. [CrossRef]
9. Abeling, S.; Ingham, J.M. Volume loss fatality model for as-built and retrofitted clay brick unreinforced masonry buildings damaged in the 2010/11 Canterbury earthquakes. *Structures* **2020**, *24*, 940–954. [CrossRef]
10. Hallquist, J.O. *Livermore Software Technology Corporation (Lstc), Ls-Dyna Keyword User's Manual Volume I*; 2018; Volume II. Available online: https://ftp.lstc.com/anonymous/outgoing/jday/manuals/LS-DYNA_Manual_Volume_I_R11.pdf (accessed on 15 March 2021).
11. Comisión Sismológica Europea. *Escala Macro Sísmica Europea EMS-98*; 1998; Volume 15. Available online: http://media.gfz-potsdam.de/gfz/sec26/resources/documents/PDF/EMS-98_Original_englisch.pdf (accessed on 15 March 2021).
12. Jalayer, F.; De Risi, R.; Manfredi, G. Bayesian Cloud Analysis: Efficient structural fragility assessment using linear regression. *Bull. Earthq. Eng.* **2015**, *13*, 1183–1203. [CrossRef]
13. De Estatística, I.N. Census Data 2011. Lisbon. 2011. Available online: http://censos.ine.pt/ (accessed on 15 March 2021).
14. Costa, C.; Arduin, D.; Rocha, F.; Velosa, A. Adobe Blocks in the Center of Portugal: Main Characteristics. *Int. J. Arch. Herit.* **2019**, *15*, 467–478. [CrossRef]
15. Correia, M.; Carlos, G.; Merten, J.; Viana, D.; Rocha, S. *Vernacular Heritage and Earthen Architecture*; Taylor & Francis Group: London, UK, 2013; pp. 111–116. [CrossRef]
16. Varum, H.; Figueiredo, A.; Silveira, D.; Martins, T.; Costa, A. Investigaciones realizadas en la Universidad de Aveiro sobre caracterización mecánica de las construcciones existentes en adobe en Portugal y propuestas de rehabilitación y refuerzo. Resultados alcanzados. *Inf. Constr.* **2011**, *63*, 127–142. [CrossRef]
17. Varum, H.; Costa, A.; Fonseca, J.; Furtado, A. Behaviour Characterization and Rehabilitation of Adobe Construction. *Procedia Eng.* **2015**, *114*, 714–721. [CrossRef]
18. Varum, H.; Costa, A.; Velosa, A.; Martins, T.; Pereira, H.; Almeida, J. Caracterização mecânica e patológica das construções em Adobe no distrito de Aveiro como suporte em intervenções de reabilitação. In *Houses and Cities Built with Earth: Conservation, Significance and Urban Quality*; CULTURA 2000—Education and Culture, European Union Program; ARGUMEMTUM: Monsaraz, Portugal, 2006; pp. 41–45.
19. Silveira, D.; Varum, H.; Costa, A.; Neto, C. Survey of the Facade Walls of Existing Adobe Buildings. *Int. J. Arch. Herit.* **2016**, *10*, 867–886. [CrossRef]
20. IPQ. NP EN 1998-1. Eurocódigo 8: Projecto de Estruturas Para Resistência Aos Sismos. In *Parte 1: Regras Gerais, Acções Sísmicas e Regras Para Edifícios*; Instituto Português da Qualidade: Caparica, Portugal, 2010; p. 230. (In Portuguese)
21. Braga, A.M.; Estêvão, J.M.C. Os Sismos E a Construção Em Taipa No Algarve. In *Sísmica 2010–8° Congr. Sismol. e Eng. Sísmica*; University of Aveiro: Aveiro, Portugal, 2010; pp. 1–13.
22. Correia, M.R. *Rammed Earth in Alentejo*; Argumentum: Lisbon, Portugal, 2007.
23. Cancela, D.C.P. Comportamento Higrotérmico e Monitorização de Construções em Adobe. Master's Thesis, Civil Engineering, Univeristy of Aveiro, Aveiro, Portugal, 2013.
24. EN 1995-1-2:2004. Part 1–2: General—Structural fire design. In *Eurocode 5: Design of Timber Structures*; CEN: Brussels, Belgium, 2004.
25. Battaglia, L.; Ferreira, T.M.; Lourenço, P.B. Seismic fragility assessment of masonry building aggregates: A case study in the old city Centre of Seixal, Portugal. *Earthq. Eng. Struct. Dyn.* **2021**, *50*, 1358–1377. [CrossRef]
26. Silveira, D.; Varum, H.; Costa, A.; Martins, T.; Pereira, H.; Almeida, J. Mechanical properties of adobe bricks in ancient constructions. *Constr. Build. Mater.* **2012**, *28*, 36–44. [CrossRef]
27. Silveira, D.; Varum, H.; Costa, A. Influence of the testing procedures in the mechanical characterization of adobe bricks. *Constr. Build. Mater.* **2013**, *40*, 719–728. [CrossRef]
28. Silveira, D.; Varum, H.; Costa, A.; Pereira, H.; Sarchi, L.; Monteiro, R. Seismic behavior of two Portuguese adobe buildings: Part I—in-plane cyclic testing of a full-scale adobe wall. *Int. J. Arch. Herit.* **2018**, *12*, 922–935. [CrossRef]
29. Sarchi, L.; Varum, H.; Monteiro, R.; Silveira, D. Seismic behavior of two Portuguese adobe buildings: Part II—numerical modeling and fragility assessment. *Int. J. Arch. Herit.* **2018**, *12*, 936–950. [CrossRef]
30. NIKER. *New Integrated Knowledge Based Approaches to the Protection of Cultural Heritage from Earthquake-Induced Risk*; Research Report; University of Padova: Padova, Italy, 2010.

31. Karanikoloudis, G.; Lourenço, P.B. Structural assessment and seismic vulnerability of earthen historic structures. Application of sophisticated numerical and simple analytical models. *Eng. Struct.* **2018**, *160*, 488–509. [CrossRef]
32. Ginell, W.S.; Tolles, E.L. Seismic Stabilization of Historic Adobe Structures. *J. Am. Inst. Conserv.* **2000**, *39*, 147–163. [CrossRef]
33. Vargas, J.; Bariola, J.; Blondet, M.; Mehta, P.K. Seismic strength of adobe masonry. *Mater. Struct.* **1986**, *19*, 253–258. [CrossRef]
34. Blondet, M.; Ginocchio, F.; Marsh, C.; Ottazzi, G.; Villa-Garcia, G.; Yep, J. Shaking Table Test of Improved Adobe Masonry Houses. In Proceedings of the 9th World Conference on Earthquake Engineering, Kyoto, Tokyo, Japan, 2–6 August 1988.
35. Illampas, R.; Ioannou, I.; Charmpis, D.C. Adobe bricks under compression: Experimental investigation and derivation of stress–strain equation. *Constr. Build. Mater.* **2014**, *53*, 83–90. [CrossRef]
36. Lourenço, P.B.; Pereira, J.M. *Seismic Retrofitting Project Recommendations for Advanced Modeling of Historic Earthen Sites*; Getty Conservation Institute: Los Angeles, CA, USA; TecMinho–University of Minho: Guimarães, Portugal, 2018.
37. Alshawa, O.; Sorrentino, L.; Liberatore, D. Simulation Of Shake Table Tests on Out-of-Plane Masonry Buildings. Part (II): Combined Finite-Discrete Elements. *Int. J. Arch. Herit.* **2016**, *11*, 1–15. [CrossRef]
38. Hazay, M.; Munjiza, A. Introduction to the Combined Finite-Discrete Element Method. In *Computational Modeling of Masonry Structures Using the Discrete Element Method*; IGI Global: Hershey, PA, USA, 2016; pp. 123–145. ISBN 1522502319.
39. Dis, C. LS-DYNA/LS-PrePost Ex. 0. *Introduction* **2011**, *31* (Suppl. S1), 1–74.
40. T.M. Inc. *MATLAB*; T.M. Inc.: Natick, MA, USA, 2018.
41. D'Altri, A.M.; Sarhosis, V.; Milani, G.; Rots, J.; Cattari, S.; Lagomarsino, S.; Sacco, E.; Tralli, A.; Castellazzi, G.; De Miranda, S. Modeling Strategies for the Computational Analysis of Unreinforced Masonry Structures: Review and Classification. *Arch. Comput. Methods Eng.* **2020**, *27*, 1153–1185. [CrossRef]
42. Pulatsu, B.; Erdogmus, E.; Lourenço, P.B.; Lemos, J.V.; Tuncay, K. Simulation of the in-plane structural behavior of unreinforced masonry walls and buildings using DEM. *Structure* **2020**, *27*, 2274–2287. [CrossRef]
43. Bala, S. *Tie-Break Contacts in LS-DYNA*; Livemore Software: Livermore, CA, USA, 2007.
44. Sousa, L.; Marques, M.; Silva, V.; Varum, H. Hazard Disaggregation and Record Selection for Fragility Analysis and Earthquake Loss Estimation. *Earthq. Spectra* **2017**, *33*, 529–549. [CrossRef]
45. Hallquist, J.O. *LS Dyna Theory Manual*; Livemore Software: Livermore, CA, USA, 2006.
46. Costa, A.; Humberto, V.; João, M.G. *Structural Rehabilitation of Old Buildings*; Springer: Berlin/Heidelberg, Germany, 2014.
47. Pomonis, A. Derivation of Globally Applicable Casualty Rates for use in Earthquake Loss Estimation Models. In Proceedings of the 15th World Conference Earthquake Engineering, Lisbon, Portugal, 24–28 September 2012.
48. Vamvatsikos, D.; Cornell, C.A. Incremental dynamic analysis. *Earthq. Eng. Struct. Dyn.* **2002**, *31*, 491–514. [CrossRef]
49. Vamvatsikos, D.; Cornell, C.A. Direct Estimation of Seismic Demand and Capacity of Multidegree-of-Freedom Systems through Incremental Dynamic Analysis of Single Degree of Freedom Approximation. *J. Struct. Eng.* **2005**, *131*, 589–599. [CrossRef]
50. Okada, S. Description for indoor space damage degree of building in earthquake. In Proceedings of the Eleventh World Conference on Earthquake Engineering, Acapulco, Mexico, 23–28 June 1996.

MDPI
St. Alban-Anlage 66
4052 Basel
Switzerland
Tel. +41 61 683 77 34
Fax +41 61 302 89 18
www.mdpi.com

Buildings Editorial Office
E-mail: buildings@mdpi.com
www.mdpi.com/journal/buildings